EFFICIENCY, JUSTICE AND CARE

INTERNATIONAL LIBRARY OF ETHICS, LAW, AND THE NEW MEDICINE

VOLUME 33

The titles published in this series are listed at the end of this volume.

Efficiency, Justice and Care

Philosophical Reflections on Scarcity in Health Care

YVONNE, DENIER
Catholic University of Leuven,
Belgium

Springer

A C.I.P. Catalogue record for this book is available from the Library of Congress.

ISBN-10: 1-4020-5213-8 (HB)
ISBN-13: 978-1-4020-5213-2 (HB)
ISBN-10: 1-4020-5214-6 (e-book)
ISBN-13: 978-1-4020-5214-9 (e-book)

Published by Springer,
P.O. Box 17, 3300 AA Dordrecht, The Netherlands.

www.springer.com

Printed on acid-free paper

In memory of my father
and to Lode
with love

ACKNOWLEDGEMENTS

In the spring of 2004, I attended the play called *Die Siel van die Mier*, written by David Van Reybrouck and marvellously performed by Josse De Pauw. It is about a professor who gives his final lecture. For 35 years he has done this, passing on his scientific knowledge to his students, in a clear and well-structured way. However, as clear and lucid lectures, articles, books and syllabi may be, the journey towards it is anything but that:

> As a student, you're served knowledge. Well-finished, polished, ready, and printed off. As a young researcher, however, you start to gather knowledge. In a cluttered, disorderly, and clumsy way. Sometimes you're tired, not in good shape. And you're constantly thinking: I am messing about; I was not made for doing this; my research should be like all those lectures I've attended – just as systematic, just as immaculate. Until you realise that this is only a facade. Until you realise that ... your ... colleagues are also busy muddling along. It is only later on that we enforce order; that we carve out our pottering and make it look as if it possessed clear and lucid logic from the very beginning. Hypothesis, methodology, and conclusion, yes indeed – but at the very moment [of research] things go off very muddled and confused, if they go off at all.[1]

Work on this book began in the autumn of 1999. What exactly is implied, I wanted to find out, by the idea of just health care? What are the philosophical categories, distinctions and arguments used in matters of just distribution of scarce health-care resources? Along the often-muddled road of analysing these questions, trying to find coherence in textual research and in formulating the various aspects of the problem while endeavouring to write in a lucid style, I have accumulated many debts. Finally, it is my pleasure to thank those who have, in many ways, helped to bring this often-confused project to a structured end.

Above all, I am grateful to Prof. Dr. Toon Vandevelde, who has given me the opportunity to begin this project by offering me a position as a research assistant at the Centre for Economics and Ethics. His unremitting support, ceaseless reading and re-reading of many versions of this text, continual offering of inspiring ideas and helpful comments have turned many vague and muddled documents into a coherent text. Above all, I came to appreciate his combination of academic directness with human kindness. I am exceptionally obliged to him in many respects.

The bulk of this work has been written in Leuven, in the amiable surroundings of the Centre for Economics and Ethics and of the Institute of Philosophy. I am grateful to past and present members of the Centre for Economics and Ethics, Prof. Dr. Luk Bouckaert, Dr. Stephan Cludts, Dr. Kurt Devooght, Bart Engelen,

[1] In: D. Van Reybrouck, Die Siel van die Mier, Gent: Het Muziek Lod, 2004, p. 23 (my translation, YD).

Sylvie Loriaux, Thomas Nys, Prof. Dr. Erik Schokkaert, Annabel Sels, Ann Van Espen and Prof. Dr. Luc Van Liedekerke, who each in their own way have contributed to this research project – by offering helpful comments, by helping me out with any practical concerns, and perhaps most importantly, by creating a warm and pleasant atmosphere filled with cheer and laughter. Among the members of the Institute of Philosophy, I am much indebted to Prof. Dr. Stefaan Cuypers and Prof. Dr. Herman De Dijn, for their kind and helpful comments on a number of preparatory articles, which expressed some ideas that are further elaborated in this text. I am also grateful to John Alexander, Helder De Schutter, Barbara Haverhals, Dr. Ronald Tinnevelt, Joris Van Damme and Katia Vanhemelryck with whom I have had many philosophical and amiable conversations.

From October 2001 until April 2002, I went to Germany. I am much obliged to Prof. Dr. Dietmar Mieth and to Prof. Dr. Marcus Düwell, for giving me the opportunity to join the Interdepartmental Centre for Ethics in the Sciences and Humanities of the Eberhard-Karls University of Tübingen, as a visiting scholar. During this stay, I have received tremendous help from the most kind and stimulating people. I especially wish to thank Prof. Mieth, Prof. Düwell, Dr. Lars Thielmann, Dr. Georg Marckmann, Prof. Dr. Hille Haker and Dr. Sigrid Graumann for many interesting conversations and helpful comments, which certainly contributed to unravelling my unstructured train of thoughts. I am also grateful to Dr. Susan Nurmi-Schomers and Dr. Birgit Leweke for kindly helping me out with many practical concerns. This project was financially supported by the Deutscher Akademischer Austauschdienst. I cannot imagine what this book would have looked like had I not had this opportunity. My stay in Tübingen undoubtedly gave my reflections on the matter a specific direction. For all this help I am very thankful.

As the scientific forum of this research project is predominantly English-speaking, there were many reasons for writing this work in English. However, not being a native speaker, this has never been an easy job. If my train of thoughts was already muddled in Dutch, my first formulations in English were even more so. I received excellent guidance in endeavouring to write in Queen's English, from Jo Clijsters, Claire Dawson, Fionnuala Gogarty, Klaartje Verbelen and Lode Waelbers, who each meticulously read parts of this text, unflagging in their help to search for better formulations in more refined English. For their time, willingness and patience I am exceptionally obliged to them.

Finally, how can one bring such a project to an end without the support of those that are dear to you? I could not have done this without the unqualified and loving support of my family and friends. I am grateful for their interest in my research, for their kind understanding for the solitude and peace I needed to work on it, and for their warm inquiries into how far it still was from completion. I owe great thanks to all of them. I wish to dedicate this work to two persons in particular.

My father, Thieu Denier, was an extraordinary warm and kind person. He showed me the value of things that cannot be learned from the books. His positive attitude, perseverance and ability to put things into their right perspective, were to his credit and earned him deep respect from anyone who was so fortunate to have known him. It is very hard to write about health, disease and finitude when someone so dear to you is suddenly stricken with the diagnosis of terminal cancer. He died on a spring-morning, 11 May 2003, at the age of 55. Bearing his belief in my work and his proud support firmly in mind intensified my ability to continue this work.

That this project ultimately came to a good end, I undoubtedly owe to a great extent, in countless ways, and more than to anyone else, to Lode. He has served as a sounding board and critic of many of my ideas and projects. He has meticulously read every page of this work and has ceaselessly relieved me of all possible IT and layout concerns. He has supported and encouraged me in times when it was very hard. For accepting in good spirit the many intrusions into our lives caused by this work, my final and dearest words of thanks are to him.

Yvonne Denier
Leuven, 31 March 2005

CONTENTS

La justice … fait partie intégrante du souhait de vivre bien … souhait d'une vie accomplie avec et pour les autres dans les institutions justes.

Paul Ricoeur, Le Juste, Paris: Le Seuil, 1995, p. 17.

We must consider justice and injustice – what sort of actions they are concerned with, what kind of mean justice is and what are the extremes between which the just is a mean

Aristotle, Nicomachean Ethics, Book V, 1129a.

GENERAL INTRODUCTION

What does just health care imply? Does it mean that people have a right to health care? Does it entail that there are rights-based social obligations to provide equal access to health care for everyone? And if so, why? Why are health care interests so important that they deserve special protection? What kind of social good is health care? What are its functions and do these make it different from other commodities? Furthermore, how much equality should there be in health care? What inequalities are morally acceptable and how should the burdens of achieving equality be distributed? Which matters of health care belong to the domain of justice, and which to the domain of charity? To what extent should we allow personal responsibility to play a role in allocating health-care services and resources, or in distributing the costs? And what does justice require with regard to long-term care for the chronically ill and irreversibly dependent?

These are all examples of topics raised in contemporary debate about the requirements of justice in health care. At first sight, the essence of what is going on in just health care seems quite evident. We generally know, for instance, *what health care is*. Understood broadly, it is a complex framework of institutions, services and policy measures, roughly organised in accordance with three goals: prevention, cure and care. In general, public health measures contain the category of services that promote the collective health status by means of *prevention* of disease and disability. Besides prevention, health care includes the familiar personal medical services of *cure* and the social support services of *care* for the chronically ill or disabled. Furthermore, we also know that by preserving health, by restoring it if possible, and by caring for the patients when cure is not or no longer possible, supporting them and easing their suffering, health-care institutions, services and measures have a major impact on our well-being. They determine 'the *level* and *distribution* of the risk of our getting sick, the likelihood of our being cured and the degree to which others will help us when we become impaired or dysfunctional'.[1] As such, health care involves issues of *social justice*. Finally, we have a fairly clear idea about what justice in health care entails. We are generally convinced that it has something to do with *fair treatment* in the light of what is owed to persons, i.e. with *equality* of access, based on medical need.

However, it is also commonly known that contemporary health care has become very sophisticated and expensive and that it has reached quantitative and

[1] N. Daniels, *Just Health Care*, Cambridge: Cambridge University Press, 1985, p. 2. A revised edition, *Just Health*, is forthcoming.

qualitative limits. Since the 1990s, issues of scarcity, priority setting and rationing lie at the centre of most current debates on health care. These are pressing issues: one way or another, limits have to be set. As such, the question of what is involved in just health care becomes much more complex. This complexity can be represented as an incompatible triad, a set of three propositions of which any two are compatible, but which together form a contradiction.[2] A classic illustration of the incompatible triad is the sign on the garage forecourt:

> We provide three kinds of services – cheap, quick, and reliable. You can have any two, but you can't have all three. If it's cheap and quick, it won't be reliable. If it's cheap and reliable, it won't be quick. And if it's quick and reliable, it won't be cheap.

In the case of health care, the three rival values are: (1) economic efficiency; (2) justice; and (3) decent-quality care. Here, it also seems to be that we can have any two but not all three. If we want to provide comprehensive and qualitative care that is economically efficient, i.e. that promotes the public interest in a cost-effective way, it is likely that we enter into a health-care system that offers such care only to those who can afford it. This raises objections based in considerations of justice. If, on the contrary, we want to provide decent-quality care on the basis of people's need rather than their ability to pay, it might turn out that the system is not efficient. And if we want to provide care equally to all those who need it, while preserving economic efficiency, the package might be very limited and of low quality.[3] Essentially, the central question is the following: how best to square the proverbial welfare circle.[4] How can resources be matched to needs, or needs to resources in socially acceptable and economically feasible ways?

This inspired me to put forward the following research question: how can *health care* be incorporated into a *theory of justice*, while realising an acceptable *balance* between efficiency, justice and care?

The first step in this research project is obviously to investigate the various elements that *constitute* the problem. This is the aim of Part I, *Just Health Care: Presuppositions and Objectives*, which consists of two chapters. Chapter 1 treats of the core issues of just health care and provides a mainly descriptive analysis of health care, as we know it. It yields an overview of the typical characteristics and

---

[2] Cf. A. Weale, 'Rationing Health Care', in *British Medical Journal* 316(1998)7129, p. 410; J. Butler, *The Ethics of Health Care Rationing. Principles and Practices*, London: Cassell, 1999, pp. 1–4.

[3] John Butler makes an interesting comparison. He holds that it is possible to see in broad terms where different countries have made the trade-off: 'Health-care systems that are organized on market lines (of which the private sector in the USA is the exemplar) offer comprehensive and high-quality care, but only to those who can afford it. By contrast, publicly funded health care in the USA, most famously through the Medicaid programme in the State of Oregon, offers high-quality care on the basis of need, but the package is limited. And examples of the third kind of trade-off are found in the systems that reflect the command and control economies of the former Soviet bloc, which offer comprehensive care on the basis of people's need, but the quality is patchy'. In: J. Butler, *op. cit.*, 1999, p. 2.

[4] V. George; S.M. Miller, *Social Policy Towards 2000: Squaring the Welfare Circle*, London: Routledge, 1994.

problems of contemporary health care, which basically revolve around one issue, i.e. the relation between the private and the public domain. In health care, this relation is expressed by the complex combination of our private and personal concerns for our health, with our public concern for the just social order and our fair treatment in it. In being mainly descriptive, Chapter 1 does not provide a normative framework within which the following questions can be analysed: 'What is the most adequate way to deal with these problems?' i.e. 'What *should* we, and what *can* we reasonably expect from the health care system?' These questions initiate Chapter 2.

In Chapter 2, I will concentrate on the various ways in which scarcity in health care can be expressed and interpreted (as a natural, ontological condition, as a social and anthropological mechanism and as an economic translation of finitude); and on the typical relation between preferences and needs (how can we distinguish basic non-volitional health care needs from needs that result from particular wishes and desires?). I will concentrate on the typical modern attitude towards the promises of contemporary medical technology, and on the specific character of authentic care (does not this *in se* presuppose a logic of abundance instead of scarcity?). Finally, I will focus on the following question: how should we think about scarcity in health care if we want to realise the goals of efficiency, justice and care in a well-balanced way? In analysing these various normative questions regarding just health care, I will argue in favour of a contemporary revaluation of the classical Socratic perspective.

In Part II, *Distributive Justice and Health Care*, I will assess whether and to what extent we can *incorporate* these various issues regarding just health care into a theory of justice. In this part, I will provide a detailed analysis of the function and significance of just health care within three contemporary theories of justice: (1) the Rawlsian theory of Norman Daniels (Chapter 3); (2) the capabilities approach of Martha Nussbaum (Chapter 4); and (3) the resource-egalitarian proposal of Ronald Dworkin (Chapter 5). The complexity of the relation between efficiency, justice and care will intersect these chapters. Put generally, it will appear that the Rawlsian approach of Norman Daniels incorporates the idea of *efficiency* in his conception of just health care, but cannot take comprehensive *care* into account. Its contractualist basis excludes care for the irreversible and long-term dependent from the domain of justice. In reaction to this, the capabilities approach of Martha Nussbaum reconciles both *justice* and long-term *care*. It is a rich account that concentrates on the terms that determine human flourishing. However, references to economic *efficiency* are mostly absent. Finally, Ronald Dworkin's proposal seems to offer an interesting solution to our problem of combining efficiency, justice and care, by giving personal responsibility a more central role. Recent literature, however, has put forward that overemphasising the role of personal responsibility not only meets difficulties in practical applicability, but also clashes with a consistent understanding of what we owe to each other at the bar of justice. Consequently, the problem returns: how can we set limits on health care while realising an acceptable balance between efficiency, justice and care?

In the final section on *Health Care and the Limits of Human Existence*, I will take general stock by formulating an answer to the following question: 'What are the desiderata for principles of just health care?' Here, I will bring the essential questions and positions together into a particular normative framework that contains the central elements of just health care. Following the line of reasoning of Nussbaum, I will think of these elements as *substantive benchmarks* that will have the requisite critical force to guide and direct social policy. They will provide a critical framework in the light of which various proposals on setting limits on health care and on issues of priority setting will be justifiable or not. Subsequently, I will specify a *general moral perspective* in the light of which these desiderata for principles of just health care are best understood.

However, I must be cautious and avoid raising wrong expectations. Although the central research question in this book may suggest otherwise, I do not aim to provide *cut-and-dried, practical* answers to solve the main contemporary problems in health care. On the contrary, in this work, I consider it as my philosophical task to *analyse* the questions and problems regarding just health care; to develop and critically assess, explain and justify or modify the various categories, distinctions, arguments and positions in this regard. As such, the arguments and positions developed here will not directly answer the most immediate and pressing questions about how to improve contemporary health-care systems. Nevertheless, they should help explain *why* these questions arise, *how* they are connected, and *what* in general should have to change for satisfactory answers to be given. Subsequently, the result will be a normative framework of *benchmarks* that serves as a critical *perspective* within which the search for an acceptable balance between efficiency, justice and care *continues*.

PART I

JUST HEALTH CARE: PRESUPPOSITIONS AND OBJECTIVES

INTRODUCTION TO PART I

How best to square the proverbial welfare circle? How to match resources to needs, or needs to resources in socially acceptable and economically feasible ways? In Part I, *Just Health Care: Presuppositions and Objectives*, I will investigate various elements that constitute the problem of reconciling efficiency, justice and care in health care.

Chapter 1 treats the core issues of just health care. It is a mainly descriptive analysis of health care as we know it. Section 1.1 begins with a short, historical outlook on the way in which caring for health has been organised throughout history. In Section 1.2 analyse the various levels of organisation, actors, institutions and decisions that characterise contemporary health care. Section 1.3 comprises an abstract analysis of the questions of distributive justice. Finally, in Section 1.4, I will end this chapter by mapping various existing objectives and paradigms of health care.

The analysis in Chapter 1 yields an overview of the main characteristics and problems of contemporary health care. However, it does not provide a normative framework within which we could tackle the questions: What is the most adequate way to deal with these problems? What *should* we, and what *can* we reasonably expect from the health care system? These questions initiate Chapter 2.

In Chapter 2, *Scarcity, Finitude and the Normative Value of Health*, I will concentrate on the various ways in which scarcity in health care can be expressed and interpreted. Firstly, it is reasonable to set a distinction between the *factum* of scarcity, and the *fatum*, i.e. the way we deal or cope with the inevitable. Section 2.1 concentrates on the common interpretation of scarcity as a contemporary *factum*. The following sections, which concentrate on the way in which we *deal* with the *factum* of scarcity, lend specific structure to Chapter 2, which is analogously determined according to the incompatible *triad* of efficiency, justice and care. Section 2.2 – the '*efficiency* section' – is organised alongside the lines of the economic interpretation of ontological scarcity, which focuses mainly on combining efficiency, effectiveness and equity. Section 2.3 – which would be referred as 'the *justice* section' – concentrates on scarcity as a circumstance of justice and analyses the various interpretations of equity in health care. Finally, Section 2.4 – the '*care* section' – concentrates on the viewpoint of interpreting scarcity as an economic translation of *finitude*. This viewpoint is, I believe, more apt to incorporate *care* in the discussion. In order to grasp a truly adequate conception of just health care, we need efficiency, justice and care to function as a trinity. This might happen more easily by a positive valuation of finitude. Here, we will meet a strange but interesting paradox: it seems that the logic of *abundance* that is implied by the concept of *care* enables us better to deal with the logic of *scarcity* because it motivates us to cope with *finitude* in a fully different way.

Having all this in hand, we will consequently move over to Part II, *Distributive Justice and Health Care*, where it assesses whether and to what extent we can *incorporate* these various issues of just health care, its presuppositions and objectives into a theory of justice. First things first, however. To begin with, let us have a close look at the various characteristics of health care, starting with an historical outlook.

CHAPTER 1

JUST HEALTH CARE: CORE ISSUES

Contemporary health care is a complex and heterogeneous framework of institutions, services and policy measures, roughly organised in accordance with three goals: prevention, cure and care. In general, public health measures contain the category of services that promote the collective health status by means of *prevention* of disease and disability. Besides prevention, health care includes the familiar personal medical services of *cure* and the social support services of *care* for the chronically ill or disabled. At first sight, these three pillars of health care are hierarchically structured: prevention is preferred to cure; and cure is preferred to care for those whose health is permanently impaired.

By preserving health, by restoring it when possible and by caring for the patients when cure is not or no longer possible, supporting them and easing their suffering, health-care institutions, services and measures have a major impact on our wellbeing. That is, they determine 'the *level* and *distribution* of the risk of our getting sick, the likelihood of our being cured and the degree to which others will help us when we become impaired or dysfunctional'.[1] As such, health care involves issues of *social justice*. Indeed, health care includes a poignant combination of our intimate, *individual* and personal concerns for our mental and physical health, with our *public* passion, our concern for the just social order and our fair treatment in it.

To get a clear view on the matter, it is interesting (1) to have a look at the history and development of health care; (2) to map the various organisational levels, actors, institutions and types of decision that characterise contemporary health care; (3) to inquire into the circumstances and characteristics of distributive justice and (4) to map the different objectives and paradigms that characterise *all* contemporary health-care systems, however much they each may differ in their various concrete forms.

1.1 PUBLIC AND PRIVATE: A HISTORICAL OUTLOOK

1.1.1 Antiquity: A Cultivated Mind in a Disciplined Body

'Your health is your wealth', declares the well-known saying. Being attentive to health and taking care of it is a very old concern. The first written expressions of this general human concern can be found in antiquity.[2] Egypt papyri of the

---

[1] N. Daniels, *Just Health Care*, Cambridge: Cambridge University Press, 1985, p. 2. A revised edition, *Just Health*, is forthcoming.

[2] R. Porter, *The Greatest Benefit to Mankind: a Medical History of Humanity*, New York: Norton, 1999.

second millennium BC contain healing rules and health regulations, encoding far older oral traditions. Mesopotamian clay tablets that date from the seventh century BC include medical instructions, diagnoses and remedies that go back a long way in oral tradition. Throughout Greek civilisation, as with the Roman later, the ideal of a cultivated mind in a disciplined body gave rise to instructions in exercise, bathing, massage, gymnastics and diet. Greek philosophers like Plato (*c*.427–347 BC) and the Stoics (between the fourth century BC and second century AD) implicated the general health concern in wider debates about human nature and the status of the body. According to these philosophers, the attainment of true health depended on temperance and wisdom, or proper self-control. Achieved through moderation in eating, drinking, sex and exercise, bodily health became the template for *sôphrosynê*, or soundness of mind. During the same period, the texts of Hippocrates of Cos (460–377 BC) announce a period of appeal to reason and rationality in understanding the workings of the body, together with concern for ethical excellence in practising the medical profession. The historical starting point of medical ethics, the Hippocratic Oath, which dates between the fifth and third centuries BC, still serves as the Magna Carta of contemporary professional ethics in medicine.

During its long history, the general health, medical and related ethical concerns used to be atomised, being mainly a matter of the individual's *private* concerns and – in cases of illness – his face-to-face relationship with the physician, or with other health-care providers, who patch up the sick individual. Although the classical urban civilisations evolved relatively sophisticated *public* health measures, implying a strong movement towards public organisation of health care, the decline of the Western Roman Empire meant a return to a rural society and it was not until the rise of towns and cities in the medieval period that public organisation of health care slowly became reinstated.[3]

1.1.2 Towards Institutionalisation of Health Care

The late Renaissance and early modern period witness two elements that helped pave the way for institutionalisation of public health.[4] Firstly, *mercantilism* counted population as a source of a nation's wealth. Fear of depopulation stimulated measures to force down high morbidity. A second element was the development of *political arithmetic*. Morbidity and mortality statistics are basic to understanding the health of a population and determining health policy. Both

[3] In the second millennium BC, the Minoans developed elaborate plumbing systems that included flush toilets. The great Roman aqueducts that were built between 312 BC and about 100 AC, sections of which still survive, are familiar to all; but what is not so well known is that the Roman water systems, at least the one for Rome, differentiated between water for common use and for drinking. See: J. Duffy, 'Public Health: History of Public Health', in W.T. Reich (ed.), *Encyclopedia of Bioethics*, 2nd edn., New York: Simon & Schuster/Macmillan, 1995, pp. 2153–2156.

[4] J. Duffy, *op. cit.*, 1995, p. 2157.

elements reflect a utilitarian concern for the common good, expressed in the form of 'statistical lives' rather than 'identifiable lives'.[5]

In the eighteenth and nineteenth centuries of industrialisation, urban and commercial explosion, and immense population growth – three factors that had devastating health effects and provoked a rain of epidemics – one acquired an enlarged social awareness, confronting the interplay of sickness, medicine and society.[6] It became clear that the living conditions are major determinants for the general health status of a population. Accordingly, the stage was set for systematic attention to the distant social determinants of disease and health such as climate, food and wages, working conditions, education and housing, poverty and public hygiene and the presence of hospitals, making medical services available to people. High tide modernity and its philosophy of progress, declaring that reason, science and technology would enhance man's control over nature and social progress, and that prosperity and the conquest of disease would follow, stimulated attention for the public duty and possibility to promote the people's health. Future medical progress, it was believed, supported by the civilisation process would extend longevity.

1.1.3 Politics of Health in the Nineteenth Century

The nineteenth century opened on a diversity of approaches to the politics of health, all dealing differently with the tension between individual and society.[7] German-speaking Europe embraced health paternalism in the form of the medical police, grounded in the view that the citizen has the duty to be healthy. The French revolutionaries argued that the new state would inaugurate a reign in which health and hygiene were integral, thus decreeing that every *citoyen* had a right to health, as well as to life, liberty and property. Free-market England and the USA supported the idea that the duties a virtuous physician owed to the patient ought to be extended to the public, while at the same time leaning towards voluntarism,

[5] D.E. Beauchamp, 'Public Health: Philosophy of Public Health', in W.T. Reich (ed.), *Encyclopedia of Bioethics*, 1995, pp. 2161–2165; T.L. Beauchamp; J.F. Childress, *Principles of Biomedical Ethics*, 5th edn., Oxford: Oxford University Press, 2001, p. 252. A classic article on this now commonplace distinction is T.C. Schelling, 'The Live You Save May Be Your Own', in S.B. Chase (ed.), *Problems in Public Expenditure Analysis*, Washington: Brookings Institution, 1966, pp. 127–176.

[6] L. Breslow, 'Public Health: Determinants of Public Health', in W.T. Reich (ed.), *Encyclopedia of Bioethics*, 1995, pp. 2153–2156; J. Duffy, 'Public Health: History of Public Health', in *ibid.*, pp. 2157–2161; R. Porter, *The Greatest Benefit to Mankind*, 1999, pp. 245–303, 397–427; The World Health Report 2000, *Health Systems: Improving Performance*, Geneva: World Health Organization, 2000, pp. 11–13.

[7] R. Porter, *The Greatest Benefit to Mankind*, 1999, pp. 405 ff.; L.B. McCullough, 'Justice and Health Care: Historical Perspectives and Precedents', in E.E. Shelp (ed.), *Justice and Health Care*, Dordrecht: Reidel, 1981, pp. 51–71. Philosophical analyses on the tension between individual freedom and public-health measures is offered in: D.E. Beauchamp, *The Health of the Republic: Epidemics, Medicine, Medicine and Moralism as Challenges to Democracy*, Philadelphia: Temple University press, 1988; G. Dworkin, 'Paternalism' in *Monist* 56(1972)1, pp. 64–84; R.M. Dworkin, *Taking Rights Seriously*, Cambridge, MA: Harvard University Press, 1977; J. Feinberg, *Social Philosophy*, Englewood Cliffs NJ: Prentice-Hall, 1973.

looking at charity to bestow health care on the poor, and leaving it to local initiative to install pumps, provide soup kitchens and remove nuisances.

1.1.4 Contemporary Health Care: A Complex Framework

During the twentieth century, medicine became more and more integral to the social and political apparatus of industrialised societies. By now, most industrialised countries have become welfare states with complex networks of various social services, facilities and institutions, acting and interacting at different levels, constituting a proverbial Leviathan that plays a central role in the ordering of society. The personal health concerns and medical relations are embedded in a complex framework of institutions; services and policy measures, all lumped together under the general term 'health care'. Today, health care is a broad and heterogeneous notion, including not only the most visible and familiar personal medical services, but also the less visible range of health-related services such as preventive medical and public health measures, including health and safety regulation, as well as certain social support services for the chronically ill or disabled. Health ethics today entails concern for issues of social justice, going far beyond the mere individual and private level, involving various levels of organisation and finance, different actors and various institutions.

1.2 LEVELS, ACTORS, INSTITUTIONS AND DECISIONS

1.2.1 Four Levels

Four different levels of organisation and finance of health care can be distinguished.[8] The *comprehensive social level* is the level of the decent minimum and regulates the allocation of social goods, including housing, education, culture, safety, defence; some of them being health-related, others not. Health is not our only goal or value, and expenditures for other goods inevitably compete for limited resources with health-targeted expenditures.[9]

The level of allocation within the health budget determines health policies and programmes for occupational safety, environmental protection, sanitation, injury prevention, consumer protection and food and drug control. Society protects and promotes health in many ways besides the provision of medical care. The term 'health resources', then, is not a substitute for 'medical resources', and the budget for health vastly exceeds the specific portion for health care.

Once society has determined its budget for health care, it still must decide at *the level of allocation within the health-care budget* by selecting certain projects and procedures for funding, e.g. whether priority should go to prevention or treatment. In many cases, preventive care – like polio vaccination, prenatal care and preventive dentistry – is more effective and more efficient in saving lives, reducing

---

[8] T.L. Beauchamp; J.F. Childress, *Principles of Biomedical Ethics* 2001, pp. 250–235; Also: J.F. Childress, 'Priorities in the Allocation of Health Care Resources', in E.E. Shelp (ed.), *Justice and Health Care*, 1981, pp. 139–150.

[9] Cf. infra on the external aspect of scarcity in Section 2.1.1.1 *The Twofold Dynamics of Scarcity.*

suffering, raising levels of health, and lowering costs. However, one-sided concentration on prevention can lead to neglect of needy persons who would benefit directly from critical care.[10] Most societies are inclined to favour 'identifiable' persons and to allocate resources for critical care, even though evidence exists that preventive care is more effective and efficient in saving 'statistical lives'. Accordingly, our moral intuitions often drive us in two conflicting directions: should we allocate more to rescue persons in medical need or should we allocate more to prevent people from falling into such a need?

Determining which categories of injury, illness or disease (if any) should receive a priority ranking in the allocation of health-care resources is another example of decisions at this level of allocation. For example, should treatment of heart disease have priority over cancer research? In trying to determine such priority questions, policymakers have to examine various diseases in terms of factors such as their communicability, frequency, cost, associated pain and suffering and impact on length of life and quality of life. From efficiency perspective, it might be justified to concentrate less on killer diseases, such as some forms of cancer, and more on widespread disabling diseases, such as arthritis.

Fourthly, there is *the level of allocation of scarce treatments or goods to patients*. This comprises not only the most visible and familiar system for the delivering and financing of personal medical services – like acute care and some individually focused preventive techniques – but also the less visible and less glamorous health-care services, responsible for social support and personal care for the mentally and physically disabled or the chronically ill. Not everyone who needs a particular form of health care, such as medicine, therapies, operations, transplantations and even space in intensive care units (ICUs), can gain access to it. These allocative decisions are more difficult when an illness is life-threatening and the scarce resource potentially life-saving. The question can become 'Who shall live when not everyone can live?' Here we meet the tension between individual and society in a form that has not so much to do with the classic freedom–paternalism problem. Rather, the difficulty lies in the seeming inappropriateness of abstract allocative principles on the level of face-to-face relationships. The general utilitarian concern of the system, which in the context of scarcity comes down to calculating and choosing between patients on the basis of abstract reasoning, collides with the doctor's Hippocratic duty of doing as much as possible for any patient.[11]

---

[10] This problem is analogous to what Amartya Sen has called the problem of distributive-aggregative tension in N. Daniels; B. Kennedy; I. Kawachi; A. Sen (eds.), *Is Inequality Bad for Our Health?* Boston: Beacon Press, 2000, pp. vii–xvii.

[11] See: J. Butler, *op. cit.*, 1999, p. 17; E.J. Cassell, 'Do Justice, Love Mercy: the Inappropriateness of the Concept of Justice Applied to Bedside Decisions', in E.E. Shelp (ed.), *Justice and Health Care*, 1981, pp. 75–82; P. Schotsmans, 'Is het ten laste nemen van de gezondheidszorg door de solidariteit begrensd? Een ethische verkenning', in J. Hallet; J. Hermesse; D. Sauer (eds.), *Solidariteit, gezondheid, ethiek*, Leuven: Garant, 1994, pp. 33–47.

1.2.2 Various Actors

Many different actors in various positions influence the selection of specific procedures or criteria to allocate health-care resources.[12] *Policymakers* shape the allocative principles, and thus the general character of the system, through their control over the resources used by institutions. At the core of the process we find the *allocative officers*, the individuals in the institution that is charged with the allocative task: the health-care providers, surrounded by managers, accountants and fund-raisers. At the same time as they are trying to handle pressure from above, institutions have to cope with claims from below, i.e. from the *patients*, asking for help and often expressing their claims in terms of 'justice' and 'fairness', an appeal that may or may not be adequate. Finally, latently present but nevertheless important, is the diffusive and pervasive force of *public opinion*, spontaneously focused on intermittently occurring scandals, often crystallised in the media. Sometimes, a public scandal about allocation patterns arises when one becomes aware that, despite wide publicity, sufficient funds cannot be raised for a person in need of an expensive medical procedure.[13]

1.2.3 Diverse Institutions

All this happens in an amalgam of diverse purpose-built institutions blessed or burdened with complex infrastructures, bureaucracies, funding arrangements and back-up facilities: government departments, research sites and lobbies, universities, hospitals doubling as medical schools, professional organisations, multinational pharmaceutical companies and insurance groups, institutions providing non-medical supporting care for the elderly and the handicapped.[14]

1.2.4 Two Types of Decision-Making

Finally, different-level decisions determine the character of the health-care system. Important is the difference between macro or social decision-making and micro or individual decision-making.[15]

---

[12] See also: J. Elster, *Local Justice: How Institutions Allocate Scarce Goods and Necessary Burdens*, Cambridge: Cambridge University Press, 1992, pp. 5 ff.

[13] As such, unique cases may accelerate public ethical reflection. See also: T.L. Beauchamp; J.F. Childress, *Principles of Biomedical Ethics*, 2001, pp. 255–259; J. Elster, *Local Justice*, 1992, pp. 5–6, 155–157, 178–183.

[14] R. Porter, *op. cit.*, 1999, pp. 629, 668.

[15] Or 'global' and 'local' justice, as referred to by Jon Elster, *op. cit.*, 1992, pp. 4–5. Next to the difference between macro and micro level, it is possible to distinguish a third level that is situated in between, viz. the meso-level. Whereas macro decisions are mainly taken by policy makers and micro decisions by health-care workers and their patients, meso-decisions are taken by people who work at the level of institutions (i.e. hospital managers, accountants, insurers and fundraisers). As this book mainly concentrates on characteristics of macro-level problems, I will not go into the specifics of meso-level decisions. From time to time, however, I will deal with micro-level problems, insofar as these may enlighten or aggravate macro issues. For more on the various levels of decisions, see: H.T. Engelhardt, 'Zielkonflikte in nationalen Gesundheitssystemen', in: H.M. Sass (ed.), *Ethik und öffentliches Gesundheitswesen: ordnungsethische und ordnungspolitische Einflußfaktoren im öffentlichen Gesundheitswesen*, Berlin: Springer, 1988, pp. 35–43. Referred to by L. Thielmann, *Ethische Grundlagen einer Prioritätensetzung im Gesundheitswesen*, Bayreuth: PCO Verlag, 2001, p. 60. More generally on first-order, second-order and third-order decisions, see: G. Calabresi; P. Bobbitt, *Tragic Choices*, 1978; J. Elster, *Local Justice*, 1992.

1.2.4.1 *Macro-Level Decisions*

Macro-level decisions shape the framework within which micro issues arise. They are, roughly speaking, characterised by three features:[16] (1) macro decisions are designed centrally, at the level of the national government; (2) they are intended to compensate people for various sorts of bad luck, resulting from the possession of morally arbitrary properties; (3) they typically take the form of transfers 'in cash'.

In sum, it amounts to what becomes available and how. Macro-allocation decisions determine the funds to be expected and the goods made available, as well as the methods of distribution. They determine what kinds of health-care services will exist in a society, who will get them and on what basis, how much equality there will be, how the burdens of financing them will be distributed and how the power and control over these services will be distributed.

With reference to the different levels of organisation and finance concerning health care, macro decisions are situated on the first three levels: the comprehensive social level, the level of allocation within the health budget, and the level of allocation within the health-care budget. On these levels, policymakers shape the general character of the health-care system by their control over the general resources to be allocated in society, thus determining the framework within which micro issues arise.

1.2.4.2 *Micro-Level Decisions*

The micro structure concerns moral problems of individual or small group decision-making. Micro distribution is designed by relatively autonomous groups or persons, which, although they may be constrained by guidelines laid down by the centre, have some autonomy to design and implement their preferred scheme. They are not compensatory, or only partially so. A scheme for allocating scarce medical resources may compensate people for bad medical luck, but not for other kinds of bad luck. Finally, micro decisions mainly concern allocation 'in kind', i.e. in kind of goods and burdens, not of money. Shortly, micro-allocation decisions determine which particular person will receive what specific services or goods. They are situated on the fourth level and are mainly taken by the health-care providers in consultation with the patient.

1.2.4.3 *Reflective Equilibrium*

It is interesting to note that since its beginning in the 1970s, medical ethics has tended to concentrate on micro issues, mostly the dramatic ones, situated on the cutting edge of life and death.[17] Should aggressive treatment be stopped

---

[16] J. Elster, *op. cit.*, 1992, p. 4.

[17] It lasted until the 1980s, until the works of Earl E. Shelp and Norman Daniels that theorists of justice coherently analysed the fundamental moral questions about the macro structure within which the micro issues arise and showed how general principles of justice can illuminate the particular problems we face in health care. See: A. Buchanan, 'Justice: a Philosophical Review', in E.E. Shelp (ed.) *Justice and Health Care*, Reidel, 1981, pp. 3–21; N. Daniels, *op. cit.*, 1985; O. O'Neill, 'Public Health or Clinical Ethics: Thinking Beyond Borders', in *Ethics and International Affairs* 16(2002)2, pp. 35–45; *Ibid.*, *Autonomy and trust in Bioethics*, Cambridge: Cambridge University Press, 2002, pp. 1–27.

for a terminal patient with metastasised bone cancer, or started for a hydrocephalic newborn with open spinal lesions? Which of several medically eligible patients should receive the life-saving liver transplant? Is the geriatric patient with Alzheimer's disease competent to agree to amputation of his foot and commitment to a nursing home? Should a physician tell one of his female patients that her fiancé, who is also one of his patients, is HIV positive?

The main reason for this is the concrete, interpersonal character and high level of visibility of these micro issues. In the words of Norman Daniels, these real dramas have become 'modern morality plays', with typical characters, dilemmas, dramatic choices, and weighing of interests, often performed and replayed in case studies, jurisdiction and the media, where they have a highly sensational character.[18] Although seemingly exotic, these dramas have a universal empathic appeal. It is easy to participate in them first hand, as a patient, a family member or physician. Moreover, these problems are forced on us by biology, technology and very general moral concerns. They might arise in any health-care system and seem to be part of our self-made human condition.

In this work, I will concentrate mainly on the *social* or *macro* level of decision-making. This might make it less exciting, since macro decisions are less visible and glamorous. They are the result of abstract reasoning on the level of the just society, regulating its basic social and public institutions. Although less dramatic, they have a greater impact on the general health status of its members. Macro decisions determine the *design* and *scope* of basic health-care institutions within which the micro problems arise.[19] For example, the decision of whether to treat a severely impaired patient aggressively or to allow him to die is possible because specialised medical technologies and ICUs now exist, which make it possible to prolong the lives of patients who previously would have died. These technological possibilities and ICUs exist because certain policy decisions have been made, which allocated certain social resources to the development of technology for sustaining severely impaired lives.[20] Since not only particular action but also policies and institutions may be just or unjust, serious theorising about just health care forces us to broaden the narrow focus of the micro approach by raising fundamental queries about the general background of social, economic and political institutions from which the micro problems emerge.[21]

However, it is best not to draw the line between micro and macro levels of decision-making too sharply since attention to individual cases can provide a

---

[18] N. Daniels, *op. cit.*, 1985, p. 1.

[19] A. Buchanan, *op. cit.*, 1981, pp. 3–21. N. Daniels, *op. cit.*, 1985, pp. 1–4.

[20] This example has been taken from A. Buchanan, 'Philosophic Perspectives on Access to Health Care: Distributive Justice in Health Care', in *The Mount Sinai Journal of Medicine* 64(1997)2, pp. 90–95.

[21] See also: Y. Denier; T. Meulenbergs, 'Health Care Needs and Distributive Justice. Philosophical Remarks on the Organisation of Health Care Systems', in R.K. Lie; P.T. Schotsmans; B. Hansen; T. Meulenbergs (eds.), *Healthy Thoughts. European Perspectives on Health Care Ethics*, Leuven: Peeters, 2002, pp. 265–297.

much-needed concrete focus for refining and assessing macro decisions. Public distress at making difficult choices at the micro level may sometimes induce society to modify its macro-allocational policies to increase supply of the resources concerned.[22] The adequacy or inadequacy of a moral theory cannot be determined by merely inspecting the principles that constitute it. Instead, rational assessment requires an ongoing process in which general principles are revised and refined through confrontation with the rich complexity of our considered judgements about particular cases, while our judgements about particular cases are gradually structured and modified by our provisional acceptance of general principles.[23] Accordingly, it will sometimes be the case throughout the text that micro-level issues appear as a critical force to judge the adequacy of various proposed abstract principles and frameworks.

1.2.5 Private Concern and Public Passion

Let us take stock. Taken together, it has become clear that today's health-care systems involve a diverse and constantly evolving set of various institutions with actors taking decisions at different levels of society; all these decisions together have a major impact on the level and distribution of our well-being, understood as the risk of our getting sick, the likelihood of our being cured and the degree to which others will help us and care for us when cure is not or no longer possible.[24] Because of this, issues of social justice are involved. Certain features, like medical need or ability to pay, will determine the distribution while others, like race or gender, will not do so. Inequalities in access to care, in utilisation of care and in risk of getting sick are then the direct results of certain distributive features in the basic structure and design of our health-care institutions. In order to be just, the distributive features determining the design and scope of our health-care system, the macro decisions that lead to them and the micro decisions that are taken within the system should all be in accordance with *acceptable* moral principles. If inequalities in health care are justifiable, it must be because *legitimate* and *valid* moral principles provide justification for them.[25] These are principles that could serve as the basis for public agreement.

By now, it has become clear that in health care there is a poignant combination of our *private and personal* concerns for our mental and physical health, with our *public passion*, our concern for the *just* social order and our *fair* treatment in it.[26] Let us take a closer look at what this public passion for justice entails.

---

[22] Cf. T.L. Beauchamp; J.F. Childress, *Principles of Biomedical Ethics*, 4th edn., New York: Oxford University Press, 1994, p. 364: 'Sometimes the public becomes alarmed by [society's] allocation patterns when it becomes aware, for example, that, despite widespread publicity, private funds cannot be raised for a leukemia victim who needs a bone marrow transplant'.

[23] Cf. Section 3.1.3 *Reflective Equilibrium*.

[24] N. Daniels, *op. cit.*, 1985, p. 2.

[25] T.L. Beauchamp, J.F. Childress, *op. cit.*, 2001, pp. 225–226; See also: O. Höffe, *Gerechtigkeit*, München: Beck, 2001; E. Tugendhat, *Vorlesungen über Ethik*, Frankfurt: Suhrkamp, 1993, pp. 364–391.

[26] N. Daniels, *op. cit.*, 1985, p. 2.

1.3 DISTRIBUTIVE JUSTICE: CIRCUMSTANCES, PRINCIPLES AND THEORIES

Although our passion for justice and concern for fair treatment is deep, the exact meaning of 'justice', 'equity' or 'fairness' often remains vague. This can be solved by a preliminary conceptual analysis, which enables us to understand the concepts more precisely and to determine which principles can be legitimately applied to which field. According to Brian Barry, a theory of justice should in principle answer three questions.[27] Firstly, what is justice? Secondly, why should one behave justly? And thirdly, how can we determine what justice demands?

1.3.1 What is Justice?

The concepts of 'fairness', 'equity', 'desert' and 'entitlement' have been used by various philosophers in attempts to explicate *justice*.[28] These accounts interpret justice as fair, equitable and appropriate treatment in the light of what is *due* or *owed* to persons. A situation of justice is present whenever persons are due benefits or burdens because of their particular properties or circumstances. One who has a valid claim based in justice has a right, and therefore is due something. An *injustice* involves a wrongful act or omission that denies people benefits to which they have a right, or fails to distribute burdens fairly. This common interpretation of justice can be traced back to the two most important points of orientation: Plato (*c*.427–347 BC) and Aristotle (*c*.384–322 BC) and their analyses of *suum cuique tribuere*. Let us take a closer look at their views on justice.

1.3.1.1 Suum Cuique Tribuere

PLATO: ESTABLISH ORDER IN THE SOUL AND IN THE CITY. In *The Republic* Plato starts his investigation of 'What is justice?' with a paraphrase of the poet Simonides: justice means 'to proshekon hekasto apodidonai'. The Roman jurist Ulpianus (*c*.170–228 AD) couched this in the formula 'justitia est constans et perpetua voluntas jus suum cuique tribuendi' – justice consists in rendering to every man his due.[29] According to

---

[27] B. Barry, *Theories of Justice*, Berkeley: University of California Press, 1989.

[28] See: A. Buchanan, 'Justice: a Philosophical Review', in E.E. Shelp (ed.), *Justice and Health Care*, 1981, pp. 3–21; M. Golding, 'Justice and Rights: a Study in Relationship', in the same volume, pp. 23–35; R.C. Solomon, M.C. Murphy (eds.), *What is Justice? Classic and Contemporary Readings*, 2nd edn., New York: Oxford University Press, 2000; E. Tugendhat, *Vorlesungen über Ethik*, Frankfurt am Main: Suhrkamp, 1993. For a specific emphasis on *entitlement*, see: R. Nozick, *Anarchy, State, and Utopia*, Oxford: Blackwell, 1974. John Rawls stresses justice as *fairness* in J. Rawls, *A Theory of Justice*, Oxford: Oxford University Press, 1971 and Brian Barry interprets justice as *impartiality* in his *Justice as Impartiality*, Oxford: Clarendon Press, 1995. For attention to justice as *desert*, see: A. MacIntyre, *Whose Justice, Which Rationality*, Notre Dame: University of Notre Dame Press, 1988.

[29] See: Plato, *The Republic*, trans. T. Griffith; ed. G.R.F. Ferrari.), Cambridge: Cambridge University Press, 2000, I, 331(e); Ulpianus, Dig. 1,1,10, in *The Digest of Justinian*, Vol. I, trans. A. Watson, eds. T. Mommsen; P. Krueger, Philadelphia: University of Pennsylvania Press, 1985, pp. 1–3. 'The Ancient Formula' is explicitly stated by W.K. Frankena in his article 'The Concept of Social Justice' in R.B. Brandt, *Social Justice*, Englewood Cliffs: Prentice-Hall, 1962, pp. 1–29. For other comments on the 'ancient formula', see: O. Höffe, *Gerechtigkeit: eine Philosophische Einführung*, München: Beck, 2001, p. 20–23, 49–53; E. Tugendhat, *Vorlesungen über Ethik*, Frankfurt am Main: Suhrkamp, 1993, pp. 364–391; G. Vlastos, 'Justice and Equality' in J. Waldron, *Theories of Rights*, New York: Oxford University Press, 1984, pp. 41–76.

Plato, this means that every part of the soul and every class in the city must *perform* its *due*, i.e. its *proper task*.[30]

This relates to his view that the social hierarchy in the city responds to the hierarchy of the soul and that the justice principle answers for the internal order of both. In the soul, Plato distinguishes among three basic powers – desire, spirit and reason.[31] These relate to the virtues of temperance, fortitude and prudence. In order that every power should fulfil its proper function and the right order would be established in the soul, a general order principle is necessary. This is performed by the harmonising virtue of justice. Together, these virtues are known as the four cardinal virtues: temperance, fortitude, prudence and justice. Because of its harmonising function, justice counts as the chief of all virtues. The just social hierarchy within the *polis* is derived from the harmony of the soul. The soul forms a mirror for the structure in society. According to Plato, three courses of action – answering the three powers of the soul – are present in all human beings: (1) commercial instinct; (2) spiritedness; and (3) love of learning.[32] In the just city every man fulfils the function that complies with his predominating talent. This establishes the class of the businessmen (e.g. craftsmen, farmers and merchants), soldiers and finally, the rulers, who are the philosopher kings. In the just society, everyone performs his own function. It is well known that Plato's interpretation of Simonides' account is embedded in a kind of philosophical psychology. Justice has to do with *being* in harmony and *performing* one's due, i.e. one's proper task. As such, *suum cuique* refers to 'being' and 'order', rather than to 'rendering' and 'giving'.

ARISTOTLE: SUBSTANTIVE DISTINCTIONS AND THE VARIOUS TYPES OF JUSTICE. When Aristotle in Book V of *The Nicomachean Ethics* comes to grips with the theme of justice, he starts with setting the difference between justice in the universal and justice in the particular sense.

UNIVERSAL JUSTICE. Universal justice or legal justice is synonymous with virtuousness and is related to the moral quality of the person and to lawful conduct.[33] Even more, justice in the universal sense is the *perfect* virtue, and 'neither evening star, nor morning star is such a wonder'.[34] Justice is the chief of all virtues because it combines all virtues as far as they are important to other people and to the happiness of the city.[35] The just man is honest, modest, courageous, moderate and wise. Justice or *dikaiosyne* is the attitude to freely fulfil the demands of law and morality. It is that moral disposition 'which disposes people to do just

---

[30] Plato, *The Republic*, IV, 434(a–c), 441(d–e). See also: O. Höffe, *op. cit.*, 2001, pp. 20–22, 52–53.

[31] In the text called: 'the desiring element', 'the spirited element' and 'the rational element' in Plato, *op. cit.*, 2000, IV, 439(d), 440(e)–441(a).

[32] Plato, *op. cit.*, 2000, IV, 435(e)–436(a).

[33] *Ibid.*, V, 1129(a). See also: B. Williams, 'Justice as a Virtue', in A.O. Rorty (ed.), *Essays on Aristotle's Ethics*, Berkeley: University of California Press, 1980, pp. 189–199.

[34] *Ibid.*, V, 1129(b). Aristotle is here thought to quote from the lost play *Melanippe* of Euripides.

[35] *Ibid.*, V, 1129(b).

actions, act justly and wish for what is just'.[36] Here, Aristotle resumes the Platonian idea of justice as the perfect virtue.

PARTICULAR JUSTICE. However, Aristotle finds it necessary to explain that, though the theme is justice, he will not be discussing justice in the universal sense, but will mainly concentrate on the various particular forms of justice.[37] This means justice as related to fairness and equality, concerning the way things have to be distributed, or corrected among members of the political community. For the unjust man is not only *paranomos*, the one who acts against the law but also *anisos*, the unfair who takes too much benefits and too little burdens.[38] With this characterisation, Aristotle moves from the domain of *being*, i.e. of the moral qualities of men, to the domain of *having* and just proportion of property. Instead of justice as every part *doing* his own, justice is now a matter of every part *having* his own. The reflection on *dikaiosyne*, justice as virtuousness, makes way for a reflection on *to dikaion*, the just in specific interpersonal relationships.[39] In the main part of Book V, Aristotle expounds the various particular types of justice that are up until now substantially relevant. The concepts we use today are mainly indebted to Aristotle's framework.

Particular justice is a part of universal justice and concerns all matters of honour, money or self-preservation, i.e. matters that are always threatened by the danger of transgressing limits and insatiability (*pleonexia*). The essence of what is going on here is to make sure that everyone is rendered his share of external goods. This is a matter of citizens related to each other, and even more of the government and the court of justice. Here we meet, more than in Plato's interpretation, the essence of *suum cuique tribuere* – justice as *rendering* to every man his due. Particular justice needs firstly, to determine the criterion of distribution of divisible goods; secondly, to correct where private transactions have gone wrong; and thirdly, to regulate reciprocity in exchange of goods and services.

DISTRIBUTIVE JUSTICE. According to Aristotle, the distribution of social goods 'that have to be shared among members of the political community'[40] is just as far as it complies with the principle of equality in *geometrical proportion*. This means that justice demands to treat equals equally and unequals unequally. To simply render the same amount to everyone would be unjust, because people are different and real equality takes into account the relevant differences between people. However, this is a purely minimal and formal principle. For who is equal

---

[36] *Ibid.*, V, 1129(a). Justice as *dikaiosyne* means righteousness, the quality of acting rightly. This means justice, understood as complete virtue in the fullest sense. We will see that Aristotle will not concentrate on this, but will go on further studying justice as equality, as *to ison* or *isotes*.

[37] *Ibid.*, V, 1130(a).

[38] *Ibid.*, V, 1129(b).

[39] Cf. M. Villey, *Leçons d'histoire de la philosophie du droit*, 2nd édn, Paris: Dalloz, 2002, pp. 23–36; Also in: T. Vandevelde, 'Het intrestverbod bij Aristoteles', in L. Bouckaert (ed.), *Intrest en cultuur. Een ethiek van het geld*, Leuven: Acco, 1994, pp. 37–58.

[40] *Ibid.*, V, 1130(b).

and who is unequal? In itself it states no particular respects in which equals ought to be treated equally and provides no specific criteria for determining whether two or more individuals are in fact equals. It merely asserts that, whatever respects are under consideration as relevant, persons equal in those respects should be treated equally. That is, no person should be treated unequally, despite all differences with other persons, unless some difference between them is relevant to the treatment at stake.

Consequently, we encounter a *criterion* problem. We need a standard that defines the relevant respects under which people are equal or unequal and thus define what one's 'due' is. For this, we need to take into account the specific *qualities* of persons. A prevalent example that illustrates issues of distribution according to relevant criteria is the example of dividing a cake. When a cake has to be divided among a group of children, we could give each child an equal share. That would be a strict egalitarian or arithmetical division. However, there could be various reasons for an unequal division. One child could claim that he is exceedingly hungry. This is the so-called criterion of *need*. Another child could say that the mother previously promised him half of the cake – the argument of attained *rights*. The third one could bring forward that he did chores for the mother. This would be the criterion of *contribution*. The fourth one could claim that he deserves the biggest piece, because he is the first-born. That would be the criterion of *status*. According to Aristotle, the generally accepted criterion that determines in what way persons ought to be treated equally or unequally is the distributive principle of assignment by *merit* (*axia*). A just distribution is a distribution *kat'axian*. However, he leaves it open how exactly 'merit' should be understood, being aware of the fact that in practice, not all defend the same sort of merit:

> everyone agrees that justice in distribution must be in accordance with some kind of merit, but not everyone means the same by merit; democrats think that it is being a free citizen, oligarchs that it is wealth or noble birth, and aristocrats that it is virtue.[41]

Corrective Justice. Aristotle's second form of particular justice, corrective justice or justice in rectification restores inequalities in private transactions between persons. Here also, justice is equality, and injustice inequality. This time, however, equality does not involve geometrical but *arithmetical* proportion. The court does not consider the quality and merits of the person, related to the quality and merits of the other person, but only the numerical quantity of the goods or services that are exchanged. The focus lies solely on the nature of the damage. The parties are treated as equal legal entities, and the judge merely asks whether one has done and the other suffered injustice, whether one inflicted and the other sustained damage: 'For it makes no difference whether it is a good person who has defrauded a bad or a bad person a good'.[42] Instead of specifically looking at the relevant qualities

[41] *Ibid.*, V, 1131(a). Here, Aristotle leaves it open. In *Politica* II, he will defend the aristocratic view.

[42] *Ibid.*, V, 1132(a).

of persons, corrective justice looks away. No specific distinguishing characteristics of persons may be taken as important. The well-known figure of corrective justice is blind and her attributes are the scales, representing the idea of balanced judgement and of equal consideration. All are considered equal and only the nature of the case determines the compensation.[43]

Consequently, the judge impartially decides which party deserves what kind of compensation. Through this impartial determination, the judge restores equality. In a situation of unbalance, the greater has gained, and has more than one's own, while the other party has less than at the outset. Hence, the unjust being the unequal, the judge endeavours to equalise it by the penalty or loss he imposes, taking away the gain:

> What is equal is therefore a mean between the greater and the less, but the gain and the loss constitute the greater and the less in contrary ways: more good and less evil constitute gain, while the contraries constitute loss. And the mean between them, as we saw, is what is equal, which we say is just. So what is justice in rectification will be the mean between loss and gain.[44]

The mean is the *suum cuique* situation in which people have their own. In this case, it means that they have, after the transaction, an amount equal to the one they had before.

JUSTICE IN THE FORM OF RECIPROCITY. In Chapter 5 of Book V of the Nicomachean Ethics, Aristotle discusses reciprocity, a third form of justice 'that fits neither distributive nor rectificatory justice'.[45] Justice as reciprocity involves people associating with one another for purposes of exchange and contains the well-known passage on money as an instrument to make comparability of commodities possible. As such, money helps establish justice in exchange of goods and services.

EPIEIKEIA. A final point, again in connection with universal or legal justice, that deserves attention is the concept of equity and what is equitable.[46] Equity is the not so adequate translation of the Greek term *epieikeia*, which according to its original meaning can also be understood as mildness or indulgence. Equity is a specific sort of justice and is – Aristotle insists to this – even superior to legal justice. Its essence is that it is a rectification of legal justice, in cases where the law is defective. The reason is that all law is general. There are, however, some things

---

[43] The well-known figure of justice, blindfolded and with a balance in one hand, represents justice as impartiality. Impartiality means proportionate equality. In the case of corrective justice, a just decision is impartial when based, and only based upon the arithmetical proportion. It is based upon impartial consideration of the deed and the question whether one person concerned has taken more than his due, and the other, through this, has suffered damage. The ideal judge is then so to speak justice personified in being himself as a mean or medium between the litigants. In judging impartially, he (re-) installs the mean, i.e. which is just (NE V, 1132(a)). In the case of distributive justice, impartiality is related to the geometrical proportion. A just allocation is based on the sole consideration of the relevant differences between persons. See also: M. Villey, *op. cit.*, 2002, p. 28; O. Höffe, *op. cit.*, 2001, p. 11.

[44] *Ibid.*, V, 1132(a).

[45] Aristotle, *op. cit.*, 2000, V, 1132(b).

[46] Also in: O. Höffe, *op. cit.*, 2001, pp. 58–59.

in which one cannot speak correctly in terms of general rules. In such cases, the virtue of *epieikeia* helps us in considering the nature of the particular case itself. This requires judgemental power and insofar, it completes the virtue of *phronêsis*, which is the practical wisdom and the intellectual insight in individual cases, guided by practical power.[47] Equity is thus one sort of justice in particular cases, which requires situational flexibility that can correct for general rule-based legal justice. So when in a particular case, equity considerations advise an exception to the general law, then this correction is good and just, because it follows from the nature of the case. Moreover, it is even better than to follow the law: 'This will be by saying what the lawgiver would himself have said had he been present, and would have included within the law had he known.'[48]

ARISTOTLE'S LEGACY. What is the value of this detailed analysis on Plato's and Aristotle's views on justice? I believe it is of great value since it provides us of better insight into the current contemporary view that justice is about treatment in the light of what is due or owed to persons. In contemporary moral, political and social philosophy, one can discern four overlapping, though still distinguishable spheres of justice: political justice, justice in law and punishment, commutative justice and distributive or social justice.[49] It is easy to see how this division in justice for the most part can be traced back to the substantive distinctions made by Aristotle.

Political justice is concerned with the justification of political authority and political order and focuses on issues that concern the constitution and legislation of the state. Theories of *justice in law and punishment* are concerned with the justification of rules of law and of social norms, with following them and with the justification of punishment of those who break the law or disregard the generally accepted moral and social norms. *Commutative justice* bears on the relations between individuals, especially with respect to the private equitable exchange of goods and fulfilment of contractual obligations. Reflections on *distributive or social justice* focus on the distributional issues of society's resources, of common property. Distributive or social justice refers to fair, equitable and appropriate distribution of goods and burdens, determined by justified norms that structure the terms of social cooperation. Its scope includes policies that allot diverse benefits and burdens, such as rights and responsibilities, property, resources, taxation, privileges and opportunities. Issues of law and punishment, of distribution and of regulation of private economic relations presuppose a strong coordinating authority. To adjust unequal distribution, correct lawless behaviour and incorrect exchange, the presence of an authority that has the

---

[47] Aristotle, *op. cit.*, 2000, VI, 1143(a).

[48] *Ibid.*, V, 10, 1137b.

[49] See: S. Hellsten, 'Theories of Distributive Justice', in R. Chadwick (ed.), *Encyclopedia of Applied Ethics*, San Diego: Academic Press, 1998, Vol. 1, pp. 815–828; O. Höffe, *op. cit.*, 2001; D. Mieth, 'Ethik der Gerechtigkeit. Ansätze, Prinzipien, Kriterien', in D. Mieth; P. Magino (eds.), *Vision Gerechtigkeit? Konziliarer Prozeß; und Kirchliche Jugendarbeit*, Dusseldorf: Altenberg, 1992, pp. 12–32.

power to enforce and implement the corresponding decisions, is a necessary condition for *real* justice. That is why we can say that even if the different spheres of justice are often overlapping, in this division political justice covers the scope of the other three spheres.

1.3.1.2 The Circumstances of Justice

Another important and common idea that is related to justice is the idea, which refers to the *circumstances* in which issues of justice arise. It holds that the question of *justice* arises when there is a conflict of interest within particular circumstances, usually called the circumstances of justice. This expression goes back to David Hume's analysis of the origin of justice and property. The *locus classicus* is Hume's Treatise:

> 'tis only from the selfishness and confin'd generosity of men, along with the scanty provision nature has made for his wants, that justice derives its origin.[50]

Rawls, taking up Hume's observations, summarises the idea as follows:

> the circumstances of justice obtain whenever mutually disinterested persons put forward conflicting claims to the division of social advantages under conditions of moderate scarcity. Unless these existed, there would be no occasion for the virtue of justice.[51]

Regarding the issue of *distributive* or *social justice*, David Miller adds three premises to these circumstances.[52]

1.3.1.3 Bounded Society

Firstly, we have to assume a bounded society with a determinate membership, forming the universe of distribution, whose present fairness or unfairness we can meaningfully assess. The bounds of this social universe are usually taken for granted, assuming it concerns the borders of nation-states. John Rawls, for instance, says that his principles of justice are worked out to apply to a society conceived as a closed system: 'it is self-contained and has no relations to other societies. We enter it only by birth and exit only by death.'[53] His assumption is that 'the boundaries of these schemes are given by the notion of a self-contained national community'.[54]

1.3.1.4 Institutions

Connected to this first additional premise – that in speaking of social justice we tacitly or openly envisage a connected body of people who form the universe of distribution – is a second, i.e. the principles we advance must apply to an identifiable

[50] In: D. Hume, *A Treatise of Human Nature*, 2nd edn., ed. L.A. Selby-Bigge, rev. P.H. Nidditch, Oxford: Oxford University Press, 1978, p. 495.

[51] J. Rawls, *A Theory of Justice*, Oxford: Oxford University Press, 1971, p. 128; rev. edn., Cambridge, MA: The Belknap Press of Harvard University Press, 1999, p. 110.

[52] D. Miller, *Principles of Social Justice*, Cambridge, MA: Harvard University Press, 1999, esp. pp. 4–20.

[53] J. Rawls, *Political Liberalism*, New York: Colombia University Press, 1993, p. 68.

[54] J. Rawls, *Theory of Justice*, 1971, p. 457; 1999, p. 401.

set of institutions whose *impact* on the life chances of different individuals can also be traced. Once again, we can refer to Rawls's work to see this assumption made explicit. According to Rawls, the subject matter of social justice is the basic structure of society, understood as the major social institutions that

> distribute fundamental rights and duties and determine the division of advantages from social cooperation. By major institutions I understand the political constitution and the principal economic and social arrangements ... [that] taken together as one scheme ... define men's rights and duties and influence their life-prospects, what they can expect to be and how well they can hope to do.[55]

Rawls assumes that we can understand the basic structure well enough to regulate it by principles of justice.

1.3.1.5 Agency

The third additional premise follows naturally from the second that there is some *agency* capable of *changing* the institutional structure in more or less the way our favoured theory demands. It is no use setting out principles for reforming the basic structure if we have no means to implement these reforms.[56] The main agency here is the state: theories of social justice propose legislative and policy changes that a well-intentioned state is supposed to introduce. I do not mean to imply that the theories in question are exclusively addressed to legislators and other state officials. Very often, agreement and cooperation of citizens is needed to make the reforms work, so that we can say that the theory put forward is a *public* doctrine that ideally every member of the political community is supposed to embrace.[57]

Nevertheless, given that the theory is meant to regulate the basic structure, and given that the structure is a complex of institutions with its own internal dynamics, an agency with the *power* and *directing* capacity that the state is supposed to have is essential if a theory of justice is to be more than a utopian ideal.[58] Included in this idea of an agency being capable of changing the structure is that we can distinguish the unfair from the unfortunate. To describe a state of affairs as just or unjust, it must be a state of affairs, which has either resulted from the actions of societies or individuals, or is at least capable of being changed by such actions. Miller uses the example of rain. Although we generally regard rain as burdensome and sunshine as beneficial, a state of affairs in which half the country is drenched by rain while the other half is bathing in sunshine cannot be discussed in terms of justice, unless we believe that Divine intervention has

---

[55] *Ibid.*, 1971, p. 7; 1999, pp. 6–7.

[56] Cf. supra on the strong coordinating authority as a necessary condition for real justice. That is why we can say that even if the different spheres of justice are often overlapping, *political justice* covers the scope of the other three spheres. Cf. Section 1.3.1.1 *Suum Cuique Tribuere*.

[57] Agreement and consensus is an essential element in contemporary liberal theory. Cf. Part II.

[58] Cf. David Miller: 'There has to be a culture of social justice that not only permeates the major social institutions, but also constrains people's behavior even when they are not formally occupying an institutional role'. In D. Miller, *op. cit.*, 1999, p. 13.

caused this state of affairs, or that meteorologist could *alter* it. As long as a state of affairs is regarded simply as a product of natural causes, questions about its justice or injustice do not arise. This conclusion is important for my argument on just health care, for it implies that *not all health inequalities are per se unjust.*[59]

Let us take stock. Together, the three premises – bounded society, identifiable set of institutions and an agency with directing power – give a more refined definition of the circumstances of social or distributive justice:

> If we do not inhabit bounded societies, or if people's shares of goods and bads do not depend in ways we can understand on a determinate set of social institutions, or if there is no agency capable of regulating that basic structure, then we no longer live in a world in which the idea of social justice has any purchase.[60]

1.3.1.6 Formal and Material Principles of Justice

A theory of justice proposes principles that evaluate people's claims within the circumstances of justice. Eventually, this will lead to an impartial division of advantages and burdens acceptable for all. Because of the importance of impartiality, the virtue of justice is to be distinguished from sentiments of benevolence, charity, care and friendship, which all create warm but partial feelings. Justice, on the other hand is rather cold.[61]

No single principle can address all problems of justice. Relative to the *context* and the *distribuendum*, several principles of justice appear in common morality and merit acceptance. Common to all theories of justice is the principle of *formal* equality, traditionally attributed to Aristotle: equals must be treated equally, and unequals must be treated unequally.[62] This principle is formal because it states no particular respects in which equals ought to be treated equally and provides no criteria for determining whether two or more individuals are in fact equals. Therefore, it needs to be refined. In addition, *material* principles specify the relevant

---

[59] See also: H.T. Engelhardt, 'Health Care Allocations: Responses to the Unjust, the Unfortunate, and the Undesirable', in E.E. Shelp (ed.), *Justice and Health Care*, 1981, pp. 121–137. I will come back to this in detail in Section 3.3.2.2 *Which Health Inequalities Are Unjust?* and in Section 4.4.6 *The Tragic and the Unjust*.

[60] D. Miller, *op. cit.*, 1999, pp. 6, 17–20.

[61] Cf. Justice is 'a cautious, jealous virtue' in D. Hume, *Enquiries Concerning Human Understanding and Concerning the Principles of Morals*, 3rd edn., ed. L.A. Selby-Bigge, rev. P.H. Nidditch, Oxford: Clarendon Press, 1975, p. 184. In Aristotle, we find reference to the flip side of this idea, which holds that 'when men are friends, they have no need of justice'. In Aristotle, *The Nicomachean Ethics*, Book VIII, 1155a. In the same line of reasoning, the lady of corrective justice is blind, being impartial, objective and not personally concerned. This cold character of justice is one of the main points to which the critique of the early feminists and defenders of an ethics of care has been addressed. In Part II, I will come back to this discussion in somewhat more detail and I will defend the idea that justice is necessarily cold and must remain as such, if it wants to continue offering true equal protection to all individuals, impartially guaranteeing a safety net – in which care is included – to *all* citizens. A very illuminating article on this subject is: J. Waldron, 'When Justice Replaces Affection: The Need for Rights', in *id.*, *Liberal Rights*, Cambridge: Cambridge University Press, 1993, pp. 370–391.

[62] Cf. Section 1.3.1.1 *Suum Cuique Tribuere*.

characteristics for equal treatment by providing answers to the question: 'Equality of what?' Instances are the principle of distribution according to need, according to effort, to contribution, to merit, or to free-market exchanges, or otherwise, the principle that defends distribution of an equal share for each person.

Most societies invoke several of these material principles in different spheres and contexts. Many unemployment subsidies, welfare payments and health-care programmes are distributed on the basis of need; jobs and promotions in many sectors are awarded on the basis of demonstrated achievement and merit; the higher incomes of some professionals are allowed and often encouraged on grounds of free-market wage scales, superior effort, merit or potential social contribution; and, at least theoretically, the opportunity for elementary and secondary education is distributed equally to all citizens. In itself, abstract principles alone provide only rough guidelines for forming specific policies or concrete actions. We need further moral argument that justifies, specifies and balances principles and assesses competing claims in order to determine which concrete aspects of a situation are morally relevant and decisive in forming a reasoned judgement and which are morally arbitrary. In order to resolve possible conflicts among the above-listed principles, they need to be integrated into a general and coherent framework of principles, a theory of justice.

In Chapter 2, I will sketch various material principles that are applicable to health care by inquiring into possible answers to the question 'Equality of What?' Part II will provide detailed analysis of the task and meaning of just health care within three contemporary theories of justice: the Rawlsian theory of Norman Daniels, the capabilities approach of Martha Nussbaum and the resource-egalitarian proposal of Ronald Dworkin.

1.3.2 Why Behave Justly?

Theories of distributive justice differ, depending on their answer to the second and third question posed by Brain Barry, i.e. the question of the motivation to behave justly and the question of how to determine what justice demands. At the one extreme of the different motivations to behave justly, some theories assume the willingness to accept *impartial* arrangements even if this goes against one's own interests. At the other extreme, *self-interest* prevails over impartiality. Impartial arrangement is only acceptable as a form of reasonable self-interest. Of course we can imagine intermediate positions, in which both concerns are, to some extent, taken into account.[63]

---

[63] An interesting consequence of the distinction between reasonable self-interest and impartiality is that both may be logically connected respectively to the *insurance principle* and the *principle of solidarity*. Although I will not analyse these connections in detail, it is interesting to bear them in mind. See also: K. Bayertz, *Solidarity*, Dordrecht: Kluwer, 1999; T. Vandevelde, 'Ideologie en sociale zekerheid', in M. Despontin; M. Jegers (eds.), *De sociale zekerheid verzekerd*? Brussel: VUB Press, 1995, pp. 151–176; E. Schokkaert, 'Warm en koud. Solidariteit en verantwoordelijkheid in de ziekteverzekering', in *Ethische Perspectieven* 8(1998), pp. 135–146.

1.3.3 How to Determine What Justice Demands?

Concerning the question of how to determine what distributive justice demands, one may, on the one hand, focus on the *procedure*, or on the other hand, fully concentrate on *outcomes*. In the first case, the just is the outcome of a process, which is in itself procedurally acceptable. In the second case, principles of justice act as more direct constraints on acceptable outcomes of the political process. Again, most theories take both aspects into account.

1.3.4 The Just and The Good

In contemporary social and political philosophy, there is a broad liberal consensus that justice has to do with providing the general framework of conditions (e.g. rights, liberties and goods) that have to be fulfilled so that people consequently can live their life according to their *own* comprehensive conception of the good life. Therefore, it is reasonable to say that in an indirect but very important way personal well-being, the good of mankind and of society, together understood as 'the good life', is the general goal of justice; not with respect to development of the *content* of the good, but more importantly, with respect to providing the *framework* in which the good – whatever it is – can be realised.[64] In the same line of thought, we have to understand Paul Ricoeur's idea that I have quoted at the very beginning: 'Justice ... is an integral part of the wish to live well ... the wish for a fulfilled life for and with others in just institutions.'[65] Through this function of providing and organising the conditional framework of the good, the principles of justice greatly affect our personal lives, prospects and chances, 'what [we] can expect to be and how well [we] can hope to do'.[66]

The same can be concluded with regard to our problem of just health care. Various organisational frameworks are possible, depending on the respective weight attributed to the different objectives and organisational paradigms. Accordingly, these various frameworks differently affect the level and distribution of health and related well-being of our personal lives. Below I will illustrate this in somewhat more detail.

---

[64] Critique on the liberal consensus about the relation between 'the just' and 'the good' can be found in: A. MacIntyre, 'The Privatization of Good: an Inaugural Lecture', in *Review of Politics* 52(1990), pp. 344–361.

[65] P. Ricoeur, *The Just*, trans. D. Pellauer, Chicago: University of Chicago Press, 2000, p. xv; translation of 'La justice ... fait partie intégrante du souhait de vivre bien ... souhait d'une vie accomplie avec et pour les autres dans les institutions justes'. In P. Ricoeur, *Le Juste*, Paris: Le Seuil, 1995, p. 17.

[66] J. Rawls, *op. cit.*, 1971, p. 7; 1999, pp. 6–7. This idea is also expressed in D. Hume, *Enquiries Concerning Human Understanding and Concerning the Principles of Morals*, 3rd edn., ed. L.A. Selby-Bigge, rev. P.H. Nidditch, Oxford: Oxford University Press, 2000, p. 192; and in: Aristotle, *The Politics*, trans. B. Jowett, rev. trans. J. Barnes, ed. S. Everson, Cambridge: Cambridge University Press, 1996, VII. 1–2, where he discusses the idea that a political arrangement has as its task the securing to its people of the necessary conditions for a good human life. It is to create a context in which one may choose to function in the ways that are constitutive of a good human life. Analysed in M.C. Nussbaum, 'Nature, Function, and Capability: Aristotle on Political Distribution', in *Oxford Studies in Ancient Philosophy*, Suppl. Vol. I (1988), pp. 145–184.

Just health care			
Decent-quality care	Equal access	Freedom of choice	Economic efficiency
Public ↔ Private			

Figure 1. Just health care: objectives and organisational paradigms

1.4 FOUR OBJECTIVES, TWO PARADIGMS

Let us take stock by recapitulating the broad definition of health care understood as a complex and heterogeneous framework of institutions, services and policy measures, roughly organised in accordance with the goals of prevention, cure and care. By preserving health; restoring it if possible; and caring for the patients when cure is not or no longer possible; by supporting them and easing their suffering, health-care institutions, services and measures have a major impact on our well-being. That is, it determines 'the *level* and *distribution* of the risk of our getting sick, the likelihood of our being cured and the degree to which others will help us when we become impaired or dysfunctional'.[67]

How can this complex framework be realised, organised and, very important, be maintained? From a general and abstract perspective, it is possible to distinguish four objectives and two organisational paradigms. Accordingly, they enable us to understand and compare various health-care systems, as they exist in the real world.

1.4.1 Four Objectives: The Internal Health-Care Trade-Off

Four objectives are involved in just health care. Firstly, a good health-care system should provide *decent-quality care*. Secondly, a just health-care system should find an acceptable balance between equality and liberty. It should guarantee *equal access* to care while at the same time maintaining sufficient *freedom of choice* on the part of health-care provider and consumer. Finally, for the sake of its continued existence, it should be *economically efficient* and control the costs through cost-containment programmes.[68] In reality,

[67] N. Daniels, *Just Health Care*, 1985, p. 2.

[68] In this I follow T.L. Beauchamp; J.L. Childress, *op. cit.*, 2001, p. 231, where they distinguish four goals in health care: superior care, equality of access, freedom of choice and social efficiency. Since balancing freedom and equality is a basic problem of distributive justice, one could take these elements together under the justice goal. See: H. Pauer-Studer, *Autonom Leben. Reflexionen über Freiheit und Gleichheit*, Frankfurt a.M.: Suhrkamp, 2000. However, I strongly believe that *all* four elements are distinguishing aspects of the general justice goal in health care. Otherwise, the typical problems between efficiency and justice, and between justice and care are unavoidable, as I will demonstrate respectively in Chapter 2 of Part I and in Part II. As most economists believe that speaking of efficiency without considerations about justice is nonsensical, I shall argue that the same goes for care. As efficiency must function as a value *within* the framework of justice, care must also function as a value *within* the framework of justice. As such, justice functions, Aristotelian-like, as a harmonising umbrella value, keeping all the other values in health care in right balance. I will come back to this in detail in Section 4.2.1 *Other Virtues?* Finally, it is important to note that I do not believe that providing *superior* care can be considered as a general objective of just health care. The goal of *decent-quality* care is more applicable to the demands of justice.

scarcity of resources demands that choices and trade-offs are made between these goals. Social and political situations, social and cultural values and the predominant sense of justice determine emphasis on one goal, rather than another.

For reasons of clarity, I call this the *internal* health-care trade-off because it is situated on the third level, i.e. the level of allocation *within* the health-care budget.[69] As such, it has to be distinguished from the *external* health-care trade-off (in which the size of the health-care budget has to be weighed against other *social* goals (e.g. housing, education, culture, safety and defence – on level 1) and against other health goals (e.g. occupational safety, environmental protection, sanitation, injury prevention, consumer protection and food and drug control – on level 2).

1.4.2 The Private and the Public Paradigm

Furthermore, existing health-care systems can be characterised according to their position between the two earlier mentioned paradigmatic models of organisation and finance: the private and the public paradigm.[70] The *private paradigm* is inspired by libertarian theory and comes down to a pure free market system of health care. On this view, health care is seen as a private commodity to be traded on the market in exchange for other private goods. Redistribution is just only when the individuals freely and explicitly choose for a redistribution of goods through private insurance or through charity. Thus, the other objectives (i.e. economic efficiency, decent-quality care and equal access) are subordinated to the primary goal of individual freedom. Equality is minimalistically understood as equal negative freedom and equal political rights.[71] Good quality care is accessible for those who are voluntarily insured. Furthermore, it is believed that this system will be most efficient, for individuals will engage in cost containment by simply not purchasing too expensive health-care services when the cheaper alternative is just as well.[72]

In the *public paradigm*, the trade-off is decided in favour of egalitarianism in health care, which comes down to equal access for equal need. Here, everyone

---

[69] Cf. Section 1.2.1 *Four Levels*.

[70] See: The World Health Report 2000, *Health Systems: Improving Performance*, Geneva: World Health Organization, 2000; U.K. Hoffmeyer; T.R. McCarthy (eds.), *Financing Health Care*, 2 vols, Dordrecht: Kluwer, 1994; E. Jakubowski; R. Busse; G.R. Chambers, *Health Care Systems in the European Union: a Comparative Study* (European Parliament Working Paper), Luxembourg: European Parliament, 1998.

[71] Note that this involves equal *political* rights. Access to politics, suffrage and equality before the law, social class, religion and ethnicity should not be allowed to produce inequality. It does not involve equal *social* rights such as equal access to education or health care. See: S. Hellsten, 'Theories of Distributive Justice', in R. Chadwick (ed.), *Encyclopedia of Applied Ethics*, 4 vols, San Diego: Academic Press, 1998, Vol. 1, pp. 815–828.

[72] H.T. Engelhardt, *The Foundations of Bioethics*, New York: Oxford University Press, 1986, p. 357; H.T. Engelhardt, 'Freedom and Moral Diversity', in *Social Philosophy and Policy* 14(1997) pp. 180–196.

would receive the same standard of care, presumably one that offers at least a decent minimum, and there is no possibility to purchase additional care on the private market. To realise this goal, both finance and facilities in health care are public. This underlies an inclusive health-care system, which attempts to encompass all health-care providers and health-care recipients. The system is financed through general compulsory taxation and the organisation and facilities are under strict governmental budgetary control.[73] In an inclusive or unified health-care system, freedom of choice, quality of care and economic efficiency are subordinated to the goal of equal access for all.

It is important to note that the two organisational paradigms, as well as the four objectives, function as *benchmarks*. Only as such are they significant. For in reality health systems are not exclusively private or public, fully efficient, or perfectly egalitarian, but two-tiered or mixed. A *two-tiered system* is a compromise between a free market and a strict egalitarian system of health care: a decent minimum is provided independently of the ability to participate in the market, and in addition, one can purchase additional care if one has the resources, and if someone is willing to sell the services. This combines both concepts of freedom. The first tier makes the egalitarian concept of positive freedom operational by offering people a decent minimum of health care that opens prospects for their personal development. Enforced social coverage for basic and catastrophic health-care needs constitutes the first tier. This would presumably cover at least public health protections and preventive care, primary care, and acute care, as well as special social services for those with disabilities The second tier acknowledges the libertarian concept of negative freedom since consumers are not obstructed to use private resources to provide themselves with whatever care they want for other health-care needs and desires.[74] In this two-tiered conception, society's obligations are not limitless but restricted to providing a basic safety net for everyone.

Different countries implement the two-tiered model differently. This results from various combinations of the four goals, while leaning more towards one paradigm than to another. In the USA, for example, the market system remains the prevailing paradigm for health policy.[75] Patients are free to subscribe to a private health insurance or can obtain health insurance through their employer. This results in lack of adequate insurance for many people, either because of *cream skimming* or because they are employed by companies offering no health

---

[73] An interesting analysis of three genres of systems (pure free market, two-tiered or mixed and egalitarian) is provided in H.T. Engelhardt, 'Health Care Allocations: Responses to the Unjust, the Unfortunate, and the Undesirable', in E.E. Shelp (ed.), *Justice and Health Care*, 1981, pp. 121–137.

[74] H.T. Engelhardt, *op. cit.*, 1986; p. 361; *Id., op. cit.*, 1981, p. 121. T.L. Beauchamp; J.F. Childress, *op. cit.*, 2001, pp. 244–245.

[75] Consequenlty, the USA has no universal compulsory health insurance that gives equal access to basic health-care services for all. T.L. Beauchamp; J.F. Childress, *op. cit.*, 2001, p. 231; see also: H.T. Engelhardt, *op. cit.*, 1981, p. 128; *op. cit.*, 1986, p. 356.

insurance benefits.[76] This problem of unequal access is partially overcome by charity on the one hand and by public health plans on the other. Federal and local governments encourage religious communities and private hospitals to do charitable work to fill the urgent health-care needs of the underinsured, the uninsured and the uninsurable.[77] Public health plans like Medicare and Medicaid cover, respectively the elderly (above 65 years), and the poor and disabled who are ineligible for the private insurance schemes.[78] In general, high-quality care is offered to those who can afford it. Medicaid and Medicare offer high-quality care on the basis of need, but the package is limited. Whether the system is efficient, is contestable. The US health-care expenditure largely exceeds the average health-care expenditure of other industrialised countries, less predominantly organised according to market forces.[79] A broad consensus appears to be emerging that all citizens should be able to secure equitable access to health care. The problem with this consensus is its thinness. There is wide disagreement on a range of political solutions proffered to improve access, on the role of government in these solutions, and on methods of financing them.

Britain's National Health Service (NHS) is an example of a health system leaning close towards the public paradigm, with the goal of providing equal care for all on top. Initially, the NHS was set up as a strictly egalitarian health-care system. Both finance and facilities were public. The NHS was financed mainly by state funds from general taxation. The actual level of health-care funding is, then, under strict government budgetary control.[80] Apart from the health-care

---

[76] *Cream skimming* (also called *cherry picking*) is a typical insurance problem. It happens when unusually low-cost people select an insurance plan. Cream skimming occurs when an insurer comes to know more about consumer's expected costs than the consumers themselves by using marketing techniques or insurance plan designs to enroll a healthier-than-usual population. For example, an insurance plan that offers excellent obstetric care but poor oncology or elderly care will probably attract a younger and healthier population than one that offers the opposite. A common criticism of low-cost health plans is that they keep their costs low by enrolling healthier people (or encouraging unhealthy people to leave the plan) rather than by treating their enrollees more efficiently.

[77] During the 1990s, the number of uninsured Americans increased from 36.3 million in 1990 to 43.9 million in 1998. S.A. Schroeder, 'Health Policy 2001: Prospects for Expanding Health Care Insurance Coverage', in *The New England Journal of Medicine* 344(2001)11, p. 847.

[78] Taken together, more than 60% of the US population has employer-based health insurance coverage. Another 25% has either private health insurance unconnected to employment or some form of publicly supported health insurance. The remainder of the population has no health insurance. See: T.L. Beauchamp; J.F. Childress, *op. cit.*, 2001, p. 240.

[79] In 1997, the USA's estimated total (public and private) expenditure on health was 13.7% of the gross domestic product (GDP). During the same period, the average expenditure on health in the EU countries was 7.3% of the GDP. See: The World Health Report 2000, *Health Systems: Improving Performance*, Geneva, World Health Organization, 2000, pp. 192–195. Cf. Table 1 in Section 2.2.1 *Increasing Health Care Costs*.

[80] This system of health care funding is referred to as the Beveridge Model according to the Beveridge Report (1944) that cleared the way for the NHS to be established in 1948. See also: R. Porter, *op. cit.*, 1999, esp. pp. 628–667.

funding, the central government also had control over the health-care facilities. On the level of access, the NHS guaranteed universal access to health-care services on the condition of the general practitioner as the entry point for medical care. At the beginning of the 1990s, however, Britain's NHS underwent fundamental reform including among others more internal competition (internal markets) and greater hospital autonomy as a response to problems of efficiency and quality of care, for instance, long waiting lists for non-urgent medical interventions. In addition, countries with NHS's have been introducing more and more market-oriented elements, like private contributions and co-payments.[81] Analogous circuits of state-owned hospitals and private health institutions appear and in primary care general practitioners now often function in both public and private contexts.[82]

This illustration shows that neither the 'egalitarian' nor the 'free market system' of health care are static and pure. All systems move, evolve within the field, demarcated by the four objectives and two paradigms, seeking a balanced equilibrium. The NHS has incorporated a number of free market elements in order to eliminate problems of efficiency and quality of care, without touching the goal of basic and affordable care for all. The USA example shows with Medicare and Medicaid the creation of a safety net to guarantee minimum access for the least well-off, while the extensive network of private facilities leaves enough room for private purchase of care.

Another type of two-tiered system can be found in the so-called social insurance or Bismarck model.[83] As the counterpart of the Beveridge Model, funding under the social insurance model is organised through intermediary insurance funds that are largely independent from the government and often aligned with political or religious institutions or ideologies. In a social insurance system, the role of the state is limited and there is more freedom of choice for both patients and providers than in the original NHS. Patients have the freedom to choose what health-care insurance they subscribe to while an additional number of health-care institutions are privately owned. At the same time, the first tier, guaranteeing equal access to basic health care, remains assured because health insurance is compulsory. Furthermore, the private insurance funds are not driven by profit seeking but by solidarity motives. The aim is to share the costs for medical care between the sick and the healthy, and to adjust for different levels of ability to pay.[84] And finally, social insurance focuses not only on freedom of choice and basic and affordable care, but

---

[81] Alongside the UK, Ireland, the Scandinavian countries as well as Southern European countries (Italy, Portugal and Spain; partly in Greece) have nationalised health-care systems.

[82] E. Jakubowski et al., *op. cit.*, 1998, p. 22.

[83] Referring to the German chancellor Otto von Bismarck who introduced the first system of social insurance in 1883. See: R. Porter, *op. cit.*, 1999, pp. 353–354; 630, 638. Following the German blueprint, different modes of social insurance systems are operational throughout Europe. Among the countries that essentially have a system of social insurance are Austria, Belgium, France, Greece, Luxembourg and The Netherlands.

[84] E. Jakubowski et al., *op. cit.*, p. 10.

also on efficiency and cost containment. Private funds are merged and the role of the state is growing because of budget control.[85]

The above illustrations of the different ways in which particular health-care systems try to shape and reshape an acceptable balance between the four objectives and the two organisational paradigms show that the tension between individual and society, present in all thinking about political, social and ethical affairs, also determines the health-care scene. How can we reconcile freedom and equality in providing good quality care that remains cost-effective? What is the society's obligation in these matters? To what extent is limiting individual freedom legitimate? And what limits do autonomy and individual liberties of physicians and patients place on the social distribution of health care? Where does individual responsibility start? The underlying tension between individual and society causing these and other questions has been present from the beginning of attention for the public aspect of health and medicine.[86] This has not changed and never will. It is the *archê*, the continuing source of our thinking about it.

In some situations and some cases, however, this tension between individual and society, between the private and the public, may be *intensified*. This is the case in contemporary health care. However different the various health-care systems may be in reality, they all have to deal with the same problem: that of cost increase, and the subsequent necessity of rationing and priority setting. This problem in turn originates from the problem of *scarcity*. How can we deal with scarcity of health-care resources in ethically acceptable ways? This question opens in Chapter 2.

---

[85] *Ibid.*, p. 5.

[86] D.E. Beauchamp, 'Public Health: Philosophy of Public Health', in W.T. Reich (ed.), *Encyclopedia of Bioethics*, 1995, p. 2165.

CHAPTER 2

SCARCITY, FINITUDE AND THE NORMATIVE VALUE OF HEALTH

The relevance of the analysis made in Chapter 1 originates from the condition of scarcity and competition. In order to obtain a clear view on what we owe to each other at the bar of just health care it is important to have a closer look at this condition of scarcity. I will do this in a fourfold way.

Firstly, I will define scarcity as a contemporary *factum*, i.e. as a factual presence in contemporary societies. Subsequently, I will characterise two ways in which it turns up in health care, i.e. the external and internal aspect of scarcity of health-care resources. The first is the aspect, which refers to scarcity as a *natural, ontological* condition. It expresses the idea that questions of justice in health care would not arise if we had complete abundance of goods and services. We are not in the Garden of Eden. The second aspect refers to scarcity as an *anthropological* or *social* construction. It holds that even if we would hypothetically face abundance of resources, we would still experience scarcity due to anthropological or social intensification of our needs and desires.

In Section 2.2, I will focus on the way in which the issue of scarcity in health care is usually dealt with in academic literature and social policy debate, i.e. on the discussion of the causes of cost increase and on the development of various mechanisms of rationing and priority setting aiming at enhancing efficiency and effectiveness of care while trying to secure the ideal of equity. I will call this course the economic interpretation of scarcity because the development of such mechanisms is mainly carried out in economic research.

In Section 2.3, I will look at the various ethical implications of this interpretation of scarcity by trying to grasp what the ideal of equity in health care implies. Here, scarcity turns up as one of the circumstances of justice and forces us to answer the question: 'Equality of What?' Careful analysis of various possibilities will show that equal access for equal need is probably the most adequate equalisandum. Consequently, it is necessary to inquire into the structure of the concept of need and its relation to preferences. How are they related? Or put differently: can we define a set of characteristics to which basic needs, and consequently, basic health-care needs should answer? In this section, arguments using the objective, truncated scale of well-being, the ideas of normal functioning and corresponding opportunity range prove to be central. I will conclude this section by providing an answer to the question: what does a human right to health care imply?

To conclude this chapter, I will put these reflections into a broader philosophical perspective by offering a view on scarcity that understands it as an economic translation of *finitude*. This viewpoint is, I believe, more apt to incorporate *care* in the discussion. I will argue that in order to grasp a truly adequate conception of just health care, we need efficiency, justice and care to function as a trinity. This might happen more easily by a positively refined valuation of finitude. Here, we will meet a strange but interesting paradox: it seems that the logic of *abundance* that is implied by the concept of *care* enables us better to handle the logic of ontological *scarcity* because it motivates us to deal with *finitude* in a different way.

2.1 ON SCARCITY OF HEALTH-CARE RESOURCES

2.1.1 Context: Scarcity as a Contemporary Factum

Quite a lot of problems that *all* contemporary health-care systems face originate from scarcity of health-care resources. Scarcity means that we do not have sufficient resources, and that there are limits to our technology and management skills in producing enough to meet the existing demand. Consequently, it implies that not all of our own or society's goals can be realised at the same time; that it is necessary to make choices as to how the scarce goods and services are allocated and used; and thus, that we must *trade off* various goods against others.

2.1.1.1 The Twofold Dynamics of Scarcity

To begin with, it is illuminating to distinguish two different aspects of the dynamics of scarcity. The first is the *external* aspect and refers to the fact of *opportunity cost*. This is a term used in economics to mean the cost of something in terms of an opportunity foregone (and the benefits that could be received from that opportunity). For instance, if I give up the option of going for a nice dinner and a theatre visit in order to continue writing this manuscript, the cost of writing is the enjoyment I would have received by having a night out. The point is that we are not living in the Garden of Eden; I cannot realise everything at the same time, and my choice for one thing implies giving up something else. This means that I need to consider consciously how to use my limited resources (time, money, attention, etc); I need to consider my goals in the short and long term; I need to think about what I want to realise and what I am willing to give up. This external aspect refers to scarcity as it is a *natural* condition of resources being limited.

Next to this first aspect, it is important to bear in mind the second, *internal* aspect of scarcity. This refers to scarcity as a modern anthropological or social construction.[1] It holds that even if we would hypothetically face abundance of resources, we would still experience scarcity due to anthropological or social intensification of our needs and desires. It refers to the idea that scarcity also originates in the impossibility to fulfil *unlimited* subjective wants. Even if my resources were endless, my wants or needs for other resources would also be endless; and I would

[1] Cf. I. Illich, *Medical Nemesis: the Expropriation of Health*, London: Calder & Boyars, 1975; G. Calabresi; P. Bobbitt, *Tragic Choices*, New York: Norton, 1978; H. Achterhuis, *Het rijk van de schaarste*, Baarn: Ambo, 1988; N. Xenos, *Scarcity and Modernity*, London: Routledge, 1989.

still have to choose. As such, scarcity is a perpetual condition that exists because people have unlimited wants and needs, but limited resources. For instance, even if I were endlessly rich, the amount of time I have would still be limited. Furthermore, there are some resources – like good health or loving relationships with friends and family members – that cannot be bought, not even with the greatest amounts of money. Finally, there is the fact that 'wanting more' is always conceivable. The internal aspect of scarcity as such refers to the *tendency towards infinity* that is internally connected to obtaining and using resources. I will always lack time or money to realise all of my projects, and I can always imagine wanting to have more of it, or wishing I had made more efforts to realise this or that.

This dual aspect (of both nature and of human attitude) is also implicitly present in Hume's definition of the circumstances of justice. Justice arises he says, because of 'the *scanty* provision *nature* has made for [men's] *wants*'.[2] As we will see, nature and human wants, needs, desires and expectations play an important role in the experience of scarcity in health care as well.

2.1.1.2 Scarcity in Health Care

Both aspects of scarcity also apply to the public organisation of health care. Imagine that society decides to spend all its resources exclusively to health care. Even then, we encounter Kenneth Arrow's bottomless pit argument, which holds that available resources will never be enough to meet all demands because the latter are endless and drain away everything.[3] The history of medicine discloses this *internal* aspect of scarcity. It discloses the infinite capacity to do ever more – and ever more expensive – things for the patients.[4] The nature of health care is such that supply often generates its own demand; and to spend more on the provision of health care is often no more than to stoke the fires of further demand. As John Butler puts it:

> Since to conquer one peak is merely to reveal yet others to climb, we cannot assume that a doubling or even a trebling of the volume of resources allocated to [health care] would close the gap between supply and demand.[5]

The *external* aspect of scarcity reveals itself in the fact that all countries must set limits on the amount of resources they can spend on health care *in relation* to other social goals. This becomes clear when we recapitulate the four levels of allocation in health care: (1) the *comprehensive social level*; (2) the level of allocation within the *health budget*; (3) the level of allocation within the *health-care budget*; and (4) the *level of bedside decisions*, allocating scarce treatments or goods to patients.[6]

---

[2] In D. Hume, *A Treatise of Human Nature*, 1978, p. 495 (my emphasis, YD). Cf. Section 1.3.1.2 *The Circumstances of Justice*.

[3] K. Arrow, 'Some Ordinalist-Utilitarian Notes on Rawls's Theory of Justice', in *Journal of Philosophy* 70(1973)9, pp. 245–263; p. 251. Cf. Section 3.3.1 *Health Care: an Additional Primary Social Good?*

[4] Cf. R. Porter, *op. cit.*, 1999; see also J. Butler, *op. cit.*, 1999, pp. 5–39.

[5] J. Butler, *ibid.*, p. 7.

[6] Cf. Section 1.2.1 *Four Levels*.

Because of this combination of aspects of scarcity, the gap between supply and demand is inevitable. As a *factum*, it is inherently linked to our thinking of distributive justice in health care. Nevertheless, it is important to distinguish between the *factum* of the gap and the issue of *coping* with it.

2.1.2 Setting Limits: Coping with the Gap

It is possible to distinguish various ways of coping with the gap between supply and demand. First one could, utopian-like, ignore the gap and assume away the problem of scarcity. This could be called the 'Star Trek' approach. It eliminates scarcity by assuming abundance. It envisions a technological society that is so advanced that all material needs and wants can be met. In the Star Trek future this is done with the replicator. If one wants a cup of coffee or anything else, one merely goes up to the replicator, orders it, and it magically appears. Of course, this would be a technologically translated version of the Garden of Eden and I will not devote more time to this non-realistic interpretation. Another translation of presupposing abundance would be simply to keep raising arguments for increasing the public budget spent on health care without thorough regard for other social goals.

An alternative approach is one that is generally valued and developed in literature. It holds that the true challenge in health care lies not so much in increasing the public budget spent on health care but rather in optimising the allocation of resources by setting limits efficiently and fairly through adequate rationing and priority setting. It consists in developing various mechanisms that all aim at realising the best combination of efficiency, effectiveness and equity. I will call this approach the economic approach because the development of such mechanisms is mainly carried out in health economics research. However, it is important to know that analysis of the equity framework is mainly based on social and political philosophy, in particular, on philosophical theories of distributive justice.

A third approach is founded on philosophical reflections on scarcity as an expression of finitude, not so much finitude of resources as such, but rather of finitude as an essential characteristic of the *condition humaine*. This approach tries to install a different and positively refined valuation of the limits of human existence. It does not mean to eliminate scarcity in the way of Buddha, which is to eliminate want, or in the way of the Stoa, which entails an unmoved acceptance of fate and fortune. It does entail, however, more than the mechanism-development approach, critical and fundamental arguments for a thoughtful and balanced attitude towards the endless possibilities of modern medicine and medical technology, founded on being able to relate meaningfully to the inevitable and the uncontrollable, even though we find it tragic.

2.2 COMBINING EFFICIENCY, EFFECTIVENESS AND EQUITY

Let us start with the common interpretation of scarcity as a key concept in economics. In fact, neoclassical economics, the dominant school of economics today, defines its field as involving scarcity. According to Lionel Robbins's classical definition, economics is the study of the allocation of scarce goods among competing ends.[7] Economists study – among other things – how societies perform the optimal allocation of scarce resources, seeking the best combination of efficiency, effectiveness and equity.

2.2.1 Increasing Health-Care Costs

Economic literature on health-care allocation usually starts with the conclusion that since the 1960s all Organisation for Economic Cooperation and Development (OECD) countries face a skyrocketing increase of health-care costs – measured as a percentage of the gross domestic product (GDP) (cf. Table 1).

2.2.2 Various Causes

Various causes of this increase in expenditure can be specified: demographic development and the problem of ageing; the financing structure of health care (viz. the famous moral hazard problem: being well-insured may stimulate risk-taking and raise medical consumption); the organisational structure of health care (viz. how much is included in the insurance package; to what extent is the general practitioner a gatekeeper of the hospital doors?).[8]

TABLE 1. Total expenditure on health – % GDP

Country	1960	1970	1980	1985	1990	1995	2000
Belgium	3,4	4	6,4	7,2	7,4	8,7	8,7
Canada	5,4	7	7,1	8,2	9	9,1	9,1
Denmark	3,6	5,9	9,1	8,7	8,5	8,2	8,3
Germany	4,8	6,3	8,8	9,3	8,7	10,6	10,6
Ireland	3,6	5,1	8,4	7,6	6,6	7,2	6,7
Japan	3	4,5	6,4	6,6	5,9	7	7,8
The Netherlands	3,8	5,9	7,5	7,3	8	8,4	8,1
Norway	2,9	4,4	7	6,7	7,8	8	7,8
Spain	1,5	3,6	5,4	5,4	6,6	7,7	7,7
Switzerland	4,9	5,6	7,6	8	8,5	10	10,7
UK	3,9	4,5	5,6	5,9	6	7	7,3
USA	5,1	6,9	8,7	10	11,9	13,3	13

Copyright *OECD Health Data 2002*, 4th edn.

[7] 'Scarcity of means to satisfy given ends is an almost ubiquitous condition of human behavior. Here, then, is the unity of the subject of Economic Science, the forms assumed by human behavior in disposing of scarce means.' In: L. Robbins, *The Nature and Significance of Economic Science*, London: Macmillan, 1932, p. 15.

[8] For a detailed discussion and an extensive overview of references to the academic literature see: K. Kesteloot; M. Marchand, 'Hervormingsvoorstellen in de gezondheidszorg: een micro-economische benadering', in M. Despontin; M Jegers (eds.) *De sociale zekerheid verzekerd?* Brussel: VUB Press, 1995, pp. 371–412; K. Kesteloot, 'Hervormingen in de gezondheidszorg: economische achtergrond en ethische implicaties', in B. Raymaekers; A. Van de Putte; G. Van Riel; G. Verstaen (eds.), *Moeten, mogen, kunnen: ethiek en wetenschap*, Leuven: Universitaire Pers, 2001, pp. 159–185.

Even though all these factors obviously influence the level of health-care expenditure, they only provide a limited explanation of the skyrocketing cost increase. The lion's part, it appears, should be attributed to the progress and improvement of modern medical technology.[9] Since the 1960s, medicine has been revolutionalised by technological advances in diagnosis and treatment of illnesses; an advance that has stimulated increasing levels of spending. In this, two factors function as catalyst.

2.2.2.1 Medical Technology: Exponential Increase

Firstly, the simple contradiction between endless clinical possibilities of diagnosis and therapy, and economic affordability that underlies the idea of an increasing gap between supply and demand in health care is inherently linked to scientific and technological progress: to conquer one peak is merely to reveal yet others to climb. The history of medicine shows an exponential increase of diagnostic capabilities and of related therapeutic possibilities (for instance in AIDS and cancer research). Furthermore, it happens that advance in medicine creates new needs that did not exist until the means of meeting them came into existence, or at least into the realm of the possible (like the various possibilities in assisted reproduction).

2.2.2.2 Modern Medicalisation of Life

Secondly, the cost increase in health care is also catalysed by the increasing *medicalisation* of life since the 1960s.[10] With this, I mean that we live our lives – in the personal as well as in the social context – within a medical framework (in personal *care* and *cure*) or against a medical horizon (of public health measures aiming at *prevention*). This phenomenon of medicalisation of life is fostered by people's general expectations of medicine and health care that are based in the idea that life and health are the *summum bonum* or, as René Descartes put it, chief among all goods that need to be protected and promoted by all means. It is also fostered by the traces of high tide modernity and its philosophy of progress

---

[9] W.B. Schwartz, 'The Inevitable Failure of Current Cost-Containment Strategies – Why They Can Provide Only Temporary Relief', in *Journal of the American Medical Association* 257(1987)2, pp. 220–224; J. Newhouse, 'Medical Care Cost Increases: How Much Welfare Loss?' in *Journal of Economic Perspectives* 5(1992)3, pp. 3–21; R.P. Ellis; T.G. McGuire, 'Supply-Side and Demand-Side Cost Sharing in Health Care', in *Journal of Economic Perspectives* 7(1993)4, pp. 135–151.

[10] Cf. I. Illich, *Medical Nemesis*, London: Calder & Boyars, 1975, pp. 31–60. *Id.*, 'Body History', in *Lancet* 11(1986), pp. 1325–1327. In this latter article, Illich modifies his earlier statement: it is not so much the medical establishment that is the biggest problem, but the general contemporary attitude of striving for full and healthy well-being. The concept of medicalisation of life is also analysed by M. Foucault in his work *Naissance de la clinique*, Paris: Presses Universitaires de France, 1963. See also: M. Karskens, 'Biopolitiek en de gezonde mens', in J. Rolies (ed.), *De gezonde burger: gezondheid als norm*, Nijmegen: Sun, 1988, pp. 71–88. It is important to bear in mind that although both Foucault and Illich concentrate on the phenomenon of medicalisation, their respective analyses differ. Unfortunately, I cannot go into further detail on this matter here.

declaring that reason, science and technology enhance medical advances, and through this, man's control over nature, health and disease:

> the preservation of health is ... without doubt *the first good and the foundation of all the other goods of this life*; for even the mind depends so much on the temperament and on the disposition of the organs of the body that, if it is possible to find some means that would render men more wise and more skilful than they have been up until now, I believe that it is in *medicine* that one ought to be searching for it. It is true that the medicine that is now practiced contains few things of which the utility be so remarkable; but not that I had any intention of disparaging it, I am sure that there is no one, even among those who make a profession of it, who would not admit that all that which one knows in medicine is almost nothing in comparison with that which remains to be known there, and that *one might rid oneself of an infinity of maladies*, as much of the body as well as of the mind, and maybe even of the enfeeblement of old age, too, if one had *sufficient knowledge* of their causes and of all the remedies that nature has provided us.[11]

This modern optimism finds contemporary expression for instance in general expectations with respect to the promises of genetic research and technology. A final stimulating factor in this regard is the time-honoured doctor's instinct to save the life of the patient under his immediate care whatever may be the opportunity cost of doing so. Taking these elements together, it should not surprise us that medicine is often called a 'new religion'.[12] In this context, there is no reason to bridle medical and technological progress.

Taking the determining role of medical technology into account it is reasonable to believe that changing the ways in dealing with the demographic evolution, the problem of ageing, moral hazard, and financial and structural organisation of health care may cause only a *partial* curb of the cost increase, but it will not alter the *general evolution* of this increase.[13] Therefore, we must focus on the domain of medical technology.

---

[11] In: R. Descartes, *Discourse on the Method*, ed. and trans. G. Heffernan, Notre Dame: University of Notre Dame Press, 1994, p. 87 (my italics, YD). 'la conservation de la santé ... est sans doute le premier bien et le fondement de tous les autres biens de cette vie; care même l'esprit dépend si fort du tempérament, et de la disposition des organes du corps que, s'il est possible de trouver quelque moyen qui rende communément les hommes plus sages et plus habiles qu'ils n'ont été jusqu'ici, je crois c'est dans la médecine qu'on doit le chercher. Il est vrai que celle qui est maintenant en usage contient peu de chose dont l'utilité soit si remarquable; mais, sans que j'aie aucun dessein de la mépriser, je m'assure qu'il n'y a personne, même de ceux qui en font profession, qui n'avoue que tout ce qu'on y sait n'est presque rien, à comparaison de ce qui reste à y savoir, et qu'on se pourrait exempter d'une infinité de maladies tant du corps que de l'esprit, et même aussi peut-être de l'affaiblissement de la vieillesse, si on avait assez de connaissance de leur causes, et de tous les remèdes dont la nature nous a pourvus'. In: R. Descartes, *Discours de la Méthode, Sixième Partie*, in A. Bridoux, (ed.), *Descartes: Oeuvres et Lettres*, Paris: Gallimard, 1952, pp. 168–169.

[12] J. Rolies, 'Gezondheid: een nieuwe religie?' in *id.* (ed.), *De gezonde burger: gezondheid als norm*, Nijmegen: Sun, 1988, pp. 11–30; K. Dierickx, *Genetisch gezond?* Antwerpen: Intersentia, 1999, pp. 159–166. See also L. Reuter, *Modern Biotechnology in Postmodern Times?* Dordrecht: Kluwer, 2003, esp. pp. 125–160, for an interesting analysis of the views of French critical thinkers as J.P. Sarte, G. Deleuze, F. Guattari, J. Ellul, M. Foucault and B. Latour on technology beyond modernity.

[13] W.B. Schwartz, *op. cit.*, 1987, pp. 220–224; J. Newhouse, *op. cit.*, 1992, pp. 3–21; J.M. Poterba, 'A Skeptic's View of Global Budget Caps', in *Journal of Economic Perspectives* 8(1994)3, pp. 67–74; K. Kesteloot, *op. cit.*, 1995.

Below I will concentrate first on the catalyst that originates in the contradiction between endless clinical possibility of diagnosis and therapy and economic affordability, and inquire into the various strategies of enhancing efficiency and effectiveness, while preserving equity. In Section 2.4, I will concentrate on the second catalyst: the phenomenon of medicalisation and our modern attitude towards life, health and finitude. Note, moreover, that both catalysts are not to be understood as separate phenomena. They are intrinsically related, whereby the second catalyst functions as a more fundamental dynamics, underlying the first catalyst.

2.2.3 Enhancing Efficiency and Effectiveness…

A possible strategy for managing the gap between supply and demand in publicly financed health care lies in the various efforts that have been made, and continue to be made, to enhance the efficiency and effectiveness of the health-care system, also known as 'rationing'.[14] To walk this road, it is argued, is to generate more beneficial care from a given budget with no diminution in its quality. The more beneficial care that can be generated from the budget, the narrower will become the gap between supply and demand.

Efficiency (or, to be more precise, productive efficiency) is about maximising the quantity and quality of what is achieved from a given quantum of resources. It is about getting the best result for the money available, not about getting away with the lowest expenditure. *Effectiveness*, unlike efficiency, is less contaminated by the pressure of resources and it therefore relates to the problem of cost-containment in a slightly different way. Treatment A is clinically effective if it alters the development of a disease for the better; and A is more effective than B if it does so more thoroughly, or with fewer side effects. Although the cost of A and B is in itself unrelated to the question of their effectiveness, it may be relevant to their affordability. Today, scientific evaluation of medical treatments resulting in an accumulating body of evidence about therapies that work and the conditions under which they work best, is widespread.

Various strategies are open to the actors in the health-care field (politicians, planners, managers and clinicians) in bridging the gap between supply and demand in publicly financed health-care systems. A distinction is commonly drawn between

[14] On *efficiency* and *effectiveness*, see: A.L. Cochrane, *Effectiveness and Efficiency*, London: The Nuffield Provincial Hospitals Trust, 1972. The concept of *rationing* might best be explained as follows: whenever valued things are in short supply, ways must be found of resolving the dilemma: who will get what they want and who will have to go without? Aside from contexts of anarchy, where people grab and hold what they want by sheer brute force, the share-out of valued goods and services takes place within rules or customs that are socially negotiated and sanctioned. In broad terms, these arrangements can be thought of as belonging to one of two types: *market* distributions and *rationed* distributions. The market is a way of regulating distribution of goods and services according to the principle of *willingness to pay*. Some commodities, however, are distributed within a socially negotiated framework that takes little or no account of price. It is for this allocationary mechanism that the notion of rationing is best reserved. When goods are rationed, it is something substantially other than willingness to pay that determines the share of the cake. Cf. J. Butler, *op. cit.*, 1999, pp. 20–30.

those that affect the allocation of resources or services, which are taken at a fairly high level in the system (*macro rationing*), and those that affect the allocation of concrete care, which operate at a lower level (*micro rationing*).[15] This distinction is somewhat artificial because although the language may vary, the idea is similar, and also because, as I have mentioned before, measures taken at the macro level have repercussions at the micro level and vice versa. Let us take a brief look at some examples of existing rationing mechanisms. These can be thought of as belonging to one of two types of rationing: supply-side or demand-side rationing.

2.2.3.1 *Reduce Supply*

At the supply-side, various mechanisms can be determined. There is the possibility of reducing the number of beneficiaries (only the elderly above 65 years (Medicare in the USA), or only the poor and disabled who are ineligible for the private insurance schemes (Medicaid). Also known is the application of Quality Adjusted Life Years (QALY) and Disability Adjusted Life Years (DALY). Both function as complementary tools in cost-effectiveness research to measure and compare different types of health states or health outcomes. In particular, they make it easier to include the burden caused by disability and chronic diseases in cost-effectiveness studies. DALY may be called a modification of QALY. Both approaches multiply number of years lived by the 'quality' of those years. This process is called 'quality adjustment' in QALY and 'disability adjustment' in DALY. QALY are years of healthy life lived – DALY are years of healthy life lost. Consequently, whereas DALY's represent a loss and should be minimised, QALY represent a gain and should be maximised. Yet another possibility is that of reducing the public health-care package by way of *evidence-based policy* and *evidence-based medicine.* Only care that is proven to be effective will be publicly covered. This path can be supported by various techniques of *health technology assessment* (like cost-benefit analysis, cost-effectiveness analysis, and cost-utility analysis).

The general idea of the supply-side strategy lies in a sharper definition of the range of services that the public system provides, thereby taking services off the public menu that have hitherto been available, leaving more resources for providing those that remain. Put bluntly, it is a process of rationing by *exclusion*, in which those who require the excluded services must either buy them privately, or go without.[16] As we will see in Part II, a hypothetical approach to rationing by exclusion has been proposed by Ronald Dworkin in his *prudent insurance principle.* Against the background of some core assumptions (we all have equal income and wealth, full information of the effectiveness of different treatments, and know nothing about the genetic, cultural and social influences

[15] G. Calabresi; P. Bobbitt, *op. cit.*, 1978; R. Klein; P. Day; S. Redmayne, *Managing Scarcity. Priority Setting and Rationing in the National Health Service*, Buckingham: Open University Press, 1996; K. Kesteloot; M. Marchand, *op. cit.*, 1995, pp. 371–412; J. Butler, *op. cit.*, 1999.

[16] R. Klein; P. Day; S. Redmayne, *op. cit.*, 1996; B. New; J. Le Grand, *Rationing in the NHS: Principles and Pragmatism*, London: King's Fund Publishing, 1996; J. Butler, *op.cit.*, pp. 5–39.

on our health), Dworkin invites us to ponder the following question: which treatments would we choose to be insured for and which would we regard not worth the cost of insurance? Consequently, Dworkin identifies a number of treatments that might become candidates for omission (e.g. expensive life-saving treatment above 85 years, or in case of persistent vegetative state or expensive heroic technology of doubtful effectiveness, ...) and treatments that most prudent people would want to buy insurance for (e.g. immediate and expert treatment for handicapping conditions in childhood, custodial care in conditions of dignity and adequate comfort in the case of very old age, irreversible dementia or persistent vegetative state).[17]

2.2.3.2 Reducing the Demand for Care

Another assortment of strategies is designed to bear down upon the level of demand for health care, thus trying to stimulate 'proper' use of public health care. If less use can be made of services, especially in circumstances that are thought to be inappropriate, resources will be conserved for deployment elsewhere. From the patient side, this could be done by increasing the share of private pocket contribution to the general health-care costs (through prescription charges or a higher personal share in hospital costs), or by installing waiting lists. From the side of health-care provider, this can be done by stimulating economical management of care (by means of denying questionable treatment or by dilution, i.e. by reducing the comprehensiveness of care, using cheaper materials or hiring cheaper staff).[18]

The general case for improving efficiency and effectiveness of health care cannot seriously be faulted. To fail to do so is to be reckless in the stewardship of public resources. Yet enthusiasm for ever more efficiency and effectiveness may not ignore the ethical responsibility for equity in health care. Savings in one area of health care must not be claimed as a virtue if they have been achieved at the cost of greater expenditure elsewhere, i.e. at the expense of quality (*quality-skimping*) and accessibility of care (*cream skimming* or *cherry-picking*).[19] In such cases, the poor, the weak, the sick and the dependent, i.e. the least-advantaged

---

[17] R. Dworkin, 'Justice and the High Cost of Health', 1994, repr. in *id.*, *Sovereign Virtue: the Theory and Practice of Equality*, Cambridge, MA: Harvard University Press, 2000, pp. 307–319; R. Dworkin, 'Justice in the Distribution of Health Care', 1993, repr. in M. Clayton; A. Williams (eds.), *The Ideal of Equality*, Basingstoke: Palgrave Macmillan, 2002, pp. 203–222. For critical analysis of Dworkin's model, see: J. Butler, *The Ethics of Health Care Rationing*, London: Cassell, 1999, pp. 21–23; M. Powers, 'Hypothetical Choice Approaches to Health Care Allocation', in J. Humber; R. Almeder (eds.) *Allocating Health Care Resources, Biomedical Ethics Reviews*, Totowa: Humana Press, 1995, pp. 55–84; R. Smith, 'Being Creative About Rationing', in *British Medical Journal* 312(1996), pp. 391–392. I will come back to Dworkin's approach in detail in Part II, Chapter 5 *Setting Limits: Dworkin's Proposal*.

[18] J. Butler, *op. cit.*, 1999, pp. 16–36.

[19] *Quality skimping* means that the quality of care is put at risk, for instance by exploiting health care staff by paying them less than they merit or causing them undue stress through the imposition of extremely heavy workloads, or other such measures. See: K. Kesteloot; M. Marchand, *op. cit.*, 1995, p. 388; J. Butler, *op. cit.*, 1999, pp. 31–32 *Cream skimming* and *cherry picking* have been defined previously. Cf. fn. 76 in Section 1.4.2 *The Private and the Public Paradigm*.

suffer the most from the expense. Put generally, it must not be forgotten that enhancing efficiency and effectiveness is to be praised not as an end in itself but only as it is directed towards sustaining equitable health care.

2.2.4 *While Preserving Equity*

Next to efficiency and effectiveness, there is another criterion upon which exclusion can be founded. If costs, as well as effectiveness are added into the equation, treatments might be dropped because they are too costly, or ineffective. But this leaves open the fact that some services may be dropped not because they are costly, or ineffective, but because they are simply judged *not* to form part of the *core services* that ought to be provided by a just public health-care system.[20] The flip-side of this idea is that some things are just *too important* to be left to private pocket payment, because the consequences of lacking them due to inability to pay are unacceptable. A good deal of health care is important and special, and should therefore be protected in a publicly-funded health-care package that can guarantee its availability to all who need it. This gives rise to the following question: what is it in health care that is so important that it should be equally guaranteed to everyone? What should and what can we reasonably expect from just health care to guarantee to all? These questions require us to think about the essence of equity in health care.

2.3 THE ESSENCE OF EQUITY IN HEALTH CARE

The core question is: does the health-care system produce an allocation that meets society's requirements for justice? Clearly, this is a normative issue: the decision made depends on a certain valuation of health and health care.

To begin with, we can say that equity concerns fairness and justice, the idea of balancing legitimate, competing claims of individuals in society in a way that is seen as impartial or disinterested.[21] In matters of distributive justice, as in the

[20] Examples could be for instance in vitro fertilisation and a range of cosmetic procedures. The idea is then that if people want to achieve assisted fertility or to have their bodies reshaped, then that is a matter for their personal budgets, not for the public purse.

[21] Note again that it concerns a cold virtue. *Equity* arguments, which serve as a basis for redistribution of resources within society, can be distinguished from arguments for redistribution based on caring feelings, or even compassion. Because it is grounded in notions of *fairness* and *justice*, equity appeals more explicitly to *reasoned* arguments about what is right and just, and therefore what ought to be done as a matter of *principle*. Bear in mind, furthermore, the Aristotelian distinction between equity and justice. Whereas justice as *dikaiosyne* and *to dikaion* refers to the demands of general law and morality, equity or *epieikeia*, is a specific form of justice, considering the nature of the particular case itself, which is, in this case, health care. Cf. Section 1.3.1.1 *Suum Cuique Tribuere*. See also J. Hurley, 'An Overview of the Normative Economics of Health Care', in A.J. Culyer; J.P. Newhouse (eds.), *Handbook of Health Economics*, 2 vols, Amsterdam: Elsevier, 2000, Vol. 1, pp. 55–118; A.J. Culyer, 'The Normative Economics of Health Care Finance and Provision', in *Oxford Review of Economic Policy* 5(1989)1, pp. 34–58; G. Mooney, *Key Issues in Health Economics*, New York: Harvester Wheatsheaf, 1994; E. van Doorslaer; A. Wagstaff; F. Rutten, *Equity in the Finance and Delivery of Health Care: an International Perspective*, Oxford: Oxford University Press, 1993; A. Wagstaff; E. van Doorslaer, 'Equity in Health Care Finance and Delivery', in A.J. Culyer; J.P. Newhouse (eds.), *op. cit.*, 2000, pp. 1803–1862; A. Williams; R. Cookson, 'Equity in Health', in A.J. Culyer; J.P. Newhouse (eds.), *op. cit.*, 2000, pp. 1863–1910.

case of health care, equity concerns the fair distribution of health-care resources or services. This determination of equity, however, is not yet very helpful, for it does not provide us with a relevant notion of equity. If we take it a step further, it is reasonable to say that equity or justice is mostly related to *equality*, and to reducing inequality in some way.[22] In this line of reasoning, Amartya Sen has argued that virtually every reasonable theory of justice has at its core the proposition that justice demands *equality* in the distribution of something, which he calls the 'focal variable'.[23] The same goes for theories of justice in health care. Because health is an important component of well-being, too great inequalities are generally considered to be unacceptable; most people approve of some form of equality in health care.

This, however, still does not answer the question of the essence of equity in health care because theories differ on what – in Sen's language – the 'focal variable' is. If we want equality in health care, we must determine what it is we want to equalise. Liberty? Welfare? Health? Use of health care? Access to health care? More specifically, the choice of the focal variable is critical because achieving equality with respect to the focal variable implies inequalities in other dimensions (particularly, in other competing focal variables). In this respect, the difference between horizontal and vertical equity becomes relevant.[24] *Horizontal equity* calls for equal treatment of equals, i.e. of those who are similarly related to the focal variable. Hence, horizontal equity in allocation of health-care resources may call for equal treatment for equal need. Horizontal equity in financing health care may call for those with the same income to pay the same amount. *Vertical equity* calls for unequal treatment of unequals, i.e. of those who are differently related to the focal variable. In particular, it calls for unequal treatment according to the extent to which those receiving treatment are unequal. Concerning allocation of health-care resources, vertical equity may call for greater resources for those with greater needs, just as, with regard to financing health care, it may call for a person with a higher income to pay a higher tax than those with lower income (like a progressive income tax).[25] Many different focal variables are possible for health

---

[22] G. Vlastos, 'Justice and Equality', in J. Waldron, *Theories of Rights*, New York: Oxford University Press, 1984, pp. 41–76. With regard to health care, see also: T.L. Beauchamp; J.F. Childress, *Principles of Biomedical Ethics*, Oxford: Oxford University Press, 2001, 5th edn., pp. 225–282; R. Veatch, *A Theory of Medical Ethics*, New York: Basic Books, 1981, pp. 250–287; A. Williams; R. Cookson, *op. cit.*, 2000.

[23] A. Sen, *Inequality Reexamined*, Cambridge, MA: Harvard University Press, 1992.

[24] J. Hurley, 'An Overview of the Normative Economics of Health Care', in A.J. Culyer; J.P. Newhouse, *Handbook of Health Economics*, 2 vols, Amsterdam: Elsevier, 2000, Vol. 1, pp. 55–118. See also: G. Mooney, *Key Issues in Health Economics*, New York: Harvester Wheatsheaf, 1994, p. 75.

[25] See: J. Hurley, *ibid.*, 2000, p. 92; A. Wagstaff; E. van Doorslaer, 'Equity in Health Care Finance and Delivery', in A.J. Culyer; J.P. Newhouse (eds.), *op. cit.*, 2000, p. 1819; A.J. Culyer; A. Wagstaff, 'Equity and Equality in Health and Health Care', in *Journal of Health Economics* 12(1993), pp. 431–457, esp. p. 433. It is reasonable to relate this distinction between horizontal and vertical equity to Aristotle's distinction between equality in *arithmetical* and *geometrical* proportion. Cf. Section 1.3.1.1 *Suum Cuique Tribuere*.

care. Let us take a look at five possibilities that are generally suggested by most literature on the subject.[26] Again, it is our philosophical task to assess, explain and justify, or modify the various analyses that are made about the importance of these different focal variables.

2.3.1 *Equality of What? Defining the Focal Variable*

Several possibilities for defining equity in health care are: equal liberty, equal welfare, equality of health, equality of use and equality of access to health care.

2.3.1.1 *Equal Liberty?*

Equal liberty or equality of freedom is defended by the libertarian position as the proper goal of equity in health care. This viewpoint is based on the first principle of libertarian theory, which is the right to freedom and self-ownership. Consequently, according to libertarianism, an absolute respect for the right to private property and for negative freedom constitutes the basis and guideline for the legitimate role of the state, and for the basic principles of individual conduct.[27] Political institutions should only serve to protect this basic right.

The predominant version of libertarianism is the Entitlement Theory, developed and presented by Robert Nozick in *Anarchy, State and Utopia*.[28] Nozick understands justice in terms of the protection of rights or entitlements, in particular, rights to liberty and private property.[29] The absolute respect for

---

[26] For reference to the health economic discussions of this subject, I am much indebted to the suggestions given by Erik Schokkaert. His lecture on the Leuven Meeting of the *European Masters in Bioethics*, 7 March 2001, stimulated this analysis of the various interpretations of equity in health care. In the same way, I am indebted to Katrien Kesteloot's paper, 'Hervormingen in de gezondheidszorg: economische achtergrond en ethische implicaties', in B. Raymaekers; A. Van de Putte; G. Van Riel; G. Verstaen (eds.) *Moeten, mogen, kunnen: ethiek en wetenschap*, Leuven: Universitaire Pers, 2001, pp. 159–185; and to A.J. Culyer; A. Wagstaff, 'Equity and Equality in Health and Health Care', in *Journal of Health Economics* 12(1993), pp. 431–457.

[27] The libertarian concept of freedom signifies negative freedom, i.e. freedom from interference. One possesses negative freedom to the extent that in one's action one is not interfered with by others. Positive freedom means a claim to something, the possibility to be someone, to pursue and realise a private goal in a certain manner, the possibility to be aware of choices and to be able to explain them in relation to ideas and goals. See: P. Dasgupta, *An Inquiry into Well-Being and Destitution*, Oxford: Clarendon Press, 1993, pp. 40–46, explaining Isaiah Berlin's distinction between positive and negative freedom as presented in I. Berlin, 'Two Concepts of Liberty', in *id.*, *Four Essays on Liberty*, Oxford: Oxford University Press, 1969. Important concerning the difference between both concepts of liberty is that the former is individualistically understood. Negative freedom implies that the self-sufficient individual may not be hindered in his personal development. The starting point of the second conception of freedom is that an individual needs society in order to be truly free. Note furthermore that this dualist view on freedom is challenged by the triadic view of MacCallum. Cf. fn. 49 in Section 3.1.5.4 *Real Equality of Opportunity*.

[28] R. Nozick, *Anarchy, State and Utopia*, Oxford: Basic Books, 1974. See also: A. Wagstaff; E. Van Doorslaer, 'Equity in Health Care Finance and Delivery', in A.J. Culyer; J.P. Newhouse (eds.), *Handbook of Health Economics*, Vol. 1, Amsterdam: Elsevier, 2000, pp. 1803–1862, esp. pp. 1808–1809.

[29] R. Nozick, *ibid.*, 1974, pp. 149–182.

individual freedom and property right leads to a form of strict procedural justice. Justice is not a matter of just *results* but of just *procedures*: 'Whatever arises from a just situation by just steps is itself just.'[30] A focus on the results would lead to systematic violation of individual freedom and the right to private property: 'any principle of justice which demands a certain distributive end state or pattern of holding will require frequent and gross disruptions of individuals' holdings for the sake of maintaining that end state or pattern'.[31] Consequently, Nozicks theory is strikingly anti-redistributive. Coerced redistribution would be an unjust redistribution of private property, by illegitimately considering it as public property. This would mean sacrificing basic rights and liberties for the public interest:

> attempts to force anyone to contribute any part of his legitimate holdings to the welfare of others is a violation of that person's property rights, whether it is undertaken by private individuals or the state. On this view, coercively backed taxation to raise funds for welfare programs of any kind is literally theft.[32]

Against the argument that an absolute respect for property rights would create immense poverty and a growing gap between rich and poor, libertarians argue for the difference between justice and charity. Morality comprises more than non-violation of rights. Redistribution is legitimate only when it is voluntarily organised through charity. Consequently, those in need cannot claim fulfilment of their needs as a right: 'While justice demands that we not be *forced* to contribute to the well-being of others, charity requires that we help even those who have no *right* to our aid.'[33]

This means that there can be no such thing as a moral or human right to health care:

> A basic human right to the delivery of health care, even to the delivery of a decent minimum of health care, does not exist. The difficulty with talking of such right should be apparent. It is difficult if not impossible both to respect the freedom of all and to achieve their long-range best interest.[34]

According to libertarian theory, rights only involve avoiding violation by interference and thus involve no more than the right to be left alone. Rights have nothing to do with providing services or goods that improve the well-being of all, that meet the basic needs and that as such offer protection to the poor and vulnerable. Libertarianism is a 'hands-off' theory. In this, libertarianism does not deny the existence of basic needs. However, meeting these needs and taking care for the distressed is not a matter of justice and thus not a legitimate obligation of society:

---

[30] *Ibid.*, p. 151.
[31] *Ibid.*, p. 13.
[32] Allen Buchanan, explaining the libertarian perspective in A. Buchanan, 'Justice: a Philosophical Review', in E.E. Shelp (ed.), *Justice and Health Care*, Dordrecht: Reidel, 1981, pp. 3–21, esp. p. 12.
[33] R. Nozick, *ibid.*, 1974, p. 14.
[34] In H.T. Engelhardt, *The Foundations of Bioethics*, Oxford: Oxford University Press, 1986, p. 336.

> It may well be unfeeling or unsympathetic not to provide such help [i.e. health care for the individuals injured], but it is another thing to show that one owes others such help in a way that would morally authorise state forces to redistribute resources, as one would collect funds owed in a debt. The natural lottery creates inequalities and places individuals at disadvantage without creating a straightforward obligation on the part of others to aid those in need.[35]

As such, compulsory income taxation in order to organise a health-care system is a libertarian injustice. Even if one could show that health care contributes to individual freedom, it would still be irrelevant. Although the importance of health care is not ignored by definition, it is only relevant within the framework of charity and moral virtuousness. Besides, nothing prevents an individual from taking private health insurance. With this, only a private health-care system, i.e. a voluntary system of redistribution, can be legitimate. Hard-core libertarians support a health-care system that is based only on the free market ideal: distribution of health-care services and goods are best left to the market operating through the principle of willingness to pay.[36] According to this view, health-care services are no goods of special importance, but commodities like any others: things we agree to buy and sell in a market.[37]

The typical problem with such a viewpoint is that *willingness* to pay does not take *ability* to pay into account. Consequently, the concept of liberty is a merely *formal* concept, without reference to actual ability to be free. In the case of health care, the freedom to insure voluntarily ultimately boils down to the ability to insure, which leaves, as we have seen, a lot of people uninsured.[38] In itself, it does not provide a safety net or minimum floor below, which no one would be allowed to fall. The question is: should we allow inequalities in access to health care to vary with whatever economic inequalities are permissible? Or is health care special to such an extent that we should not allow it to be put on a par with other commodities – like cars or personal computers – we purchase on the market? Most societies, it seems, do think that health care is of a different kind because not even the most libertarian-inspired countries organise health care *fully* according to the free market.

2.3.1.2 Equal Welfare?

Would it be possible to defend an idea of equity that takes respect for personal choice and preferences into account, as well as *ability* to pay, thus incorporating some kind of floor-conception in the idea of equity in health care? This would involve distancing from the mere procedural route and taking a more *outcome-oriented* path. Perhaps a welfarist perspective that is adjusted to equity by guaranteeing everyone an equal starting position could do this.

Consider the *argument from fair shares*, which is intended to show that we can develop an equitable conception of distribution in health care that is based on

---

[35] In H.T. Engelhardt, *op. cit.*, 1986, p. 340.

[36] See also T. L. Beauchamp; J.F Childress, *op. cit.*, 2001, p. 231.

[37] I will come back to the discussion on the specific character of health care in Part II.

[38] Cf. Section 1.4.2 *The Private and the Public Paradigm*.

personal preferences and strength of preferences.[39] The argument goes as follows. Suppose we could guarantee each person a fair income share according to a particular theory of justice. Consequently, individuals's willingness to pay is less a function of their ability to pay. Such subsidisation would allow individuals to decide the extent to which they wanted to consume health care and they could, with respect to their fair income share, decide how much they wished to purchase.[40] That way, equity in health care – understood as equality of individual welfare – would be arrived at.[41]

Suppose further then, that there is a competitive medical market in which a variety of insurance schemes could be bought. Consequently, every person could protect himself against the risk of needing health care through voluntary purchase of insurance. Everyone would be responsible for buying insurance at a level he desires. Each would be entitled only to what he was prudent enough to purchase from his fair income share. Of course, this presupposes that individuals are competent and capable of making informed decisions about using the system – i.e. they have information about alternatives in therapies or insurance schemes. In this market, demand, understood as our preferences for different insurance packages, will be efficiently matched to supply, understood as health-care resources and services. Consequently, no one could claim a right to have preferences for health care met, however basic or extravagant they are, unless he had already bought appropriate insurance. Additionally, charity might intervene to keep people from dying in the streets.

Two categories of problems, however, crop up. The first category is based on the classic analysis made by Kenneth Arrow on reasons why medical markets

---

[39] Developed by Charles Fried in his work *Right and Wrong*, Cambridge, MA: Harvard University Press, 1978, pp. 126 ff. See also N. Daniels, *op. cit.*, 1985, pp. 20–23. At first sight, there seems to be an analogy to the 'prudent insurance ideal' developed by Ronald Dworkin in R. Dworkin, *op. cit.*, 2000/1994; *op. cit.*, 2002/1993. However, it is important to note in advance that in spite of this initial similarity, Dworkin's profound critique on welfare egalitarianism shows plenty of reason to think that he certainly does not approve of the equity conception of equal welfare in health care. Cf. R. Dworkin, 'What is Equality? Part 1: Equality of Welfare', in *Philosophy and Public Affairs* 10(1981)3, pp. 185–246; repr. in: R. Dworkin, *Sovereign Virtue: the Theory and Practice of Equality*, Cambridge MA: Harvard University Press, 2000, pp. 11–64. In Chapter 5, furthermore, we will see that his proposal is different from Fried's in a threefold way: (1) Dworkin's proposal concerns a *hypothetical* insurance mechanism instead of factual individual insurance; (2) through this, he intends to aggregate individual preferences for insurance; and (3) through this, reach a general outcome, which is fully different from Fried's proposal, since there will be no individual differences in access to health care. I will come back to Dworkin's view in detail in Chapter 5.

[40] See also G. Mooney, *Key Issues in Health Economics*, 1994, pp. 74–75.

[41] Equality of *individual* welfare as it is used here has to be distinguished from the *aggregate* utilitarian perspective, which aims at maximising the sum of individual utilities or welfare. The latter has much in common with the notion of efficiency as allocating resources according to the likelihood of medical success, but it also has problematic implications for the concept of individual rights, because it tends to the scapegoat mechanism where individuals may be sacrificed in order to ensure 'the greatest happiness for the greatest number'. See also: Y. Denier; T. Meulenbergs, *op. cit.*, 2002, pp. 265–297; A. Wagstaff; E. van Doorslaer, *op. cit.*, 2000, pp. 1803–1862.

are not efficient. The second is based on the argument that a fair income share is fair only if it is adequate to permit the purchase of a reasonable health insurance package.

Firstly, in his classic paper 'Uncertainty and the Welfare Economics of Medical Care', Kenneth Arrow analyses the reasons why medical markets are not efficient.[42] In short these reasons come down to the fact that the existence of *uncertainty* in a number of characteristics of medical care leads competitive markets to generate an inefficient allocation of resources and contributes to the emergence of non-market institutions (like trust, objective norms and social obligations) that compensate for these market failures. There is firstly, for instance, the fact that a person's *demand for care* is not steady in origin as for instance, food and clothing are, but irregular and unpredictable. The incidence of disease is not equally present for all. Furthermore, the related health risks cannot be guaranteed by sufficient income, as food and clothing can. In addition, health-care providers face uncertainty with regard to what is expected from them in their *caring behaviour*. The behaviour of a physician 'is supposed to be governed by a concern for the customer's welfare which would not be expected from a salesman'.[43] This distinguishes health-care provision from business, where self-interest on the part of the participants is the accepted norm. Furthermore, the amount of uncertainty as to the *quality of the product* is much greater for medical care – greater even in severe cases – than for other commodities like houses or automobiles. Moreover, there is great *informational inequality* on all sides of the medical relationship. Firstly, medical knowledge is so complicated that the information possessed by the health-care provider as to the consequences and possibilities of treatment is necessarily much greater than that of the patient. Secondly, the patient knows much more about his own situation, health behaviour, and preferences than the physician and insurer do. Lack of informational transparency may induce moral hazard on all sides. Overall, the central problem is that of departure of the medical market from the ideal of a truly competitive market on many points.

The second problem with regard to the argument from fair shares comes down to the fact that to know whether income shares are *fair*, we must know that people can buy *reasonable* coverage. However, to know what coverage is reasonable we need to know what kind of health-care needs it is *prudent* for people to insure against.[44] Just income distribution presupposes income to adequately meet reasonable needs. In order to find out which health-care needs are reasonable, we need a more *objective* and *independent* standard of justification than the one of equal welfare, which can be assessed by inquiring after individual preferences for insurance.

---

[42] K. Arrow, 'Uncertainty and the Welfare Economics of Medical Care', in *American Economic Review* 53(1963)5, pp. 941–973. See also: J. Hurley, *op. cit.*, 2000, p. 67–68; 79–80.

[43] *Ibid.*, 1963, p. 949.

[44] See also N. Daniels, *Just Health Care*, Cambridge: Cambridge University Press, 1985.

2.3.1.3 Equal Health?

Let us look at an *outcome*-oriented proposal that is strongly related to *objective* health results. It might be argued that since the objective of health-care services is to promote health, the equity goal of health care must be conceived in the same terms, i.e. in terms of equality of health.[45] As health care is purely instrumental, equality of health is the true goal. This concept of equity is concerned with the consequences of health policy and judges the goodness of the distribution according to its *objective outcome*.

However, what does equality of health mean? It surely cannot mean an identical health status for everyone, since there are natural and biological differences in health between people according to their age, gender, genetic endowments, handicaps, etc. It could, however, imply a strong tendency to try to secure an, as good as possible, health status for everyone within these categories. Consequently, as Culyer suggests, it implies a levelling-up process, pursuing equality of health in an upward direction: no one's health should decline as part of the process of pursuing equality of health.[46] On closer view, however, equality of health seems too strong and unrealistic an objective.

Unrealistic. Firstly, it is unrealistic because health cannot be as directly distributed or traded as health care can. Because of reasons grounded in the question of what *can* be distributed, together with the fact that *ought* implies *can*, equal health is a logically inappropriate object of equity, whereas for instance equal access to health care is appropriate.[47]

Furthermore, research shows that health is affected by many factors outside the direct scope of health care. The above illustrated *external* aspect of scarcity has shown that there are *limits* to what can be reasonably expected from the health-care system. Health is affected also by social class, level of education, quality of food, housing and environment, etc.[48] Taking the general health effects of social policy, and health policy into account, it seems that setting such a high standard on health care is unjustifiably overrated, seemingly ignoring the effect of the other social determinants of health. How should we allocate resources among the various determining levels of health in the pursuit of equity? Should they all be incorporated in the general ideal of health equality? This almost seems heroic. Or should the other levels be determined by another idea of equity, and if so, by what idea? This is not clear.

---

[45] The World Health Organization, for instance, appears to be primarily concerned with health as equity goal. Its main objective is to try to equalise health not just across different groups within a society but also across different societies around the world. A.J. Culyer; A. Wagstaff, 'Equity and Equality in Health and Health Care', in *Journal of Health Economics* 12(1994)4, pp. 431–457.

[46] A.J. Culyer; A. Wagstaff, 'Equity and Equality in Health and Health Care', in *Journal of Health Economics* 12(1994)4, pp. 431–457. See also G. Mooney, *op. cit.*, 1994, pp. 70–71.

[47] See also N. Daniels, *Just Health Care*, 1985, pp. 4–9.

[48] The major part of general health improvement is indebted to social policy and public health measures (like better nutrition and sanitation). Cf. Section 1.1 *Public and Private: a Historical Outlook*. See also: N. Daniels, *Just Health Care*, 1985, pp. 12–13; N. Daniels; B. Kennedy; I. Kawachi; A. Sen, *Is Inequality Bad for Our Health?* Boston: Beacon Press, 2000; Y. Denier, 'Public Health, Well-Being, and Reciprocity', in *Ethical Perspectives* 12(2005)1, pp. 41–66.

Secondly, the equity concept of equal health collides with the *internal* aspect of scarcity, i.e. we might wonder how far health care should go in pursuit of equity to equalise health or to stop someone's health status declining. For there are, firstly, a lot of internal factors as well; such as genetic endowment, personal response to treatment, personal preferences, and values with respect to health and health care (like being risk-averse and risk-loving).[49] It is difficult to see how health services could cope with different effectiveness of care across different individuals. Furthermore, we have seen that there is a contradiction between clinical possibilities in trying to improve health status and economic affordability, which underlies the idea of an inescapable gap between supply and demand.[50] To attempt to equalise health as much as possible would presumably require substantial investment, especially in those people born with poorer levels of health, and will most probably not be wholly successful because some poor levels of health can never be fully rectified.[51]

Too Strong. But suppose hypothetically we could provide an acceptable solution to both problems of external and internal scarcity, we then still might wonder whether health equality is a *desirable* equity goal. On a more fundamental level, it seems to be too strong a notion. For it would require that people would have the same or similar lifestyles. Mooney identifies two problematic factors here.[52] The first is that individuals would need to have the same *opportunities* to have that same lifestyle, which is problematic because some occupations may have more serious health consequences than others – for instance, police officers, fire fighters, members of the armed forces, miners, etc.[53] The second problem is more fundamental and refers to the fact that everyone would need to have the same *preference* for this lifestyle. This could be realised to some extent by encouraging healthy lifestyles and in some way discouraging unhealthy ones. However, with regard to the ideal of equal health, this strategy is limited and unlikely to realise the ideal in an acceptable way. For not to allow individuals to have different behaviour patterns reflecting different values for health and health care would seem to go rigidly against the sorts of principles of autonomy and personal liberty that are common to most societies.[54]

All in all, it seems that equality of health is too costly in terms of securing equal health and too intrusive in terms of respect for individual values. Although

---

[49] G. Mooney, *op. cit.*, 1994, pp. 69–73.

[50] J. Butler, *op. cit.*, 1999, pp. 5–11.

[51] This refers to the bottomless pit problem as developed by Arrow in K. Arrow, 'Some Ordinalist-Utilitarian Notes on Rawls's Theory of Justice', in *Journal of Philosophy* 70(1973)9, pp. 245–263; p. 251. Cf. Section 3.3.1 *Health Care: an Additional Primary Social Good?* See also N. Daniels, *op. cit.*, 1985, pp. 43–44.

[52] G. Mooney, *op. cit.*, 1994, pp. 71–73.

[53] See also E. Anderson, 'What's the Point of Equality?' in *Ethics* 109(1999) pp. 287–337, esp. pp. 296–297.

[54] E. Anderson, *ibid.*, 1999; Y. Denier, 'On Personal Responsibility and the Human Right to Health Care', in *Cambridge Quarterly of Healthcare Ethics* 14(2005)2, pp. 224–234. I will come back to this issue in detail in Section 5.5.3 *On Personal Responsibility and the Right to Health Care.*

health is a very important *bonum*, it is not the *summum bonum* that has to be realised at all costs. Individual persons should have an equal chance to choose a healthy lifestyle and to choose health and health care without actually being forced to be healthy. Let us have a look, then, at a less stringent conception.

2.3.1.4 *Equal Use for Equal Need?*

The idea that health-care resources should be allocated in line with health-care needs has a strong intuitive appeal. The concept of equity which aims at equal use for equal need does, however, need to give a definition of what is meant by 'need' and what is meant by 'use'.

NEED. Generally, need is seen as being determined by the objective viewpoint of a third party, usually a health-care provider; whereas preferences or demands are based on the subjective and personal wishes of the individual patient. In health economics literature, three definitions of need for health care have received much attention.[55] The first equates need for health care with *ill-health* and the degree of need with the severity of illness – the most severely ill have the greatest need for health care.[56] In this case, it is interesting to question whether one in case of severe illness can be in *need for health care* when the specific health care or treatment that is necessary to relieve this need is not available, or does not exist.[57] Can one be in need for something that does not exist?

Consequently, it might be more adequate to concentrate on the presence of effective treatment. The definition of need would then be strongly consequentialist or instrumental, centering on *effectiveness*. It implies, firstly, that need can only be defined with respect to a certain goal: x is needed to achieve y. In the case of health care, the central goal would be effectiveness or, put another way, health improvement. Consequently, the need only exists when x has been demonstrated to be effective in achieving y. In this case, i.e. if those most in need are also those who can *benefit* most from health care, then, under the efficiency objective of maximising health gain, equity and efficiency are not in conflict. The same amount of resources advances both efficiency and equity.[58] However, this would lead to the strange conclusion that no matter how ill a person is, if there is no effective treatment there is no need for health care:

> it is hard to see why someone who is sick can sensibly be said to need health care, irrespective of the latter's ability to improve the person's health. A person cannot surely be said to need health care if no technology is available to improve their health or prevent its avoidable

---

[55] J. Hurley, *op. cit.*, 2000, pp. 90–92; A.J. Culyer; A. Wagstaff, *op. cit.*, 1993, pp. 433–436. See also G. Mooney, *op. cit.*, 1994, pp. 73–81.

[56] Cf. B. Williams, 'The Idea of Equality', in P. Laslett; G. Runciman (eds.), *Politics, Philosophy and Society*, Oxford: Blackwell, 1962; repr. in B. Williams, *Problems of the Self*, Cambridge: Cambridge University Press, 1973, pp. 230–249; R. Gillon, *Philosophical Medical Ethics*, Wiley: Chichester, 1986.

[57] J. Hurley, *op. cit.*, 2000, p. 91; A.J. Culyer; A. Wagstaff, *op. cit.*, 1993, p. 434.

[58] A.J. Culyer, 'The Normative Economics of Health Care Finance and Provision', *Oxford Review of Economic Policy* 5(1989)1, pp. 34–58. *Id.*, 'Commodities, Characteristics of Commodities, Characteristics of People, Utilities, and Quality of Life', in S. Baldwin; C. Godfrey; C. Proppers (eds.), *Quality of Life: Perspective and Policies*, Routledge: London, 1990, pp. 9–27.

deterioration. They may need medical research, they may need comfort, and they may, most fundamentally of all, need health, but they surely do not need health care.[59]

Although I agree with the fact that (certain) needs originate from the presence of a specific framework that can fulfil them – in the sense, as I argued previously, that certain needs did not exist until the means of meeting them came into existence – I do not agree with the general equalisation of health-care needs with the presence of effective treatment, with *the capacity to benefit*. For this would lead to a narrow definition of health care, including only the pillars of prevention and cure. Care and social support for the chronically ill and the long-term disabled would then not be included, since in such cases, health improvement, understood as realising or restoring healthy normal functioning, is very often, if not mostly, not an option. Although health care in its strictly *conceptual* sense, i.e. care for health, might *logically* lead to equalising health-care needs with effective treatment, I do believe that taking the narrow route would be a true waste.[60]

A third proposal holds that a workable definition of health-care needs must be able to establish how *much* health care is needed. This is expressed, for instance, in the suggestion that 'If we say that A has greater needs than B, we simply mean that A needs a larger set of primary goods than B in order to achieve the same level of welfare.'[61] Consequently, it might be interesting to define need as *expenditures a person ought to have*.[62] This can be measured by the amount of resources (expenditure) required to effect maximum possible health improvement. Problems however, arise from a double conflation related to this interpretation.

Firstly, it conflates *extent of need* and the *amount of resources* required to meet the need. After all, by this definition a person suffering from a severe allergic reaction to a bee sting, who requires a simple and inexpensive antitoxin to prevent sure death, would have less need than a person with a moderate cataract who requires eye surgery to exhaust benefit.[63] Although the latter needs more health care (measured by expenditure), would we say that he has a greater need for health care? Probably not. The distinction between extent of need and the amount of resources required to meet the need becomes relevant when priorities must be set regarding the use of health-care resources either at the individual

---

[59] A.J. Culyer; A. Wagstaff, *op. cit.*, p. 434.

[60] I will express my objections against a one-sided emphasis on effective treatment more specifically in Section 2.4.3 *The Normative Value of Healthy Normal Functioning* and Section 2.4.4 *A Contemporary Socratic Perspective*.

[61] A. Weale, *Equality and Social Policy*, London: Routledge, 1978, p. 68. Note that Sen's critique on Rawls's theory of primary goods, referring to the latter as being fetishistic, is raised in the same spirit. Justice demands that we take individual differences into account. A handicapped person needs more goods (a wheelchair or a guide dog, for instance) than a non-handicapped person in order to lead a similarly independent life. Cf. Section 3.2.4 *The Inflexibility Critique*. This is also similar to Aristotle's idea of equality in geometrical proportion, and to the idea of vertical equity: unequal treatment of unequals. Cf. Section 1.3.1.1 *Suum Cuique Tribuere* and Section 2.3 *The Essence of Equity in Health Care*.

[62] A.J. Culyer; A. Wagstaff, *op. cit.*, 1993, pp. 435–436.

[63] I borrow this example from Hurley. Cf. J. Hurley, *op. cit.*, 2000, p. 91.

level or at population level. At the individual level, the principle of *triage* dictates that the person with the allergic reaction receives priority.[64] An analogous principle holds at the population level when allocating resources among regions is constrained so that each region must leave some needs unmet. One region's needs may be more urgent than another's, even if it requires fewer resources to meet them. Priority should be given to meeting all of these urgent needs before funding lesser needs in the second region. The point is that urgency does not necessarily equalise amount of resources.

The second conflation is that the concept of need is used here in two senses – a *technical* sense and a *normative* sense.[65] In a technical sense, one could hold that health care is needed as long as it has a positive marginal effect. But in a normative sense, one could deny that health care is needed, simply because society takes the view that a person in question *ought* not to have it. Within systems of public funding of health care, not just any need will be considered thus important that it has to be fulfilled by means of public resources. This means that even if certain health-care services or goods would turn out to be technically effective in meeting a certain health-care need, it might be that society does not consider it to be a need a person ought to have. Within the public framework, certain objectives are considered to be important and worthwhile, while others are not.[66] This refers to the social structure of public recognition of needs.[67] Let us put the discussion on the interpretation of needs aside for the moment and turn to the issue of use.

USE. Let us concentrate on the notion of use, whereby 'use' equals actual consumption.[68] It might be objected that the definition of equity as equal use for equal need leads to a problem similar to the previous definition of equity as equal health: equal use for equal need seems to ignore the possibility that individuals have preferences for health care and that the exercise of their preferences with respect to the use of services might lead to differences in the way in which services are used, as compared with what would occur if the service used were based solely on need. Strictly speaking, this formulation of the principle of

[64] Triage is a system used by medical or emergency care personnel to ration limited medical resources when the number of injured needing care exceeds the resources (medical supplies, time, doctors, nurses, etc.) available to perform care so as to treat the greatest number of patients possible. In most situations, patients are divided in groups like for instance, the deceased who are beyond help, the severely-injured who are unlikely to survive, the severely-injured who can be helped by immediate action and the lightly-injured who need help less urgently. See also: J. Butler, *op. cit.*, 1999, pp. 86–89.

[65] A.J. Culyer; A. Wagstaff, *op. cit.*, 1993, p. 436.

[66] This is often strongly related to the distinction between needs and mere wants or desires. I will come back to this distinction in detail in Section 2.3.2 *Needs and Preferences in Health Care*.

[67] As such, the needs discussion is much less objective and value-neutral as it seems to be at first sight. See also G. Mooney, *op. cit.*, 1994, pp. 73–81. Even more, the paradoxical use of the concept of need has led Brian Barry to say that the usage of the concept of need, can either become 'pathological' or can lead to the claim for instance that 'the disabled do not need wheelchair ramps'. In B. Barry, *Political Argument*, Hemel Hempstead: Harvester Wheatsheaf, 1991, p. lxviii. I will come back to this in detail in Section 2.3.2 *Needs and Preferences in Health Care*.

[68] G. Mooney, *op. cit.*, 1994, pp. 73–80.

equity seems – at least at the individual level – to be too coercive, implying that a person should receive health care even if he does not want it.[69] As such, it rides roughshod over personal autonomy, over heterogeneity of personal valuations of health, and of individual preferences for health care and health improvement. It seems a very patronising or paternalistic attitude not to allow individuals to choose whether to accept the medical advice. Consequently, unless one takes a very optimistic view of the medical relationship – in which the physician, or medical practitioner more generally, indeed knows best what serves the patient's well-being – and unless one believes in a paternalistic health service, it seems desirable to allow the patient's various preferences and ideas of the good life to be very relevant in the consumption of health care.[70]

People differ interindividually in their preferences and decisions to use health care. The hypochondriac is more easily inclined to visit the doctor than the person who is always 'fine'. Furthermore, people differ in their risk preferences. Some take the risk of undergoing an operation, while others do not. The point is that if we should take these preferences into account, the idea of equal use for equal need cannot count as an ideal of equity.

2.3.1.5 Equal Access for Equal Need?

ACCESS. Equality of access is another much-defended standard for health care.[71] Equal access to health care implies that everyone in society is equally able to obtain or make use of health care. As such, it respects *personal valuation* of health and preferences for using health care. It pertains to the ability or capacity to do something, and not whether it is actually done. The ethical basis for equality of access does not derive, then, from a necessary relation with its ultimate *effects* on the distribution of health, but is, on the other hand, intimately linked to the notion of *equal opportunity*.[72] As such, equality of access for equal need seems to have more affinity with what people generally believe to be the essence of equity in health care. If we take the principles of horizontal and vertical equity into account, it implies that those with equal needs should have equal access to services, whereas those with unequal needs should have differential access. Although everyone has

---

[69] I explicitly say 'individual level', because a lot of public health measures (hygiene, sanitation and vaccination) are necessarily coercive. See also: Y. Denier, 'Public Health, Well-Being, and Reciprocity', in *Ethical Perspectives* 12(2005)1, pp. 41–66.

[70] See also: Y. Denier, 'Autonomie en afhankelijkheid. Het subject in de medische ethiek', in R. Devos, A. Braeckman; B. Verdonck (eds.), *Terugkeer van het subject? Recente ontwikkelingen binnen de filosofie*, Leuven: Universitaire Pers, 2002, pp. 105–116.

[71] J. Hurley, *op. cit.*, 2000, pp. 89–90; G. Mooney, *op. cit.*, 1994, pp. 81–85.

[72] G. Mooney, *Key Issues in Health Economics*, 1994, pp. 81–85; E. Nord; J. Richardson; A. Street; H. Kuhse; P. Singer, 'Maximising Health Benefits vs Egalitarianism: an Australian Survey of Health Issues', in *Social Science and Medicine* 41(1995)10, pp. 1429–1437. From its affinity with equality of opportunity arises the relation between equality of access for equal need and the theories of Norman Daniels (fair equality of opportunity), and Amartya Sen and Martha Nussbaum (equality of capability). Cf. Chapters 3 and 4, respectively.

access to emergency care, priority is given to those most in need. The principle also justifies equality of access to primary care, but unequal access (based on private pocket payment) to other or higher-level forms of care.

NEED. The conclusion is that we best work with the concept of equal access for equal need. Nevertheless, it remains to be analysed which interpretation of needs can best be used. It is necessary to develop a consistent understanding of *health-care needs*. In what follows, I will leave the health economics forum and start the analysis of various philosophical viewpoints on basic needs, health-care needs, primary goods and basic capabilities. But before doing this, let me, for the sake of completeness, mention one more interpretation of equity in health care.

2.3.1.6 Equal Choice Set?

According to the definition of equity as providing equal choice set, *personal responsibility* is a crucial criterion. It holds that individual persons have equal access if they face the same choice set.[73] This means that they have the same level of income and wealth, the same information on health-related behaviour, etc., so that they would have equally informed access to health care, and thus equal opportunity to be healthy. Individuals would have the same opportunity to trade-off different goods and services to such an extent that differences in health, and health consumption would reflect nothing but differences in personal preferences. Consequently, if ill health results from factors beyond the person's control, the situation is inequitable. If, however, it results from voluntarily undertaking health-harming activities, the situation is equitable because this person has made choices under equal constraints. This differs from the situation in which the person did not undertake such behaviour because of the health risk.

Questions regarding the role of personal responsibility in health-care allocation regularly appear in contemporary debate and play an important role in discussions regarding the possibility and acceptability of rationing by responsibility.[74] It is an

---

[73] J. LeGrand, 'Equity, Health, and Health Care', in *Social Justice Research* 1(1987)3, pp. 257–274; *Id.*, *Equity and Choice*, London: Harper Collins Academic, 1991. See also: J. Hurley, *op. cit.*, 2000, pp. 89–90.

[74] R.M. Veatch, 'Voluntary Risks to Health', in *Journal of the American Medical Association* 243(1980)1, pp. 50–55; *Id.*, 'What is a "Just" Health Care Delivery?' in R.M. Veatch; R. Branson (eds.), *Ethics and Health Policy*, Cambridge, MA: Balinger Publishing Company, 1976, pp. 127–153; J. Butler, *The Ethics of Health Care Rationing: Principles and Practices*, London: Cassell, 1999, pp. 5–49; G. Dworkin, 'Taking Risks, Assessing Responsibility', in *Hastings Center Report* 11(1981), pp. 26–31; H. Leichter, 'Public Policy and the British Experience', in *Hastings Center Report* 11(1981), pp. 32–39, incorporated in H. Leichter, *Free to Be Foolish: Politics and Health Promotion in the United States and Great Britain*, Princeton, NJ: Princeton University Press, 1991; L.B. Russell, 'Some of the Tough Decisions Required by a National Health Plan', in *Science* 246(1989), pp. 892–896; R.L. Schwartz, 'Life Style, Health Status, and Distributive Justice', in A. Grubb; M.J. Mehlman, *Justice and Health Care: Comparative Perspectives*, New York: Wiley, 1995, pp. 244–248; Y. Denier, 'On Personal Responsibility and the Human Right to Health Care', in *Cambridge Quarterly of Healthcare Ethics* 14(2005)2, pp. 224–234. Cf. Section 5.5.3 *On Personal Responsibility and the Right to Health Care*.

important view held by many, translating the common idea that one pays for the consequences of one's own choices. As such, it definitely deserves an argumentative place in the determination of the scope of the moral right to health care. It plays a role in *helping* to set *limits* on the scope of the moral right to health care, but for various reasons that I will discuss in Part II, it cannot be the *basis* on which our conception of just health care is built. The basis of it has, as I will argue in the following sections, to do with equal access for equal need. Responsibility only comes in at a later stage. Let us have a closer look at the significative structure of health-care needs.

2.3.2 Needs and Preferences in Health Care

What exactly are health-care needs? Are *all* health-care needs so important that they need to be taken into account in the public health-care package? If not, can we in some way set a difference between morally important health-care needs and those that cannot adequately serve as a basis for moral claims on resources? To answer these questions, it is reasonable to inquire first into the structure of the concept of needs as such, before concentrating on what it implies in the field of health care.

2.3.2.1 The Principle of Precedence

To begin with, we can say that in present-day moral and political discourse the assertion that something is 'needed' plays a powerful role, especially in matters of resource allocation and argumentations, which support demands, influence the existing order of priorities or aim at establishing rights. In general we respond to attributions of need with special respect and concern. The assertion that something is needed tends to have a substantially greater moral impact than the assertion that something is desired or preferred. Needs make a claim on us more compelling than mere preferences do. Consequently, needs are likely to be treated with greater urgency and priority than desires and preferences. This priority of needs over desires is also described as the 'principle of precedence'.[75]

It is, however, important not to claim too much for needs and draw attention to the exceptions to the principle of precedence since not *every* need complies with the principle. This has to do with the way in which needs and preferences work together. Although they differ top and bottom from one another, they are at the same time strongly related. If we want to reach a clear view on the distributive duties of society, we will have to reach an understanding of how needs and preferences are related. For instance, it is reasonable to think that only in rare cases social policy would be directed at meeting needs without any regard for heeding preferences. When the preferences are compatible with the needs, social policy will most likely try to meet the needs with forms of provision suited

[75] See: D. Braybrooke, *Meeting Needs*, Princeton, NJ: Princeton University Press, 1987, pp. 60–75; H.G. Frankfurt, 'Necessity and Desire', in *id.*, *The Importance of What We Care About*, Cambridge: Cambridge University Press, 1988, pp. 104–116; R.E. Goodin, 'Priority of Needs', in *Philosophy and Phenomenological Research* 45(1985), pp. 615–625.

to the preferences. But when people's preferences run contrary to their needs, the project of meeting their needs is suddenly much less clear-cut. Social policy might then either ignore the preferences, and – because of very good reasons – continue to meet the needs anyway, or it might have reason enough to not ignore the preferences and back away from meeting the needs. Why do some needs comply with the principle of precedence and others do not? Let us have a closer look at the significative structure of the relation between needs and preferences.

2.3.2.2 Needs in Philosophical Disrepute?

First of all, as the principle of precedence has suggested, it seems to be the case that needs possess great normative force. At first sight, needs are likely to possess an *objective* moral importance, which as such gives it a more *compelling* character than can be attached to preferences. This seems to originate in the fact that a person has no *control* over what he needs and that he will suffer fundamental and crucial *harm* in case of failure of meeting the need.[76] On closer view, however, this is not evident. For various reasons the normative force of needs is not unsuspected.

VARIATION. First of all, needs vary enormously from person to person, between different cultures, and throughout history.[77] This variation is so great that they seem to overshadow the idea that there are common, universal needs. Within time and culture, people may not be free to choose between differences; they may (without feeling any loss of freedom) be strictly governed by convention. But if anyone is ever in a position to choose between conventions, is not the choice then just a matter of preference instead of need? This may lead us to conclude that needs are so largely determined either by conventions or by preferences that it is misleading to pretend that there are natural and essential grounds for them to have such objective normative force.[78]

CUT TWO WAYS. In general, the charges against the normative use of the concept of needs come down to the fact that needs-based perspectives cut two ways. Either they give us too small a claim on social resources providing only a floor on deprivation, or they imply too great a claim on resources, justifying 'needs-based'

---

[76] D. Braybrooke, 'Let Needs Diminish That Preferences May Prosper', in N. Rescher (ed.) *American Philosophical Quarterly*, Monograph Series, No. 1, Oxford: Basil Blackwell, 1968, pp; 86–107; *Id.*, *Meeting Needs*, Princeton, NJ: Princeton University Press, 1987; D. Copp; *Morality, Normativity, and Society*, Oxford: Oxford University Press, 1995, pp; 167–177; H.G. Frankfurt, 'Necessity and Desire', in *id.*, *The Importance of What We Care About*, Cambridge: Cambridge University Press, 1988, pp. 104–116; H.J. McClosky, 'Human Needs, Rights, and Political Values', in *American Philosophical Quarterly* 13(1976)1, pp. 1–11; D. Wiggins, *Needs, Values, Truth*, 3rd edn., 1998, Oxford: Oxford University Press, pp. 10–22; C. Wolf, 'Theories of Justice: Human Needs', in R. Chadwick (ed.), *Encyclopedia of Applied Ethics*, San Diego: Academia Press, 1998, pp. 335–345.

[77] D. Braybrooke, *op. cit.*, 1968, pp. 99–102; *op. cit.*, 1987, pp. 11–14; H.J. McClosky, *op. cit.*, pp. 8–9; C. Wolf, *op. cit.*, 1998, pp. 337–338.

[78] I. Illich, *op. cit.*, 1975; H. Achterhuis, *op. cit.*, 1988; N. Xenos, *op. cit.*, 1989.

social hijacking of resources by persons who simply have expensive tastes.[79] The concept seems both too weak and too strong to get us very far toward a theory of distributive justice. Too many things become needs, or too few.

Too Many Things. It is easy to see why too many things become needs, because need is an instrumental concept.[80] Without abuse of the language, we refer to the means necessary to reach any of our goals as 'needs'. To write something down, I need a pen. To play tennis, I need a racket. To go to Ferrals-les-Corbières for a week, I need to book a room. Taking the *instrumental* function of needs into account, together with its *normative* force, which installs a standard of comparison of relative well-being and relative sacrifice for various persons, it is clear that needs might easily become opportunistic.[81] For emphasis we often refer to things we simply want or desire as things we need: I can insist that I most certainly *need* to buy the latest album of my favourite musician or that I absolutely *need* to have this beautiful pair of shoes. Given the chance, needs – just like wants or desires – seem ready to expand indefinitely and in a fully unstructured way.

Abuse of the Concept of Social Justice. Indeed, the concept of need attracts abuse. Much hangs on abusing it, for wary as sophisticated people may be about demands put forward by it, to get something accepted as a need still makes a substantial difference to getting it attended to as an object of social policy.[82] Acceptance as a need may have been arrived at through unscrupulous stretching. In this connection, abuse of the concept of need facilitates abuse of the concept of social justice, leading people to claim under the heading of justice either things that concern relatively peripheral matters or things that essentially do not belong under that heading, only in order to lend them special moral weight.[83] As such, the case of social hijacking by persons with expensive tastes transforms into the problem of social hijacking by needs.[84] What people themselves claim to

---

[79] Cf. N. Daniels, *op. cit.*, 1985, pp. 22–23: 'Egalitarians use [needs-based theories] to criticize the failure of inegalitarian systems to meet basic human needs. Inegalitarians use them to justify providing only minimally for basic needs while allowing significant inequalities above the floor'. See also: R.E. Goodin, *op. cit.*, 1985, pp. 621–265; H.G. Frankfurt, *op. cit.*, 1988, p. 105.

[80] C. Wolf, *op. cit.*, 1998, p. 336; H.G. Frankfurt, *op. cit.*, 1988, pp. 106–109.

[81] Cf. H.G. Frankfurt, *ibid.*, 1988, p. 104: 'The language of need is used extensively in the representation of our personal and social lives. Its role in political and moral discourse is especially conspicuous and powerful. People commonly attribute needs to themselves and to others in order to support demands, or to establish entitlements or to influence the order of priorities; and we are often inclined to respond to such attributions with a rather special respect and concern'. See also: T.M. Scanlon, 'Preference and Urgency', in *The Journal of Philosophy* 72(1975)19, pp. 655–669; and David Braybrooke about the problem of overextension of the concept of needs in D. Braybrooke, *op. cit.*, 1968, pp. 100–107.

[82] D. Braybrooke, *op. cit.*, 1987, pp. 18–19.

[83] For a detailed analysis of this problem with regard to the techniques of assisted reproduction, see: Y. Denier, 'Behoefte of verlangen? De betekenisstructuur van de kinderwens' in *Ethische Perspectieven* 10(2000)3, pp. 163–174. For a revised English version of this paper, see: Y. Denier, 'Need or Desire? A Conceptual and Moral Phenomenology of the Child Wish', in *International Journal of Applied Philosophy* 20(2006)1, pp. 81–95.

[84] The case of social hijacking by expensive tastes is analysed in J. Rawls, 'Social Unity and Primary Goods', 1999/1982, pp. 359–387. See also N. Daniels, *op. cit.*, 1985, pp. 36–37.

need for their subsistence is always considerably more than the basic minimum required to sustain life. To avoid *needs-inflation*, one might invoke a distinction between noun and verb uses of 'need', so that not everything we say we need counts as *a* need.[85] This, however, does not answer the question of which asserted needs are *true* needs, and which are not.

PATERNALISM. Another problem that we must be careful to avoid is that of illegitimate paternalism, because the variability of the concept of need, together with its normative force, aggravates the problem of paternalism that haunts it.[86] We must be aware of the fact that individual stretching of the concept of needs paradoxically enough can be nothing else than a counter-reaction under the banner of 'rights' to contest patronising rhetoric of needs used by paternalistic authorities to improve the condition of the poor and dependent disadvantaged.[87] Arrangements that embody paternalism are suspect on practical grounds as well as on the moral ground of threatening personal autonomy and liberty. People must be able to claim the liberty to make their own decisions about the relative weights that they are going to give to needs, disputed or not, as against their preferences.

TOO FEW THINGS. On the other hand, it is also possible that too few things become accepted as needs. Since the concept of needs either is unabashedly expansive or stimulates paternalism, one might think it better to reduce the problem of importance of needs to the problem of importance of preferences or wants.[88] The concept of needs is then understood as a superfluous concept, being unnecessarily and misleadingly rigid. It misleadingly claims something like lexicographical priority, which would imply that at least until all needs have been met at accepted minimum standards, the tiniest increment of provision for needs is to be preferred to any provision for anything else. However, since the minimum standards are so far conventional or individual that this kind of priority is not plausible, the concept will soon break down. Would it therefore not be better to speak of preferences instead of needs, since needs can be conceived of being entirely subordinate to preferences? Statements about needs would thus, in principle, better be replaced by statements about what informed people prefer. Consequently, the difficulties that come with the problem of variability and with the problem of expansion cease to be difficulties, since there is no problem about preferences varying in all these ways. This is the move made by the argument from fair shares.

On the other hand, the argument countering this move has shown that there are profound reasons to believe that the concept of needs *does* play an important role

---

[85] H.J. McClosky, *op. cit.*, 1976, pp. 2–4; N. Daniels, *op.cit.*, 1985, p 26, fn. 5.

[86] I will come back to the difference between legitimate and illegitimate paternalism in Section 4.2.5.4 *Paternalism*.

[87] See also: Y. Denier, 'Public Health, Well-Being, and Reciprocity', in *Ethical Perspectives* 12(2005)1, pp. 41–66.

[88] N. Daniels, *op. cit.*, 1985, pp. 19–23; D. Braybrooke, *op. cit.*, 1987, p. 17.

in moral discourse. With regard to our problem in hand, this concept is necessary to express the essence of equity in health care. However, in trying to avoid the above problems of social hijacking on the one hand, and of illegitimate paternalism on the other, it is necessary to determine a typology of needs. *How* can we distinguish morally important needs from less important needs? Let us have a look at the criteria of relative well-being that are being used in moral philosophy to function as a standard on the basis of which distributive claims are to be judged.

2.3.2.3 Scanlon's Proposal: Urgency

Arguments in moral philosophy frequently concern themselves with appeals to some *standard* on the basis of which the benefits and sacrifices of different people can be compared. In applying principles of *justice*, we must appeal to some standard of this kind as a ground for measuring *equality*, and similar appeals must be made in defending corresponding *rights*, understood as institutionally defined prerogatives and protections.

CRITERIA OF RELATIVE WELL-BEING AND RELATIVE SACRIFICE. According to Thomas Scanlon, criteria of well-being should have three properties if they are to play the role commonly assigned to them in moral argument.[89] Firstly, if they are to serve as a basis for criticism and justification of institutions, criteria of well-being must represent a kind of *consensus*. Secondly, adequate criteria must allow for the fact of individual *variations in taste*. Finally, an adequate criterion of well-being will have to be *outcome-oriented*; i.e. it will provide an evaluation of the ways in which individuals may be affected by having these goods. This is necessary to guarantee sensitivity of the criterion to *variations* in needs, e.g. variations arising from physical disabilities.[90]

SUBJECTIVE AND OBJECTIVE CRITERIA. Furthermore, it is possible to distinguish subjective from objective criteria of well-being. Scanlon defines a subjective criterion as:

> a criterion according to which the level of well-being enjoyed by a person in given material circumstances or the importance for that person of a given benefit or sacrifice is to be estimated by evaluating those material circumstances or that benefit or sacrifice *solely* from the point of view of that person's *tastes* and *interests*.[91]

Taking the above-listed desiderata into account, we can say that subjective criteria certainly comply with the condition of recognition of *sovereignty* of individual tastes. Subjective criteria may vary in the way in which they identify individual preferences; and it is a person's preferences that constitute the ultimate standard for judgements about that person's well-being. Secondly,

---

[89] T.M. Scanlon, *op. cit.*, 1975, pp. 655–669.

[90] H.J. McClosky, *op. cit.*, 1976, p. 3.

[91] T.M. Scanlon, *op. cit.*, 1975, p. 656 (my italics, YD). An example is hedonistic utilitarianism, as well as the utilitarian position of Harsanyi in J.C. Harsanyi, 'Cardinal Welfare, Individualist Ethics, and interpersonal Comparisons', in *Journal of Political Economy* 63(1955), pp. 3.9–3.21; *Id.*, 'Can the Maximin Principle Serve as Basis for Morality? A Critique of John Rawls's Theory', in *American Political Science Review* 69(1975), pp. 594–606.

subjective criteria are *outcome-oriented* and can clearly accommodate context-sensitivity of social, cultural and physical variations in need.

That subjective criteria can be held to represent a *consensus* on which moral criticism and social policy can be based is less apparent. Most probably such consensus will entail that all individuals agree with institutions that guarantee the sovereignty of individual tastes but no more than that (e.g. the libertarian perspective). Whether or not some interests should be favoured at the expense of others in the design of distributive institutions or in the allocation of other rights and prerogatives, is merely determined by the strength of the subjective preferences they represent. There are no objective, independent standards of importance involved.

Despite the fact that subjective criteria appear to be a very attractive basis for moral appraisal and policy – for how, after all, would any other criteria of relative well-being be defended if not, ultimately, by appeal to individual preferences? – it is clear that the criteria of well-being that we actually employ in making moral judgements are objective.[92] Why is that so? Is it for mere practical reasons – because individual preferences are too varied and too many to be taken into account – or is it because they truly provide a better account of human well-being? Scanlon defines an objective criterion as:

> a criterion that provides a basis for appraisal of a person's level of well-being which is *independent* of that person's tastes and interests, thus allowing for the possibility that such an appraisal could be correct even though it conflicted with the preferences of the individual in question ....[93]

Firstly, objective criteria can be *outcome-oriented* and admit various kinds of variation – i.e. social, historical, cultural and individual – in need. Furthermore, an objective criterion of well-being need not deny the relevance of *subjective preference* altogether. A high objective value may be attached to providing a framework of conditions, fulfilment of which is necessary to allow individuals to develop their own preferences and interests and to make these felt in the determination of social policy.

The crucial advantage of using objective criteria of well-being lies in its *basis* for *consensus*. This is the idea that, insofar as we are concerned with moral claims that some interests should be favoured at the expense of others in the design of distributive institutions or in the allocation of other rights and prerogatives, it is an objective evaluation of the objective *importance* of these interests, and not merely the *strength* of the subjective preferences they represent, that is decisive. The point is that a subjective criterion seems insensitive to differences between preferences that refer to something that is *of great relevance* when these preferences are taken as the basis for moral claims, and preferences that do not. On an objective criterion, Scanlon argues, we can allow for individual variation in preference and give individual autonomy an important place without making relative

---

[92] T.M. Scanlon, *op. cit.*, 1975, p. 658.

[93] *Ibid.* (My italics, YD).

strength of subjective preferences the foundation of our distributive theory. The point that Scanlon makes is that even if it is the case that in assessing moral claims we discriminate among equally 'strong' preferences, it requires to be explained on what *basis* we do this and how the use of this basis is to be supported.[94]

URGENCY AND IMPORTANCE. When we compare conflicting interests with the aim of supporting moral judgements as to which of them should be allowed to prevail, what we do is not compare with how *strongly* the people in question *feel* about these interests, but rather inquire into the *reasons* for which they are considered desirable. Why are some interests of greater relevance than others? In this regard Scanlon proposes the criteria of *urgency* and *importance*. The urgency of a benefit will not only depend on the reason for which it is desirable. It is also relevant to judge how well off the person concerned would be with respect to this category without the benefit, or in particular, with respect to what alternatives are available and what sacrifices would be involved in shifting to one of these alternatives. Within such a picture some concerns will appear as relatively peripheral and others will appear as being more central, more important, thus installing a rough hierarchy of relative urgency. This depends of course on the question we are addressing.

This proposal, however, does not yet sufficiently solve our problem, for urgency is in itself also a two-faced notion. I may still believe that some things – whatever they are – are very urgent. The notion of urgency alone does not give us a reason why a person may not assign an unusually high value to what would normally be considered a trivial concern. It is, on the other hand, quite another thing to say that urgency is the basis of legitimate moral claims in case of sharing out scarce social goods.

We need a further characterisation. As Scanlon suggests, the legitimacy of moral claims that rightfully influence distribution has something to do with whether the need in question is voluntary or not. A strong argument against letting the subjective criterion rule is the argument that there is a strong *voluntarist* connotation related to preferences.[95] Something does not count as one of our desires unless we *identify* with it.[96] This seems to be a necessary condition for preferences. Desires involve concerns that a person need not have developed. Things could have been otherwise; he could have taken up some other pursuit. This makes some concerns peripheral and other central. Consequently, it is interesting to question how voluntary needs can be. Thomas Scanlon does not elaborate this suggestion of the connotation with voluntariness, but Harry Frankfurt does. Let us have a look at his typology of needs.

---

[94] This is necessary to avoid hijacking of social resources by opportunistic overextension of the concept of needs. See also: N. Daniels, *op. cit.*, 1985, p. 24.

[95] T.M. Scanlon, *op. cit.*, 1985, pp. 664–667.

[96] This is related to the age-old question, which goes back to the classical discussion between Socrates and the Sofists: do we find something important because it *is* good, in itself; or is something good because *we* find it important?

2.3.2.4 *Frankfurt's Characterisation: Volition and Harm*

Why do some needs comply with the principle of precedence and others do not? How can we set the difference between needs that are truly urgent and needs that merely function as intensification of a strong preference? In order to set the difference and to identify the needs that comply with the principle of precedence, it is necessary, Harry Frankfurt, argues, to concentrate on three issues, i.e. firstly, on the categorical character of the need in question; secondly, on the nature of its instrumental status; and thirdly, on its relation to harm.[97]

CATEGORICAL CHARACTER. Frankfurt distinguishes between three types of need: free volitional needs, constrained volitional needs and non-volitional needs. A *free volitional need* is a need that derives from a free and voluntary desire for a certain end, like for instance the need for a racket when one wants to play tennis. A *constrained volitional need* is caused by an involuntary desire, an inclination that emerges and continues to exist independently of the person's free will. An example of such a desire would be falling in love or jealousy. A *non-volitional need* is a need that exists fully independent from any desire whatsoever – such as the basic need for food or water. The need is always present, irrespective of whether one wants it or not.

Because of their involuntary character, Frankfurt argues, only the last two types of need comply with the principle of precedence. A free volitional need enjoys no moral priority according to the principle of precedence because it contains too little necessity in two respects. First of all the individual concerned has *control* over having the need. In other words, he does not really 'need' what he desires. Secondly, as a consequence it cannot be assumed that the need *necessarily* must be met. A need, however, which does fulfil the twofold necessity, is a *categorical need*: the individual concerned has no control over the need; hence we can assume that it necessarily has to be met. Constrained volitional needs as well as non-volitional needs are categorical needs. As such, both needs have to be met and have moral priority over corresponding desires.

INSTRUMENTAL STATUS. Secondly, Frankfurt argues in the line of Scanlon, it is important to bear in mind the instrumental status of the need.[98] Needs are always of *instrumental* value for the projects in living a human life: 'All necessities are in this respect conditional: nothing is needed except in virtue of being an indispensable condition for the attainment of a certain end.'[99] David Braybrooke calls this *the relational formula* that is typically attached to the concept of need.[100] When something is needed, it must therefore always be possible to specify what it is needed *for*. As such, meeting a particular need is not important in itself, but only as a necessary means to reach a certain *end*. The moral importance of the need thus also always depends on the moral importance of the end. This means

[97] H.G. Frankfurt, *op. cit.*, 1988, pp. 104–116.
[98] See also: C. Wolf, *op. cit.*, 1998, p. 336.
[99] H.G. Frankfurt, *op. cit.*, 1988, p. 106.
[100] D. Braybrooke, *op. cit.*, 1987, pp. 29–32.

that, again, some needs are morally more important than others, depending on the moral importance of the end. Or, as Braybrooke puts it:

> N needs *x* in order to *y*. Once this formula has been filled in, one can judge the importance of the need by the importance of the end *y*, or, more exactly, by the importance of N's attaining that end.[101]

Which needs are important and which are trivial depends to a considerable extent on social, cultural, economic and historical determinants. In some cultures, for instance one can experience a need to have a large family because of the social status attached to it while this does not play a significant role in other cultures and societies. Some needs evolve within the historical, technological and economic development of a society. In times when there was no organised national social security system, the size of a family was of far greater impact on personal well-being. With the development of a social security system the size of a family has lost its meaning in terms of the need for social security.

Nevertheless, *some* needs are simply independent from time and space, because they refer to things that are generically necessary for the fundamental projects of *every* person. They are universal and basic and differ, as such, from particular needs that originate from specific wants and desires. In the same line of reasoning, David Braybrooke distinguishes between 'basic course-of-life needs' and 'adventitious needs'.[102] The basic needs or course-of-life needs are the things that are necessary for the most fundamental projects, involved in living a human life, and which are essential to living or functioning normally. Adventitious needs are less urgent, non-basic needs that originate from particular desires, values and preferences, and which need not be associated with normal functioning in the same way.

HARM. This brings us to the third factor that helps us answer the question why some needs have to be met while others do not; why some needs answer to the principle of precedence and others do not. This is the relation to harm. As Braybrooke puts it, it is often regarded essential or criteriological of the concept of basic course-of-life needs that it is associated with the concept of harm, such that if one's basic needs are not satisfied, one will be harmed in some crucial or fundamental way. Adventitious needs are less *urgent* simply because they are not associated with *harm* in this way.[103]

It is all but simple, however, to state unambiguously when people are harmed because of non-fulfilment of needs. Frankfurt tries to solve this by differentiating between actual harm, mere failure to obtain the desired benefit and actual harm caused by failure to obtain the desired benefit.[104] In addition, three elementary observations on the relationship between benefits and harm should offer more clarity.

---

[101] In: D. Braybrooke, *op. cit.*, 1987, pp. 29–32.
[102] D. Braybrooke, *op. cit.*, 1987, pp. 29–33; *Id. op. cit.*, 1968, pp. 86–107.
[103] C. Wolf, *op. cit.*, pp. 336–337; D. Braybrooke, *op. cit.*, 1987, p. 48.
[104] H.G. Frankfurt, *op. cit.*, 1988, pp. 109–110.

Firstly, being actually harmed amounts to becoming worse off than before, whereas failure to obtain a desired benefit is merely a matter of not becoming better off. Respective examples are the consequences of not receiving proper medical care after an accident and not being able to move to a house with garden when living in an apartment, provided that is what one desires. Secondly, Frankfurt puts it that there is sometimes no other way to avoid a worse-off situation than by making it better. In this case failure to obtain a change for the better is identical with being harmed. Lack of proper education would be a good example. Thirdly, it is reasonable to hold that the life of a person whose conditions of living are poor gets worse as long as the situation does not improve, merely because the longer one lives in bad conditions, the worse it is. This way a person can be harmed even though his condition does not deteriorate. Chronic illness is an example.

Let us take stock. Applied to the different types of need this implies that non-fulfilment of a *free volitional need* amounts to a situation of failure to obtain improvement, because the context of the need is a voluntary desire for improvement. In originating from a voluntary desire, moreover, the free volitional need is a personally created necessity. One has control over it and one can get rid of it. As such the need is not *inextricably* linked to harm and cannot claim a distinctive moral quality.

The situation is quite different in the case of constrained volitional needs and non-volitional needs. Non-fulfilment of a *non-volitional need* always means actual harm since meeting the need is situated in the context of avoiding deterioration. Moreover, the non-volitional need is a categorical need, completely independent from whichever desire. It remains present, even though one does not want it at all and is aching not to have it.

An unfulfilled *constrained volitional need* entails harm too, although we have to interpret this within the context of an involuntary desire which, as a desire, aims at improvement, but whose involuntary character at the same time implies deterioration when not fulfilled. The constrained volitional need is also categorical. The desire that generates the need escapes the control of the person concerned; it cannot be suspended; it persists and remains ineradicably present.

Note however, that it may be right to add the observation that the term 'categorical' has different applications here. Whereas non-volitional needs are *universally* categorical, constrained volitional needs are *particularly* categorical. The former apply to everybody, without distinction; the latter, however, apply only to those who have the involuntary desire. As a result the inextricable link with harm is universally categorical in the first case, particularly categorical in the second. Frankfurt himself does not add this specification. In cases of priority setting, however, this specification does install a useful additional hierarchy in the typology of needs.

2.3.2.5 Objective Truncated Scale of Well-Being

The essence of what is going on here – differing subjective and objective criteria of well-being, pointing out urgency, the categorical character, its instrumental status and the issue of harm – is that in contexts of distributive justice we do not only appeal to *objective* criteria of well-being, but in fact to a *truncated* or *selective* scale of well-being.[105] What is distinctive about the scale we use is that it does not include or reflect the full or complete range of an individual's needs, preferences and their fulfilment. The goal of issues of distributive justice is not to secure complete subjective happiness. Furthermore, *not all* things we find *objectively* important may count as relevant enough to give them a weightier moral claim in contexts involving the distribution of resources. This reflects the fact that the difference between both scales might not be in the extent but in the *content* of the scale.

The *objective truncated scale of well-being* reflects as such two important properties of universal needs. Firstly, they are *objectively ascribable*, i.e. we can ascribe them to someone even if he does not realise he has them or even if he denies he does because his preferences run contrary to his needs. Secondly, and of greater interest, is that they are *objectively important*: we attach special weight to claims based on them in a variety of moral contexts. Consequently, our task is to characterise the class of things we need, which has these properties and to do so in a way that explains their importance. In this regard it is possible to identify *three strategies* in the characterisation of universal needs; i.e. of needs that posses the special character that gives them an important moral claim in contexts involving the distribution of resources.[106]

2.3.3 Basic Needs, Primary Goods and Basic Capabilities

Which needs can function as a standard for social policy? Which are the needs that are crucial to assessing social policy? The first strategy in the characterisation of universal needs is the anthropological strategy. Based on the generic characteristics of human nature – the limits of human existence, the idea of man as a deficient being, a 'Mängelwesen', the first strategy argues that we all need certain goods (adequate nutrition, shelter, medical care, etc.).[107] The second strategy entails a Rawlsian translation of this anthropological basis in the language of primary social goods. The third strategy is a translation in the language of basic human capabilities (Sen and Nussbaum). Let us turn to the first strategy.

2.3.3.1 Basic Needs: An Antropological Foundation

Which needs count as non-volitional course-of-life needs; i.e. which needs are such that all human beings may be expected to have them all through their lives or at certain stages of life through which all must pass and that a deficiency with respect to them is a deficiency of great importance; that is, a deficiency that causes *actual harm*?

---

[105] N. Daniels, *op. cit.*, 1985, pp. 23–26.

[106] C. Wolf, *op. cit.*, 1998, pp. 337–338; L. Thielmann, *op. cit.*, 2001, pp. 146–149.

[107] A. Gehlen, *Der Mensch, seine Natur, und seine Stellung in der Welt*, Frankfurt: Klostermann, 1993, p. 16.

Searching for the point where the relational formula stops coming up again may provide an answer.[108] It means that we have to try to find the point at which further information is unnecessary to justify the need. It turns out, Braybrooke argues, that questions about whether needs are genuine, or well-founded, come to the end of the line when the needs are connected with *life*, *health* and *normal functioning* as a *human being*. For instance, a person needs food and water in order to live; and needs exercise in order to function normally.

Although these answers are to a certain extent hardly illuminating and almost banal, they indeed have the advantage of expressing the point at which we reach the limits of the relational formula. One does not need a further explanation or justification why aiming to live, being healthy and functioning normally is important. Being essential to living or functioning normally may thus be taken as a criterion of a basic need. A deficiency with respect to them 'endangers the normal functioning of the subject of need considered as a member of a natural species'.[109] A related suggestion can be found in McClosky's discussion of the human and personal needs we appeal to in political argument. He argues that basic or course-of-life needs 'relate to what it would be detrimental to us to lack, where the detrimental is explained by reference to our natures as men and specific persons'.[110]

Note furthermore, that the relational formula stops at the point where it expresses the *limits of human existence*, i.e. our *finitude*. The immortal gods of the Greeks had passions, pleasures and all sorts of motivations, and they even suffered occasional agony, especially from the interferences and obstructions practised by their fellow gods. However, they did not have to meet any conditions to go on living and acting, and were consistently conceived as not having to. Human beings on the other hand are not, like the gods, immune to bodily impairment.

Based on the characterisation of finitude that is essentially attached to human existence, David Braybrooke proposes a minimal list of needs.[111] It is generated by considering what human beings must have if they are to live and function normally. The list has two parts, of which the first part is strongly coloured by notions about *physical* functioning. It includes

> [The need] 1. ... to have a life-supporting relation to the environment; 2. ... for food and water; 3. ... to excrete; 4. ... for exercise; 5. ... for periodic rest, including sleep; and 6. ... for whatever is indispensable to preserving the body intact in important respects.

---

[108] For note that the relational formula 'N needs x in order to y' in most cases keeps coming up again, over and over: 'N needs fresh bat's urine to finish his experiment on biological clocks; but does he need to finish his experiment on biological clocks? And if he needs to do that to keep his job as an experimental zoologist, does he need to keep his job? Perhaps he is growing stale in it,' etc. in: D. Braybrooke, *op. cit.*, 1987, p. 30.

[109] D. Braybrooke, *op. cit.*, 1968, p. 90.

[110] H.J. McClosky, *op. cit.*, 1976, p. 2.

[111] D. Braybrooke, *op. cit.*, 1987, p. 36.

The second part has to do with functioning as a *social* being:

> [The need] 7. ... for companionship; 8. ... for education; 9. ... for social acceptance and recognition; 10. ... for sexual activity; 11. ... to be free from harassment, including not being continually frightened; and 12. ... for recreation.

This minimal list of matters of need generates minimum standards of provision for each such matter within a framework of social policy; thus reflecting the principle of precedence that gives these course-of-life needs priority over preferences. In general, we can define two strategies of further specifying these standards of provision as an objective of social policy, one referring to the primary goods perspective of John Rawls, and the other referring to Martha Nussbaum's list of basic human capabilities.[112]

2.3.3.2 Primary Social Goods: the 'Thin' Strategy

In setting the standards of provision for the basic needs of free and equal people, John Rawls proposes a list of primary social goods: a set of basic rights and liberties, powers and prerogatives of offices and positions; income and wealth; and the social bases of self-respect.[113]

These goods are allocated by the basic social, political and economic framework of society and should be distributed justly, so that every person can form, revise and pursue his own conception of the good life. They count as all-purpose means for achieving directly or indirectly a wide range of ends, whatever these happen to be. As such, they answer to Scanlon's proposed property of allowing for the fact of individual *variations in taste, wishes and desires*.[114]

The primary social goods are *primary* goods because they are things that persons need in their status as free and equal citizens, and as normal and fully cooperating members of society over a complete life.[115] They are *social* primary goods in view of their connection with the basic structure: liberties and opportunities are defined by the rules of major institutions and the distribution of income and wealth is also regulated by them; taken together they form the social basis of self-respect.[116] Rawls acknowledges the presence of other kinds of primary

---

[112] C. Wolf, *op. cit.*, 1998, pp. 337–338; L. Thielmann, *op. cit.*, 2001, pp. 146–149.

[113] J. Rawls, *A Theory of Justice*, 1971, pp. 62, 440; 1999, pp. 54, 386.

[114] T.M. Scanlon, *op. cit.*, 1975, pp. 655–656.

[115] *Theory*, 1999, p. xiii. It is important to note that this account of primary goods replaces the famous original account according to which primary goods were said to be 'things that every rational person is presumed to want, whatever he may hope or plan to get out of life' (*Theory*, 1971, p. 62; 1999, p. 54). The weakness of the original account was that it left ambiguous whether something's being a primary good depends solely on the natural facts of human psychology or whether it also depends on a moral conception of the person that embodies a certain ideal. With the revised account, this ambiguity is resolved in favour of the latter, including a specific political conception of the person. This account is given since Rawls's essay 'Social Unity and Primary Goods', published in 1982.

[116] *Ibid.*, 1971, p. 92; 1999, p. 79.

goods, including 'natural goods' such as health and vigour, intelligence and imagination; but the distribution of these, he argues, is not so directly influenced by the basic structure.[117] As such, they are morally arbitrary:

> The natural distribution is neither just nor unjust ... These are simply natural facts. What is just and unjust is the way that institutions deal with these facts.[118]

A just society would be one in which its basic structures were so arranged that everyone had a fair share of the primary social goods, thereby compensating for the arbitrariness of fortune.

This strategy is expressly meant to be a *thin* strategy. It is consciously and deliberately not specified in more detail because it wants to be able to serve as a basis for realisation of multiple reasonable conceptions of the good. The list of primary social goods must represent a kind of *consensus* within the pluralist society. Therefore, they do not function as a natural-based list of basic needs, like Braybrooke's list, but as a list of *social* provisions that are to cover the social *means* of meeting the needs. If the person concerned has the means in hand within a just social framework, then the needs in question have been met *in principle*. Whether or not this is in fact the case is a matter of personal responsibility. After all, a person may still waste the means.

Recapitulating Scanlon's properties of criteria of well-being, it might be questioned whether this strategy is really able to be *outcome-oriented* in the sense that it provides an evaluation of the *extent* to which individuals may be affected by having these goods. This is necessary to guarantee sensitivity of the criterion to *variations* in needs, e.g. variations arising from physical disabilities. Some people may have extraordinary and very expensive medical needs.[119] Rawls's proposal has been criticised by many in this regard for not being sensitive to these variations in need.[120] Consequently, it is interesting to turn to the second alternative of dealing with basic needs, which can be characterised as both being sensitive to these variations in need, and being based, more than Rawls's strategy, on the natural, generic characteristics of human existence. This brings us to the strategy developed by Martha Nussbaum.

2.3.3.3 The 'Thick' Conception of Basic Capabilities

Martha Nussbaum defends an alternative strategy to determining minimum standards of provision for each matter of basic need within a framework of social policy. Instead of concentrating on primary social goods to determine the level of well-being, she urges that it is better to concentrate on basic human

---

[117] *Ibid.*, 1971, p. 62; 1999, p. 54.

[118] *Ibid.*, 1971, p. 102; 1999, p. 87. I will come back to this in detail in Part II for this assertion is not evident at all.

[119] In later works, he admits that there are people who may have unusually expensive medical needs. J. Rawls, 'Social Unity and Primary Goods', 1999/1982, pp. 368–371; *Id.*, *Justice as Fairness: a Restatement*, ed. E. Kelly, Cambridge, MA: The Belknap Press of Harvard University Press, 2001.

[120] Cf. Part II, Section 3.2.4 *The Inflexibility Critique*.

capabilities, i.e. with what people are able to do and be. Consequently, she defines a list of the basic human capabilities:

> 1. LIFE. Being able to live to the end of a human life of normal length ...; 2. BODILY HEALTH. Being able to have good health ...; 3. BODILY INTEGRITY. Being able to move freely from place to place; to be secure against violent assault ...; 4. SENSES, IMAGINATION, AND THOUGHT. Being able to use the senses, to imagine, think, and reason – and to do these things in a "truly human" way ... informed and cultivated by adequate education ...; 5. EMOTIONS. Being able to have attachments to things and people outside ourselves ...; 6. PRACTICAL REASON. Being able to form a conception of the good ...; 7. AFFILIATION. A. Being able to live with and toward others, to recognize and show concern for other human beings, to engage in various forms of social interaction ...; B. Having the social bases of self-respect and nonhumiliation; being able to be treated as a dignified being whose worth is equal to that of others ...; 8. OTHER SPECIES. Being able to live with concern for and in relation to animals, plants, and the world of nature; 9. PLAY. Being able to laugh, to play, to enjoy recreational activities. 10. CONTROL OVER ONE'S ENVIRONMENT. A. Political. Being able to participate effectively in political choices that govern one's life; having the right of political participation, protections of free speech and association; B. Material. Being able to hold property ..., and having property rights on an equal basis with others....[121]

In express opposition to Rawls, Nussbaum defends a *thick, vague* conception of the good that wishes to grasp the essence of what is generically involved in living and functioning as a human being.[122] Does her approach comply with the three properties of criteria of well-being proposed by Scanlon? Firstly, we can say that meeting the criterion of being *outcome-oriented* – i.e. being sensitive to individual variations in need – has been the starting point of her approach. Secondly, the list is formulated in such a way that it can serve as a basis for criticism and justification of institutions, i.e. that it can serve as a basis for *consensus*. Although her approach is thick, it is expressly meant to function as a list of criteria of well-being within a *liberal* society: that is, it must allow for the fact of individual *variations in taste* and conceptions of the good life.[123]

Of course, a lot more can and should be said about both strategies of Rawls and Nussbaum and I will come back to them in detail in Part II.[124] For now, suffice to say that with regard to our thinking about needs and preferences there are good reasons to take non-volitional needs or course-of-life needs as the basis of legitimate moral arguments affecting distribution in social policy. Whether this basis should best be translated in the language of primary social goods or in the language of basic human capabilities, I will leave aside for the moment. Now it is time to take the results of the above analysis a step further and concentrate on the issue of health-care needs.

---

[121] The list has undergone modification over time and no doubt it will undergo further modification in the light of criticism. This is the version as it appeared in M.C. Nussbaum, 'Beyond the Social Contract: Toward Global Justice', in G.B. Peterson (ed.), *The Tanner Lectures on Human Values*, Vol. 24, Salt Lake City: University of Utah Press, 2004, pp. 413–507.

[122] See: M.C. Nussbaum, 'Aristotelian Social Democracy', 1990, p. 216; *Id.*, 'Human Functioning and Social Justice: In Defense of Aristotelian Essentialism', 1992, p. 217.

[123] Cf. Section 4.4.1 *Rich but Liberal.*

[124] Cf. Chapter 3: *Justice as Fairness: John Rawls*, and Chapter 4: *Nussbaum's Capabilities Approach: a Non-Contractarian Account of Care.*

2.3.4 Health-Care Needs

Having analysed various characteristics of needs, what can we say now about the issue of health-care needs? Can we identify distinguishing features of health-care needs that can serve as a basis for distributive social policy? That means, can we denote certain objective, truncated hallmarks of health-care needs by which they can be identified as non-volitional course-of-life needs and by which they can be distinguished from other forms of health-care needs that do not possess this objective reference?

Answering these questions shows that arguments on preserving normal functioning, on the opportunity range open to an individual, on fair equality of opportunity or equality of capability, prove to be central.[125] Consequently, I will gather all the relevant elements of this chapter in order to formulate an answer to the question what it means to say that there is a human right to health care. This answer will provide the link between Part I and Part II, where I will analyse the meaning, role and place of health and health care within three contemporary theories of justice.

2.3.4.1 Health-Care Needs and Normal Functioning

It is possible to distinguish various kinds of health-care needs, and consequently, various uses of health care. Accordingly, it is interesting to ask whether it is helpful to apply the typologies of needs, proposed by Braybrooke (course-of-life needs and adventitious needs) and Frankfurt (non-volitional, free volitional and constrained volitional needs).[126] Which criteria should be fulfilled in order for health-care needs to comply with the objective, truncated category of non-volitional course-of-life needs?

To begin with, it is reasonable to hold that among the morally important needs are those health-care needs that are necessary to achieve or maintain *normal functioning* for *all* individuals, viewed as members of a natural species. As such, they are no personally created necessities, but universally categorical, and inextricably linked to actual harm in case of non-fulfilment.[127] Consequently, they are the needs that belong to the category of non-volitional course-of-life needs. In this line of reasoning, Norman Daniels proposes a broad and diverse set of basic health-care needs that should be adequately fulfilled by society in order to maintain, restore or provide functional equivalents (where possible) to normal functioning:

---

[125] Remember that the opportunity arguments are Rawlsian, and that the capabilities approach comes from Sen and Nussbaum. For reasons of simplicity and clarity, I will develop the arguments in the passages below solely along the lines of the opportunity concept and the idea of fair equality of opportunity. However, it is important to bear in mind that the argument can be equally translated into capability terms, which should not be a problem. At this stage of the discussion, both approaches can exist alongside each other. As from Part II, I will concentrate in detail on the various ways in which they differ.

[126] I take Braybrooke's basic needs or course-of-life needs as being identical to Frankfurt's non-volitional needs. Braybrooke's adventitious needs equalise Frankfurts free volitional and constrained volitional needs.

[127] Cf. Section 2.3.2.4 *Frankfurt's Characterisation: Volition and Harm.*

> 1. Adequate nutrition, shelter; 2. Sanitary, safe, unpolluted living and working conditions; 3. Exercise, rest, and some other features of life-style; 4. Preventive, curative, and rehabilitative personal medical services; 5. Non-medical personal, and social support services.[128]

Do these needs have the two properties noted earlier?

OBJECTIVELY ASCRIBABLE. Clearly, they are objectively ascribable, if we can come up with an objective notion of species-typical normal functioning. From the macro perspective, it is interesting to begin with the narrow and objective biomedical model of health and disease.[129] The basic idea of this model is that health is the absence of disease and that disease counts as a deviation from normal, natural functioning as human beings. Consequently, the task of characterising normal functioning falls to the biomedical sciences. Of course, as both Braybrooke's list of matters of need and Nussbaum's capabilities list have pointed out, this task of characterisation of *human* normal functioning should not only encompass references to *bodily* functioning but also to *social* and *mental* functioning. Therefore, there must be a way of objectively defining specific human functions such as acquisition of knowledge, linguistic communication, and social cooperation and attachment in the broad and changing range of the various environments we live in. Professionals in educational science, speech therapy, psychology and psychiatry may do the task of judging this.

Although the narrow biomedical model is not uncontroversial, it has one clear advantage. It allows us to draw a fairly sharp line between uses of health-care services to prevent and treat diseases and uses that meet other social goals. It allows us to develop a narrow, objective basis of fundamental health-care needs on which we can find general agreement. It will not matter if what counts as a disease is relative to some features of social roles in a given society, and thus to some normative judgements, provided that the core of the notion of normal species-typical functioning is left intact.[130]

---

[128] N. Daniels, *op. cit.*, 1985, p. 32.

[129] C. Boorse, 'On the Distinction between Health and Disease', in *Philosophy and Public Affairs* 5(1975), pp. 49–68; *Id.*, 'Health as a Theoretical Concept', in *Philosophy of Science* 44(1977), pp. 542–573. See also L. Kass, *Towards a More Natural Science*, New York: Free Press, 1985; A.L. Caplan (ed.), *Concepts of Health and Disease: Interdisciplinary Perspectives*, Reading, MA: Addison-Wesley, 1981. For a critical analysis, see: Y. Denier, 'Public Health, Well-Being, and Reciprocity', in *Ethical Perspectives* 12(2005)1, pp. 41–66.

[130] Accordingly, the model would count *infertility* as a disease, even though individuals might prefer to be infertile and seek medical treatment to render themselves so. Similarly, *unwanted pregnancy* is not a disease. *Dysfunctional noses* count as diseases, since noses have normal functional organisation. If the dysfunction or deformity is serious, it might warrant treatment as an illness. But deviation of nasal anatomy from personal or social conceptions of *beauty* does not constitute disease. This line of arguing, Daniels urges, does not preclude individuals or society to use health care technology to make a person's nose more conform to some standard of beauty – or put another way, less repulsive, may be; neither does it preclude society to subsidise non-therapeutic abortion. However, the arguments used for funding medical procedures which treat cases that are not strictly disease-related, cannot be arguments based on the idea of meeting health care needs, as defined above. Cf. N. Daniels, *op. cit.*, 1985, pp. 28–32.

The *importance* of such *other goals* may be different, taking place within a broader social framework and being founded on other bases, like in the induced infertility or unwanted pregnancy cases (for instance, ensuring that the poor and well-off women are equally able to control their bodies). Our problem in hand, starts with the question what *justice* demands in the distribution of health-care resources and services, which has been answered by the argument defending equal access for equal need. Against this background, it is relevant to take the narrow, natural, humanly generic and, as such, generally uncontroversial baseline of species-typical normal functioning as a *starting point*. If important moral considerations enter at another level, it is necessary to determine, first, what counts as *health and disease* and what are *other* relevant factors, and then, to reconcile the various relevant principles of distribution. This helps us in distinguishing various uses of health care and medical technology accordingly.

OBJECTIVELY IMPORTANT. Are basic health-care needs objectively important? Spontaneously we do treat them as such in many contexts. But what is of interest is to see *why* being in such a need gives them their special importance. Beside Braybrooke's argument that the relational formula comes to an end when the needs are connected with *life*, *health* and *normal functioning* as a *human being*, Daniels reminds us of the fact that there is another important point, namely that of *opportunity*.

2.3.4.2 Health, Disease and Opportunity

Why give health-care needs special moral importance because they are necessary to preserve normal species functioning? What is so special about normal functioning besides its relationship with life and human action? To answer the question, it is necessary to develop a remark on the relationship between normal functioning and *opportunity*, and more specifically, the notion of the *normal opportunity range*.

Impairment of normal species functioning, Daniels argues, reduces *the range of opportunities* that are open to the individual according to which he may construct and pursue his life plan or conception of the good life. We may think of a life-plan as a long-term plan in which an individual schedules activities so that he can harmoniously satisfy his wishes. The good for an individual is defined by reference to this plan and the choice of goals, projects and means for achieving them it contains.[131] Daniels puts it as follows:

> if persons have a fundamental interest in preserving the opportunity to revise their conceptions of the good through time, then they will have a pressing interest in maintaining normal species functioning … by establishing institutions, such as health-care systems which just do that.[132]

More importantly, the kinds of needs picked out by reference to normal species typical functioning are objectively important because they meet this fundamental interest that persons have in maintaining a *normal range of opportunities* open

[131] J. Rawls, *op. cit.*, 1971, pp. 90–95; 1999, pp. 78–81.
[132] N. Daniels, *op. cit.*, 1985, p. 28.

to them. Daniels defines the normal range of opportunities for a given society as 'the array of life plans that reasonable persons in it are likely to construct for themselves'.[133] This notion is relative in a twofold way:

SOCIALLY RELATIVE. From a general perspective, the normal opportunity range is of course dependent on certain key features of society (like its stage of historical development, its level of material wealth and technological development, and even important cultural factors). As such, it is to a certain extent socially relative. Nevertheless, with regard to our problem of determining the basis of just health care distribution, we can use normal species functioning as an objective parameter (one among many others) that affects the share of the normal range open to a given person.

SKILLS, TALENTS, LIFE-PLANS. Secondly, talents, skills, preferences, values and life projects also determine the share of the normal opportunity range that is open to an individual in a fundamental way. However, within a society, the normal opportunity range abstracts from individual differences in one's *effective opportunity range*. From the perspective of the individual who has particular life goals and has developed certain skills accordingly, the effective opportunity range will be a specific realisation of the normal opportunity range.

2.3.4.3 Macro and Micro Level

Measuring the impact of disease on opportunity by reference to an individual's share of the *normal* range entails, as such, abstraction from important interindividual differences in *effective* range; differences that derive from various individual's particular life plans according to their own various conceptions of the good life. It is important to note that this level of abstraction is appropriate given the fact that from the *macro* perspective we search for a measure of *social* importance, within a general framework of justice in access to health care, of health impairments. To determine the relative importance of health-care needs on the *macro* level, it is adequate to take the narrow, objective, truncated scale of well-being together with impairment of the normal opportunity range as determined by the objective biomedical model of health and disease, as criterion.

Of course, impairment of normal species functioning can diminish individual *happiness*, or satisfaction, which depends on the individual's conception of the good life. Such effects are important at the *micro* level, where we are dealing with individual decisions about using specific kinds of health care, deciding which particular services to use from among those that society is obliged to provide. In these situations, individual choices about happiness will be the final determinant of what should be done. Consequently, the objective Boorsean model of health and disease will not do on the micro level, since it is too narrow. On the micro level, we will need an additional model that respects both the expertise of the physician concerning a person's physical well-being, and the patient being the

---

[133] *Ibid.*, p. 33.

proper judge about what serves his interests best, according to his own conception of the good life. On the micro level, we rightly work with a full-range scale of well-being.[134]

On the macro level the absence of particular concern for individual happiness does not have to be a problem as long as three conditions are fulfilled. Firstly, the narrow, objective framework of measuring health-care needs can be the object of reasonable consensus, i.e. it provides a basis of just health care with which we can all agree (whatever our various personal conceptions of the good). Consequently, it can serve as a basis for answering our central problem, which concerns what the just society is obliged to provide. Thirdly, it does this in a way, which leaves enough room for individual freedom in one's valuation of health and health care. This is necessary to avoid illegitimate paternalism.

2.3.5 A Human Right to Health Care?

Having extensively analysed a range of elements that are important within the framework of equity in health care, it is time to take stock and develop a conclusive framework of health-care needs. I will do this by providing an answer to the question what is involved in saying that there is a human right to health care.[135] I will do this in a threefold way. Firstly, I will ask after the moral component of saying that there is such a right. Consequently, I will determine its basis, and finally, give some remarks about its scope.

---

[134] Note, however, that with regard to the duties of society, this nevertheless happens *within* the objective-truncated framework of provision. So using the full-range scale of well-being on the micro level does not entail that society is obliged to publicly provide people with whatever health-care services they are likely to have. It only entails reference to the property of Scanlon that at an individual level, socially workable criteria of well-being should be able to allow for individual variations in taste. Whenever people would like to make an appeal to health-care services that do not fall within the objective truncated category that represents a social consensus, this means that they would have to buy them privately or go without. A model of health and disease, which encompasses both the objective and subjective aspects, is the fact-plus-value model of Fedoryka. Cf. K. Fedoryka, 'Health as a Normative Concept: Towards a New Conceptual Framework', in *Journal of Medicine and Philosophy* 22(1997), pp. 143–160. See also: L. Thielmann, *op. cit.*, 2001, pp. 110–136.

[135] Following the authors of *Health and Human Rights*, I take the human rights approach as something that 'seeks to describe – and then to promote and protect – the societal-level prerequisites for human well-being in which each individual can achieve his or her full potential'. In: J.M. Mann; S. Gruskin; M.A. Grodin; G.J. Annas (eds.), *Health and Human Rights*, New York: Routledge, 1999, p. 2. As such, I understand a human right to hold both an ethical or moral component and a legal component of established rights and obligations. From the viewpoint of ethics, human rights serve as a statement of the aspirations of peoples and governments toward ideals that are not always attained in practice. The legal component refers to the recognition and protection of these aspirations by governments through actual domestic legislation and international agreements. This article focuses primarily on the *moral* component by concentrating on the extent to which human rights are guidelines for how governments *should* behave, rather than on various legal descriptions of how they actually *do* behave.

2.3.5.1 *What Is a Human Right to Health Care?*

From the moral viewpoint, the statement that there is a basic human right to health care means four things.[136] Firstly, it means that there is a *collective moral obligation*, that is, an obligation on the part of the society to ensure that everyone has access to some level of health-care services. Secondly, it means that this obligation is a very *stringent* one. Obligations that are implied by rights have exceptional moral force in public debate. Contemporary rights theorists such as Ronald Dworkin define a right as a 'trump' that overrides countervailing considerations. A countervailing consideration could be, for instance, the mere fact that abandoning moral obligation could increase overall utility. Therefore, rights serve as a powerful protection of important interests persons have.[137] Thirdly, a basic right to health care implies that access to health care is *owed* to those who have the right. A right holder is not kindly asking for a favour and if society fails to fulfil this collective obligation, it does injustice to all the individuals who lack access to health care. And fourthly, as a human right, it is ascribed to all individuals because they are *human*.

2.3.5.2 *The Basis of the Right to Health Care*

Why are certain interests, in this case health-care needs, so important that they deserve such special protection? What is it about health care that is so special? A possible answer is that health care is special because of its instrumental power. Health care is the means to an end that is highly valued in most cultures: good health and a long life free from pain and disability. Without lifelong access to appropriate health care, our chances of attaining this goal are likely to be impaired. Yet the high value of good health alone cannot explain the particular status of health care as a focus of moral concern. There are many things we value highly, like companionship, aesthetic pleasure, love and other such benefits to which we do not necessarily have a right. So in addition, three main arguments deserve attention: fair equality of opportunity, basic health-care needs and collective social protection.

FAIR EQUALITY OF OPPORTUNITY. Firstly, contemporary health care involves a complex and heterogeneous framework of institutions, services and policy measures that aim at *prevention* of disease and disability, *restoration* of health where possible, and personal and social *support and care* for the long-term ill or disabled. As such, health care greatly affects the risk of persons getting sick, the likelihood of being cured, and the degree to which one will receive support. Within this line of reasoning, Daniels has pointed at the way in which health care protects our level of normal functioning and consequently the range of

[136] Partially based on: A. Buchanan, 'Philosophic Perspectives on Access to Health Care: Distributive Justice in Health Care', in *The Mount Sinai Journal of Medicine* 64(1997)2, pp. 90–95.

[137] R. Dworkin, *Taking Rights Seriously*, Duckworth: London, 1977; J. Waldron, 'When Justice Replaces Affection: the Need for Rights', in *id.*, *Liberal Rights*, Cambridge: Cambridge University Press, 1993, pp. 370–391.

opportunities open to us to form, pursue and revise our life-plans.[138] Impairment of normal functioning through injury, disease and disability creates significant disadvantages and reduces a person's opportunities in life. What appears to make health care of special moral importance is its particular capacity, through prevention, restoration and support, to affect our chances of leading a full, active and morally fulfilling life. In this context, fair equality of opportunity means that all individuals are entitled to an equal opportunity for a chance to be healthy, insofar as possible.[139]

BASIC HEALTH-CARE NEEDS. Secondly, the effect of health-care services on opportunities in life is a general fact that is *common to all*. This is ultimately grounded in the concept of basic needs or, as Braybrooke calls them, 'course-of-life needs'.[140] Basic needs are the things that are *functionally necessary* for the most fundamental projects, involved in living a human life, and are essential to living or functioning normally. They apply to an entire range of interests that concern a person's physical (food, drink, shelter) and psychological existence (communication, affiliation, support). They are *basic* because they are restricted to *universally* recurrent phenomena rather than to particular individual whims or frivolous pursuits. This implies that basic needs are distinguishable from felt needs, preferences or wants. Persons simply have these needs, whether they want to or not. In the words of Frankfurt: basic needs are 'non-volitional needs'; they do not depend on what a person wants.[141] As such, they are typically assumed to be *given* rather than acquired characteristics of the human condition. That means that they are not constituted by any action for which the person is responsible by virtue of his greater effort. Consequently, essential needs are independent from merits. Where they are unequal, one thinks of them as fortuitously distributed; as part of a kind of natural or social lottery, or as the result of good or bad luck.

Likewise, basic health-care needs are those things that every person needs in order to maintain or restore normal and healthy functioning (like adequate nutrition, shelter, sanitation, unpolluted living and working conditions, preventive and curative medical services), or that a person needs to equal normal functioning as much as possible (like glasses, wheelchairs, hearing aids and guide dogs). Accordingly, health-care needs are basic needs: universal in character, necessary for the fundamental projects of every person, and *generically* originating from human *vulnerability* and *finitude*. Very often the advantages of health and the

---

[138] N. Daniels, *op. cit.*, 1985, pp. 27–28; *Id.*, 'Health-Care Needs and Distributive Justice', in *Philosophy and Public Affairs* 10(1981)2, pp. 149–179; *Id.*, 'Justice, Health, and Healthcare', in *American Journal of Bioethics* 1(2001)2, pp. 2–16.

[139] R.M. Veatch, 'Voluntary Risks to Health', in *Journal of the American Medical Association* 243(1980)1, pp. 50–55; *Id.*, 'What is a "Just" Health Care Delivery?' in R.M. Veatch; R. Branson (eds.), *Ethics and Health Policy*, Cambridge, MA: Balinger, 1976, pp. 127–153.

[140] D. Braybrooke, *op. cit.*, 1987, pp. 29–75.

[141] H.G. Frankfurt, *op. cit.*, 1988, pp. 104–116.

burdens of illness are arbitrary effects of a natural lottery (like one's genetic make-up) or social conditions (being poor), bad luck (being at the wrong place at the wrong time) or good fortune (accidental discovery of cancer at a curable stage). Although there are interpersonal differences in health-care needs to reach a normal functioning level, enjoying reasonably good health, being able to function normally, and through this having normal opportunities for a fulfilling life is of fundamental value for *every* person, and eliminating or reducing barriers that undermine this value, like disease, illness or injury, is a basic moral obligation for every just society.

COLLECTIVE SOCIAL PROTECTION. Thirdly, it would be unreasonable to expect that individuals generally should be able to gain sufficient access to health care, by relying solely on their own private resources for several reasons. Firstly, health-care needs are more unequally distributed than other basic needs like food, clothing and shelter (some people need considerably more health care than others, while people's need for food and clothing is generally the same). Secondly, health-care needs can be highly unpredictable due to the element of luck. Thirdly, the fulfilment of health-care needs has an important impact on a person's range of opportunities. And finally, health care can be catastrophically expensive. If private resources could generally cover health-care needs, there would be little point in declaring entitlements to health care.

This means that whereas it might be reasonably expected that people can adequately provide for food, clothing and shelter from their own private shares of income and wealth, this does not apply to goods like health-care services, which are an appropriate object of collective cost-sharing schemes. Private insurance alone cannot provide sufficient access to care for everyone because those who are most in need of health care, as well as those with especially high risk of ill health, will not be able to purchase affordable coverage, if they can find insurance at all. That is why we speak of a collective obligation on the part of society as a whole.

2.3.5.3 The Scope of the Right to Health Care

The final characterisation of the human right to health care concerns its scope. Historically, the right to an adequate level of health care has been classified under the second generation of economic, social and cultural rights (which also include food, work, social security, education, etc.). Societal obligations under this category differ from the obligations under the first generation of civil and political rights (which include the rights to life, liberty, freedom of movement, freedom from torture, etc.). Whereas civil and political rights must be guaranteed immediately, governmental obligation for economic, social and cultural rights involves action to ensure that these rights are progressively realised.[142] This means that whereas the first generation of rights are simply a matter of

[142] See J.M. Mann; S. Gruskin; M.A. Grodin; G.J. Annas (eds.), *op. cit.*, 1999, p. 25.

being protected as they are, the second generation deals with the problem of *to what extent* they should be promoted and protected. Consequently the question arises: should they be *endlessly* promoted? No, they should not.

To begin with, the right to health care cannot be an unlimited right. It cannot be a right of everyone to have access to whatever health-care services would be of net benefit to the individual. Rationing of health care has to be a fact of life. We must set priorities. Three reasons support the argument that the right to health care must be a limited right. Firstly, health care is not the only important good in life. This refers to what I have called the *external* dynamic of scarcity in health care.[143] Given that resources are finite, we must consider what economists call the 'opportunity costs' of providing health care for all. We must acknowledge that resources must be preserved for other social needs, like housing, education, scientific research, etc. It would not be rational for a society to devote all its resources to health care.[144]

Secondly, the right to health care is the right to an adequate level of care, not the right to all types of care that would be of net benefit. This refers to the *internal* dynamic of scarcity. Taking the continuation of technological development into account, there are virtually no limits to how much we could spend on health care. Every advance in medical science creates new needs that did not exist previously. Consequently, the nature of health care is such that supply often generates its own demand.[145]

Thirdly, health care is not about the endless provision of resources and services to increase individual happiness. Although meeting health-care needs may have a tendency to promote happiness, its moral importance is derived from the way in which it protects functioning and opportunity. As such, the notion of an adequate level of care is that of a floor, not of a ceiling.

The guidelines of basic health-care needs and of fair equality of opportunity suggests a path for giving content to the idea of a decent minimum of health care, and for setting priorities in the allocation of resources. It asserts that collective moral obligations exist to provide health care at the level needed for persons to receive a fair chance in life.

Nevertheless, unless additional qualifications were introduced, acceptance of these arguments for a right to health care would place immense burdens on society. After all, they lead to the conclusion that society is morally obligated to funnel resources toward bringing persons ever closer to the goal of fair equality of opportunity. However, a vast array of disabilities, injuries and diseases limit opportunity, and many persons are so seriously affected that they could never be brought or restored to a position of equal opportunity, even if immense sums were spent to bring them closer to that ideal. These arguments then need to be

---

[143] Cf. Section 2.1.1.1 *The Twofold Dynamics of Scarcity*.

[144] J. Rawls, *Justice as Fairness: a Restatement*, 2001, pp. 173–174; A. Buchanan, *op. cit.*, 1997, p. 92.

[145] J. Butler, *op. cit.*, 1999, pp. 5–49.

held in check by an account of allocation that avoids unreasonable demands on social resources in order to implement the right to health care.

Besides the various rationing mechanisms that were indicated in Section 2.2.3 and whose workability can most adequately be judged by health economics, it is most relevant to carry out a broader philosophical analysis of our general, modern attitude towards the promises of medical technology in health care. This opens the final section of this chapter.

2.4 CARE AND THE BOUNDARIES OF HUMAN LIFE

2.4.1 *Scarcity as Expression of Finitude*

As I said before, it is reasonable to inquire into a third, more abstract approach to scarcity. This approach is founded on philosophical reflections on scarcity as an expression of *finitude*, not so much of finitude of resources, but rather of finitude as an essential characteristic of the *condition humaine*. This abstract finitude approach aims to bring a different and more positive valuation of the limits of human existence into the discussion. Although this is not evident at first sight – what, after all, does a positive valuation of finitude, mortality, dependence, vulnerability, decline, imply? – it may provide avoidance of the anthropological and social intensification of scarcity. Taking human finitude into account in our social valuation of health and health care may make it easier to set meaningful limits on our seemingly limitless expectations for the role and task of medicine and just health care. It may slow down the rising spiral of expectations.

It is important to note beforehand that this approach does not mean to eliminate scarcity in the way of Buddha, which is to eliminate want, or in the way of the Stoa, which entails an unmoved acceptance of fate and fortune. Nor does it entail an uncritical rejection of the general economic interpretation of scarcity and its search for various rationing mechanisms. What it does entail, however, is a set of arguments for a renewed attitude towards the endless possibilities of modern medicine and medical technology, founded on being able to relate meaningfully to the inevitable fact of finitude, even though we find it tragic. As such, this approach aims to provide a middle course between *ignoring* scarcity, anthropologically and socially *reinforcing* scarcity by idolising the task of medicine and medical technology, and finally, the risk of a one-sided mechanical approach that could give the false impression of *solvability* of scarcity.

The main advantage of the abstract finitude approach is enclosed in the fact that it is more apt to incorporate *care* in the discussion on what we *can* and what we *should* reasonably expect from just health care. In order to grasp a truly adequate conception of just health care, we need efficiency, justice and care to function as a trinity. This might happen more easily by a positive and well-balanced valuation of finitude. Here, we will meet a strange but interesting paradox: it seems that the logic of *abundance* that is implied by the concept of care enables us better to cope with *scarcity* because it motivates us to deal with *finitude* in a different way. I will come back to this advantage later. Firstly, it is important to recapitulate three aspects of the above-illustrated health care discussion from which the finitude approach derives its relevance. These aspects are firstly, the exponential increase of modern medical technology; secondly, the importance attributed to health, independence and normal functioning; and thirdly, the internal dynamics of scarcity, which refers to the tendency towards infinity that is internally connected to the history of medicine and medical technology, as well as to our general conception of independent normal functioning.

2.4.2 Threefold Relevance of the Finitude Approach

2.4.2.1 Exponential Increase of Medical Technology

Firstly, we have seen that the lion's part of the skyrocketing cost increase in health care should be attributed to the progress and improvement of modern medical technology.[146] The simple contradiction between endless clinical possibilities of diagnosis and therapy and economic affordability that underlies the idea of an increasing gap between supply and demand in health care is inherently linked to scientific and technological progress: to conquer one peak is merely to reveal yet others to climb. The exponential increase of medical technology has been denoted as the first catalyst of increasing health-care expenditure and of intensification of scarcity in health care.

2.4.2.2 Focus on Normal Functioning

The second aspect from which the finitude approach derives its relevance is the importance of health care's contribution to health, independence and normal functioning. It has been shown that the most adequate interpretation of the essence of equity in health care comes down to guaranteeing equal access for equal need.[147] The argument went on to say that basic health-care needs are part of the non-volitional course-of-life needs, which are essential for living or functioning normally. Preserving normal functioning, people's normal opportunity range and contributing to fair equality of opportunity has turned out to be one of health care's major functions. The predominant importance of preserving normal functioning can also be experienced in the general intuition of a *hierarchical* relation between the three pillars of prevention, cure and care: prevention is preferred to cure; and cure is preferred to care for those whose health is permanently impaired. It is generally considered as the central task of health care to bring people to the level of healthy, independent normal functioning.

Nevertheless, the danger that is included in a one-sided focus on prevention and cure entails that the social valuation of long-term care becomes a condescending valuation of something that is *regrettable*. Care for the long-term dependent, the lifelong disabled and the irreversibly declining, becomes a social *burden*. As long as

---

146 Cf. Section 2.2.2 *Various Causes*.

147 Cf. Section 2.1.1.2 *Scarcity in Health Care*.

normal, healthy, independent functioning is the standard, long-term dependency care is not and will never be 'efficient' or 'effective'. Nevertheless, dealing with human finitude in a more positive way requires a valuation of dependence that is different from a narrow one-sided perspective of mere regret, tragedy and loss. It is reasonable therefore, to critically reflect on the various ways we *value* healthy, independent normal functioning. In Section 2.4.3, I will discuss three different ways of valuation. But before doing this, let me first go into the third and probably most important aspect from which the finitude approach derives its relevance.

2.4.2.3 Tendency Towards Infinity

All in all the most important aspect from which the finitude approach derives its relevance is the fact that both history of medicine and medical technology, as well as our general perception of healthy and independent normal functioning show a *tendency towards infinity*. As we have seen, history of medicine discloses the infinite capacity to do ever more – and ever more expensive – things for the patients. As such, health care is characterised by a *continuous relocation of natural limits*. As a result, our perception of what independent normal functioning is, is susceptible to the same development. This is most clearly noticeable in the major increase of people's life expectancy since the 1960s.[148] In many cases, it is technically less impossible to postpone death. Natural limits increasingly become a flexible matter. This is also shown in the realm of assisted reproduction. Involuntary childlessness has become less a matter of having to accept one's fate. The same goes for many other possibilities in medicine – like changing one's sex or successfully undergoing a life-saving organ transplantation. As such, healthy and independent normal functioning has become less and less a *given* fact than a gradually *movable* and *adjustable* phenomenon. The fact that normal functioning can be judged and ascribed in an objective way by the biomedical model of health and disease does not mean that it is an unchanging phenomenon. After all, medicine's seemingly infinite capacity to do ever more things for the patient gives the impression that normal functioning can be continuously relocated in the same infinite manner. The idea of fixed limits increasingly fades away. Of course, this is not to be judged as an undesirable development per se. It does, however, provide interesting food for critical reflection, especially with regard to our main question at hand: what *can* we and what *should* we reasonably expect from just health care?[149]

How should we think about the tendency towards infinity? *Why* do these elements found the relevance of the finitude approach? In the first place, because

[148] Cf. R. Porter, *The Greatest Benefit to Mankind: a Medical History of Humanity*, New York, Norton, 1999. See also The World Health Report 2000, *Health Systems: Improving Performance*, Geneva, World Health Organization, 2000, esp. pp. 27–31. Note that this major increase mainly applies to the industrialised and developed countries.

[149] For an application of this question to the issue of assisted reproduction, cf. Y. Denier, 'Behoefte of verlangen? De betekenisstructuur van de kinderwens', in *Ethische Perspectieven* 10(2000)3, pp. 163–174. Also published in revised version: Y. Denier, 'Need or Desire? A Conceptual and Moral Phenomenology of the Child Wish', in *International Journal of Applied Philosophy* 20(2006)1, pp. 81–95.

it turns out that the tendency towards infinity gives our general contemporary *attitude* to health and normal functioning and to the related rights and duties within the health-care system a specific character, i.e. one of *limitless expectations*. Consequently, modern optimism over the promises and increasingly successful possibilities of medicine and health care has changed the organisational dynamics of contemporary health care into an *accelerating and rising spiral movement*.[150] Contrary to what we can reasonably and – even more important maybe – realistically expect from medicine and health care, contemporary medicalisation of life seems to give the impression that the sky does seem to be the limit. Or put differently, that there are no limits to what we possibly can achieve; that we should keep being on the lookout for possible solutions; that there is nothing that cannot be done!

Nothing except for the thing that we might be inclined to forget: acknowledging finitude as an essential characteristic of the human condition. We cannot ignore this. After all, human persons are mortal beings. Their lives are characterised by growth, maturity and decline; they are lives in which periods of active, independent functioning alternate with periods of dependence. On closer look, it seems to be the case that in very general terms the main specific problem attached to contemporary health care comes down to having to make a choice between continuing to support the dynamics of the rising spiral in an uncritical way (by focusing on the quantity of life) on the one hand, and moderation and temperance (by taking quality of life as the main value) on the other. Formulated as such, the discussion on setting limits on health care ultimately comes down to asking about the extent to which we can reasonably and realistically ignore, deny, postpone, set aside and seemingly remove the inevitable fact of finitude. This is a very relevant question within the contemporary discussions on just health care because the tendency towards infinity contains a twofold risk.

Dealing with Mortality. Firstly, we might run the risk of becoming unable to deal with the inevitable: viz. with human mortality. In classical terms, this risk is known as *hybris*. *Hybris* is a common theme in Greek tragedies and mythology and refers to the vice of excessive pride or self-confidence often resulting in retribution. Applied to the contemporary rising spiral movement in health care the *hybris*-argument asks for awareness of the following questions: to what extent can we legitimately speak of overconfidence? To what extent can we say that all the efforts made within health care, initially set up for the benefit of man, ultimately turn itself *against* this initial beneficiary like a boomerang? Ivan Illich expresses a similar idea in his discussion of the concept of medical nemesis:

---

[150] Remember the idea expressed in Section 2.1.1.2 *Scarcity in Health Care*: 'The nature of health care is such that supply often generates its own demand; and to spend more on the provision of health care is often to do no more than to stoke the fires of further demand'. Or as John Butler puts it: 'Since to conquer one peak is merely to reveal yet others to climb, we cannot assume that a doubling or even a trebling of the volume of resources allocated to [health care] would close the gap between supply and demand'. In: J. Butler, *op. cit.*, 1999, p. 7.

> I will designate [the] self-reinforcing loop of negative institutional feedback by its classical Greek equivalent and call it *medical nemesis*. The Greeks saw gods in the forces of nature. For them, nemesis represented divine vengeance visited upon mortals who infringe on those prerogatives the gods enviously guard for themselves. Nemesis is the inevitable punishment for inhuman attempts to be a hero rather than a human being. Like most abstract Greek nouns, Nemesis took the shape of a divinity. She represents nature's response to hubris: to the individual's presumption in seeking to acquire the attributes of a god. Our contemporary hygienic hubris has led to the new syndrome of medical nemesis.[151]

The essence of applying the ideas of *hybris* and *nemesis* to our modern valuation of health and health care is the following: has it become necessary to revalue the Socratic perspective of learning to prepare oneself for dying and death because we otherwise run the risk of becoming unable to deal with the inevitable, viz. with human mortality?[152] That is, can the *hybris*-argument serve as a rightful argument for setting limits on medical technology? The reasonableness and relevance of these questions originate from the fact that we may not forget that health care derives its moral significance precisely and primarily from the horizon of finitude and from our acknowledgement of the limits of human existence. Health-care services, goods and institutions are significant, precisely because people encounter dependency, illness, disease, handicap and ageing. In such situations, we encounter human finitude. At the same time, they are generically characteristic for the human condition. As such, revaluing the Socratic perspective entails that setting limits in health care may come to mean that we have to set limits on its tendency towards infinity, i.e. its tendency to back away from our finitude as much as possible. The result is that we may be better able to take human finitude, mortality, vulnerability and dependency into account when we have to answer questions regarding allocation of scarce resources in health care. We may be better able to distance from the *hierarchical* relation between the three pillars of prevention, cure and care.

---

[151] Cf. I. Illich, *Medical Nemesis: the Expropriation of Health*, London: Calder & Boyars, 1975, p. 28. The idea of nemesis forms part of his analysis of clinical, social and structural iatrogenesis. Iatrogenesis is 'composed of the Greek words for "physician" (*iatros*) and for "origins" (*genesis*). Iatrogenic disease comprises only illness which would not have come about unless sound and professionally recommended treatment had been applied' in *ibid.*, p. 22.

[152] The Socratic answer holds that the essence of human life is strongly connected to the virtue of preparing oneself for dying and death. It entails acceptance and acknowledgement of finitude into the valuation of human life. Cf. Plato, *Crito* and *Phaedo*, trans. H. North Fowler, Cambridge, MA: Harvard University Press, 1971. See also: O. Höffe, *Medizin ohne Ethik?* Frankfurt: Suhrkamp, 2002, esp. pp. 119–142; Philippe Van Parijs puts it as follows: 'à mesure que se creuse l'écart entre le techniquement possible et l'équitablement accessible, l'éthique, dans ce domaine, consiste de moins en moins à faire tout ce qu'on peut pour offrir l'éternité à chacun. Elle consiste toujours plus à apprendre, ou réapprendre, à mourir'. In: P. Van Parijs, 'Y a-t-il des limites à la prise en charge des soins de santé par la solidarité?' In: J. Hallet; J. Hermesse; D. Sauer (eds.), *Solidarité, Santé, Ethique*, Louvain: Garant, 1994, pp. 57–68; also available in Dutch as 'Zijn er grenzen aan de ten laste neming van de gezondheidszorg door de solidariteit?' In: J. Hallet; J. Hermesse; D. Sauer (eds.), *Solidariteit, gezondheid, ethiek*, Leuven: Garant, 1994, pp. 57–67.

A WELL-BALANCED RELATION BETWEEN THE HEALTH CARE PILLARS. The second risk that is included in health care's tendency towards infinity has to do with the fact that it may give us a narrow-minded view on the role and task of health care. The hierarchical relation between prevention, cure and care is not unproblematic. Considering it as the central task of health care to bring people to a level of healthy, independent normal functioning as much as possible, entails the risk of installing a *condescending* attitude towards care for the dependent, and more specifically for the long-term and asymmetrical dependent, i.e. those who will never or no longer come to the level of normal functioning.[153] To avoid the risk of condescendence, it is important to argue for a *well-balanced* relation between the pillars of prevention, cure and care. For just health care should, in the first place, involve equal basic, respectful and decent-quality care for *all*.

A well-balanced relation between the pillars of prevention, cure and care, and related to this, between the more abstract goals of health care – efficiency, justice and care – might best be found in the attitude of complex valuation of healthy and independent normal functioning. This attitude represents a middle course between two extreme approaches, namely the instrumental valuation, and the strong intrinsic valuation. Let us have a brief look at these various attitudes to health and their related perception of the rightful task of just health care.

2.4.3 *The Normative Value of Healthy Normal Functioning*

Let us focus now on the second catalyst explaining the skyrocketing cost increase in health care, i.e. the factor referring to the phenomenon of medicalisation and to our modern *attitude* towards life, health, independence and normal functioning.[154] To begin with, we can identify two viewpoints.

2.4.3.1 *Instrumental Valuation*

The first viewpoint on the normative value of health, independence and normal functioning contains an instrumental valuation. This can be interpreted in the context of value for the society or for the individual person. In the latter context, health is valuable because it is necessary for an individual to pursue many other values in life. Disease and disability may prevent him from doing so. In the same line of reasoning, death might be defined as being bad because it causes a person to have had fewer of the goods of life than he might otherwise have had. Frances Kamm has defined this viewpoint as the *Deprivation Account*.[155]

---

[153] In this context of condescendance, one might wonder to what extent access to health care is truly democratic, and considered of equal importance for all, i.e. also for those who need long-term or lifelong supporting care. I will come back to this in detail in Section 4.1 *A Second Extension to Long-Term Care?*

[154] I have explained the two catalysts in Section 2.2.2 *Various Causes*.

[155] F. Kamm, *Morality, Mortality*. Volume 1: *Death and Whom to Save from It*, Oxford: Oxford University Press, 1993, pp. 13–24. Kamm refers to this as Thomas Nagel's view as analysed in T. Nagel, 'Death', in *id.*, *Mortal Questions*, Cambridge: Cambridge University Press, 1979, pp. 1–11.

The instrumental valuation that I wish to focus on here, however, is the valuation on the level of society, i.e. on the macro level. This kind of valuation holds that citizen's healthy normal functioning has an important productive value for society. In a general sense, this is of course the case, and it does not need to be problematic. It becomes a problem, however, when it comes down to aiming at maximisation of utility, understood in the aggregate-utilitarian sense. The significance of health-care goods and services is then understood *solely* from its focus on maintaining, restoring or providing functional equivalents to healthy and independent normal functioning. Most insolently, it implies taking cost effectiveness thus serious that every use of health care should be justified by its productive value for society, measured, for instance, by its contribution to the gross national product (GNP). In this case, every person is one utility unit, so that 'each is to count for one and no one for more than one'.[156] Each unit is equal and none deserves special protection. Consequently, this perspective can justify exclusion of certain individuals from certain health care entitlements if this would maximise utility, regardless of the value of these forms of health care for a particular individual.[157] After all, a person's dignity is derived from its socio-economic worth, measured in terms of productive contribution.[158]

In a certain sense, the general view of a hierarchical relation between the pillars of prevention, cure and care, expresses this utilitarian component. It expresses a condescending attitude towards long-term care for those persons who cannot or can no longer function in a healthy and normal way. Of course, we can think of refinements to make the utilitarian valuation of health and health care less harsh, for instance, by introducing the difference

[156] J. Bentham, *An Introduction to the Principles of Morals and Legislation*, eds. J.H. Burns; H.L.A. Hart, London: Methuen, 1982.

[157] Allen Buchanan provides a good example: 'Consider Down syndrome. Individuals with this chromosomal disorder have a high incidence of cardiac and gastrointestinal defects, in addition to varying degrees of mental retardation. It might turn out that the way to maximize overall utility is to allow infants with Down syndrome to be removed from their parents' sight at birth, before attachments could deepen. If this were the case, then Utilitarianism would not justify a right to health care for all', in A. Buchanan, 'Philosophic Perspectives on Access to Health Care: Distributive Justice in Health Care', in *Mount Sinai Journal of Medicine* 64 (1997)2, pp. 91–92.

[158] Cf. Section 1.1 *Public and Private: a Historical Outlook* where I have pointed at the historical fact that mercantilism counted population as a source of a nation's wealth. The health of citizens has an important economic value. Accordingly, the same goes for taking people to the level of independent normal functioning as much as possible. See also P. Van Parijs, 'Assurance, solidarité, équité, les fondements éthiques de l'Etat-Providence', in *Cahiers de l'Ecole des sciences philosophiques et religieuses* 12(1993), pp. 49–72. For a critical analysis of variations on this viewpoint, see *id.*, 'Au-delà de la solidarité, les fondements éthiques de l'Etat-Providence et de son dépassement', in *Futuribles* 184(1994), pp. 5–29.

between act-and rule-utilitarianism.[159] But however refined, common to all variations of utilitarianism is that the value and significance of things or acts is measured by their utility value. Hence, the social recognition of fulfilment of individual health-care needs is derived from the extent to which a health-care system maximises utility.

This tends to the scapegoat mechanism for, put bluntly, we might say that long-term care for Alzheimer patients, persons in persistent vegetative state, seriously mentally and physically handicapped people is not efficient, nor cost-effective as long as we take maximising utility as the central goal of health care. The best that we might get out of it for those people would be to keep care in these cases limited to the minimum of basic custodial care. The utilitarian reasoning behind this would merely come down to avoiding a public scandal, not to expression of respect for their human dignity. And this precisely reflects the problem for it is doubtful whether a one-sided focus on utility fits in with our spontaneous views on morality. After all, morality is also and perhaps primarily linked to the existence of moral prohibitions that limit our actions.[160] The *way* in which an end is achieved determines the morality of human conduct. The moral quality of society is to a great extent reflected in the way in which it treats the weak and the vulnerable. A society that for utility reasons neglects help to those who need it most, can hardly be called a just society. The same would hold for the imaginary situation in which the job of caring for the Alzheimer patient or the persistent vegetative state (PVS) patient would be done for instance by highly sophisticated caring machines, because these

---

[159] Act utilitarianism assesses a particular act in itself. An act is right when it maximises utility. According to rule utilitarianism an act is right when it complies with a rule, which if obeyed by all, will maximise utility over time. Mostly, both are in conflict: act utilitarianism does not maximise utility over time and following the rule can in individual cases lead to less than optimal utility results. In his article 'The Survival Lottery', John Harris proposes a scheme that would maximise total welfare by randomly selecting individuals in society, killing them and use their organs to save the life of several other individuals with organ failure. One donor person could thus save several lives and maximise utility for this act. Over time, utility will nevertheless diminish because of the reigning atmosphere of fear for selection. The rule of respect for bodily integrity then maximises utility over time by taking away that fear but results in individual cases in a lack of a sufficient amount of donor organs. In: J. Harris, 'The Survival Lottery', in *Philosophy* 50(1975), pp. 81–87; and P. Singer, 'Utility and the Survival Lottery', in *Philosophy* 52(1977), pp. 218–222. Discussed in J. Elster, *Local Justice: How Institutions Allocate Scarce Goods and Necessary Burdens*, Cambridge, 1992, pp. 190, 222. The difference between act and rule utilitarianism is important within a policy perspective since the latter offers an explanation for the justification of institutions. Institutional rules aim at maximising utility over time. A particular act that does not maximise utility at that particular moment can, thus, still be justified because it complies with the rule of an institution that maximises utility over time.

[160] See S. Hampshire, 'Morality and Pessimism', in *id.*, *Morality and Conflict*, Oxford: Basil Blackwell, 1983, pp. 82–100. See also: H. De Dijn, *De herontdekking van de ziel. Voor een volwaardige kwaliteitszorg*, Nijmegen: Valkhof Pers, 1999; *Id.*, 'Technology in Health Care: a Philosophical Ethical Appraisal', in C. Gastmans, *Between Technology and Humanity: the Impact of Technology on Health Care Ethics*, Leuven: Leuven University Press, 2002, pp. 15–34; H. De Dijn, *Taboes, monsters en loterijen. Ethiek in de laat-moderne tijd*, Kapellen: Pelckmans, 2003.

machines would be much cheaper than trained nursing staff and the patients would not be aware of the difference anyway. On the contrary, the moral quality of society resides to a large extent in its statement that not everything is taken into the utility calculus and that some values deserve absolute respect. Caring for the weak and most vulnerable should not be a matter of cost-benefit analysis but should be an inviolable starting point.[161]

2.4.3.2 Strong Intrinsic Valuation

Against the utilitarian perspective, which values human life and healthy, independent normal functioning instrumentally as a means to increase society's productivity, it might be argued that all human life is sacred, and that a person's life and health knows no price. Consequently, morality demands that we should do everything we can to prolong a person's life, or try to bring an individual's functioning as close as possible to normal species-typical functioning. Health and life is the *summum bonum* for all, and society should do all it can to maintain and preserve it. This viewpoint reflects an extreme interpretation of the Kantian injunction to treat each man as an end in himself, and never as a means only. In the field of medicine this idea returns in the Hippocratic obligation to do whatever one can for one's patient. An extreme translation of this obligation could justify therapeutic obstinacy in each particular case. Besides the fact that it might be doubted exactly how desirable this attitude is on the micro level, on the macro level it certainly follows that this strategy leads to a bottomless pit, draining away all social resources.

2.4.3.3 Complex Valuation: Reassessing Finitude

Can we think of an acceptable alternative between these two extreme forms of valuing life and health? Can we provide an acceptable characterisation of a middle course, valuing human life and healthy functioning neither as a mere function of utilitarian calculation, one-sidedly focused on preserving efficiency and effectiveness, nor as the sole soul-saving value to be safeguarded at all costs? Is there a way out of this profound inconsistency of valuations?[162]

What we are looking for here is an approach that allows us to set limits on health care and to make the necessary choices without inequitable exclusion of certain groups of people on the one hand and without wanting to manage and control the life and health of all on the other. It is my viewpoint that an acceptable valuation of human life and healthy functioning that wants to be both socially efficient and equitable for all, should start from a significant and

---

[161] Also in: Y. Denier; T. Meulenbergs, 'Health Care Needs and Distributive Justice. Philosophical Remarks on the Organisation of Health Care Systems', in R.K. Lie; P.T. Schotsmans (eds.), *Healthy Thoughts: European Perspectives on Health Care Ethics*, Leuven: Peeters, 2002, pp. 265–297.

[162] An excellent analysis of this value inconsistency as it appears in health care is provided by J. Butler, *The Ethics of Health Care Rationing. Principles and Practices*, London: Cassell, 1999 and by G. Calabresi; P. Bobbitt, *Tragic Choices*, New York: Norton, 1978.

meaningful revaluation of human finitude, i.e. from a renewed Socratic perspective of learning to prepare oneself for dying and death.[163]

In order to have a constructive critical function within the contemporary debates regarding setting limits on health care, the proposed contemporary revaluation of the Socratic attitude should have four important characteristics. Firstly, it should entail a more *positive*, *complex* and *refined* valuation of human life, its limits and its finitude; secondly, it needs critical rethinking and reassessment of the concept of scarcity; thirdly, it should be able to express *equal* respect of the human dignity of *all* persons; and finally, it should be able to explain the value of using the logic of *abundance* that comes with the idea of care for the dependent, for a discussion on setting *limits*. Below, I will analyse these four elements in more detail. Note, however, that I here solely intend to provide the atmosphere, to sketch the Socratic attitude in broad and abstract outlines. These outlines by no means intend to immediately provide concrete directions, practical strategies or solutions. Later on, I will be better able to provide a more substantive view. Let us turn now to my first prudent specification of the renewed Socratic perspective in contemporary health care.

2.4.4 A Contemporary Socratic Perspective

The complexity of adequately specifying what a renewed Socratic attitude towards health and independent normal functioning would entail within the framework of contemporary health care has to do with the *dual significance* of human finitude.[164]

On the one hand, finitude is an existential characteristic of human life. Throughout the course of human life finitude reveals itself most clearly in times of dependence; i.e. in times when people are in need of care. Illness and dependence is a manifestation of human finiteness.[165] For most people this dependence is a temporary phase, like in the case of illness, or during childhood. For some, it is a lifelong situation, like in the case of severe mental and/or physical handicap. Furthermore, dependency may reveal itself in a slowly growing manner, as can be experienced in ageing, or in the case of terminal illness. As such, human finitude is all around, as an essential characteristic of human life. Consequently, dependence has an ontological status. It is inevitably linked to human existence.

---

[163] Cf. Plato, *Crito* and *Phaedo*, trans. H. North Fowler, Cambridge, MA: Harvard University Press, 1971. See also: O. Höffe, *Medizin ohne Ethik?* Frankfurt: Suhrkamp, 2002, esp. pp. 119–142; P. Van Parijs, 'Y a-t-il des limites à la prise en charge des soins de santé par la solidarité?' In J. Hallet; J. Hermesse; D. Sauer (eds.), *Solidarité, Santé, Ethique*, Louvain: Garant, 1994, pp. 57–68.

[164] Also analysed in: Y. Denier, 'Autonomie en afhankelijkheid. Het subject in de medische ethiek', in R. Devos; A. Braeckman; B. Verdonck (eds.), *Terugkeer van het subject? Recente ontwikkelingen binnen de filosofie*, Leuven: Universitaire Pers, 2002, pp. 105–116. See also M. Warnock, *Nature and Mortality*, London: Continuum, 2003.

[165] See also: D. Callahan, 'Biomedical Progress and the Limits of Human Health', in R.M. Veatch; R. Branson, *Ethics and Health Policy*, Cambridge, MA: Ballinger, 1976, pp. 157–165.

On the other hand, however, it also has a negative connotation. Although healthy and independent, normal functioning can also be seen as being linked to human existence – just as the Heideggerian fact of 'Geworfenheit' includes the duty of personal 'Entwurf' – there remains an important normative difference between healthy independence and dependence. Being healthy and independent possesses a kind of positive normativity, whereas disease, handicap and dependence possess a kind of negative normativity. Healthy and independent normal functioning is universally seen as a value worthy of pursuit. Despite the fact that finitude reflects a basic characteristic of human life dependence is not seen as a valuable aspect of life. On the contrary, we try to back away from it, and try to retain, to realise and accomplish healthy normal functioning as long as possible *in spite of* the limits of human existence. And this is by all means generally considered as the first task of modern medicine and other health-care services. In this respect, medical intervention and health care have an important emancipating function.

Taking the analysis on contemporary health care's tendency towards infinity into account, it might be interesting to reconsider the generally negative attitude to human finitude.[166] Let us see what an alternative and more positively qualified attitude might entail.

2.4.4.1 Revaluation of the Limits of Human Existence

Firstly, in order to provide a valuable and significant contribution to the contemporary debate on setting limits on health care, a renewed Socratic attitude towards health and independent normal functioning should express a positively refined valuation of human life, its limits and its finitude.

To begin with, it is interesting to ask the question: how tragic is the fact of human finitude? Put negatively, it is indeed true that it makes certain things impossible. Being confronted with one's finitude is usually annoying and uncomfortable. In some cases, for instance in the case of terminal illness, it is harsh and raw. It messes up one's life plans and causes reactions like denial, grief, even shame or aggression and anger, behaviour like increased self-care, avoiding risks and intensive search for all medical possibilities. Against this background, it is very difficult to grasp what a positive valuation of human finitude could possibly mean.

Nevertheless, it is also significant to say that things become meaningful only against the horizon of finitude. The essence of this idea has been aptly expressed by Jorge Luis Borges. In his story *The Immortal,* Borges explores the idea of immortality. To the immortals, he writes, all undertakings are in vain and nothing is meaningful. When our lives would lack finality, they would become senseless, incomprehensible and as such, become horrible and atrocious. We would lethargically fall into complete indifference, become invulnerable to pity, insensitive to

[166] This negative attitude is also captured in the idea of the deprivation account. In: F. Kamm, *op. cit.*, 1993, pp. 13–24.

moral or intellectual merits, because whatever we would do or say today, could be done or said tomorrow. As Borges puts it:

> Death (or its allusion) makes men precious and pathetic. They are moving because of their phantom condition; every act they execute may be their last; there is not a face that is not on the verge of dissolving like a face in a dream. Everything among the mortals has the value of the irretrievable and the perilous. Among the Immortals, on the other hand, every act (and every thought) is the echo of others that preceded it in the past, with no visible beginning, or the faithful presage of others that in the future will repeat it to a vertiginous degree. There is nothing that is not as if lost in a maze of indefatigable mirrors. Nothing can happen only once, nothing is preciously precarious. The elegiacal, the serious, the ceremonial, do not hold for the Immortals. Homer and I separated at the gates of Tangier; I think we did not even say goodbye. ... I remember one whom I never saw stand up. A bird had nested on his breast.[167]

In this story, the significance of finitude is enclosed in the fact that it makes human life meaningful. Instead of disappearing in benumbed lethargy, there is the possibility of having a meaningful relation to the 'now', to the present, to what we do, and to what we choose to forego.

Another way of expressing the horror that comes with the idea of infinity is the Aristotelian critique both on *kapêlikê* or *chrêmatistikê* in the bad sense, which refers to the idea of endless increase, and on *pleonexia*, i.e. the injustice of the infinite desire to have more, especially to have more than others.[168] Without aiming to go into full detail, let me point at the essence of the argument. In *The Politics*, Aristotle sets an important difference between *oikonomikê* and *chrêmatistikê*. The first, which is also called the art of household management, refers to the natural process by which people cooperate to satisfy their needs by the use of their common human capacities. It is the art of using and spending wealth. *Oikonomikê* requires a supply of the means by which it is conducted, and if these are not found ready at hand, they must be acquired. Acquiring them is itself an art, the art of acquisition or *chrêmatistikê*, and this is a part of *oikonomikê*.[169] It is the art of acquiring wealth. There is, however, a second and quite different art of acquisition, which, because of its affinity to natural acquisition, is supposed by many people to be identical with it, but which is in fact unnatural. It is the art of trade, i.e. of endless accumulation, which he calls *kapêlikê* or *chrêmatistikê* in the bad sense.[170] *Kapêlikê* aims at wealth, considered as a quantity of money. Natural exchange, Aristotle writes, aims at the satisfaction of natural needs. It is natural because it serves to satisfy the natural requirement of sufficiency.[171] It serves a natural goal, or *telos*. Endless accumulation, on the other hand, does

---

[167] J.L. Borges, 'The Immortal', in *id.*, *Labyrinths: Selected Stories and Other Writings*, Harmondsworth: Penguin Books, 1964, pp. 135–149, esp. pp. 145–146.

[168] Aristotle, *The Politics*, trans. B. Jowett, rev. trans. J. Barnes, ed. S. Everson, Cambridge: Cambridge University Press, 1996, I.8–10; *Id.*, *Nichomachean Ethics*, trans and ed. R. Crisp, Cambridge: Cambridge University Press, 2000, Vol. 1129b1. See also B. Williams, *op. cit.*, 1980, pp. 189–199.

[169] Aristotle, *The Politics*, 1256b 27ff.

[170] *Ibid.*, 1256b40–1257a5.

[171] *Ibid.*, 1257a17–30.

no longer serve the natural requirement of sufficiency. It has no natural terminus or *telos*: 'there is no limit to the end it seeks'.[172] There is no limit, no pause in the process of seeking; it does not serve a goal except for its own endless movement. As such, it is the most unnatural act of all.

The essence of this analysis for the sake of my argument lies in the fact that according to Aristotle all human activities, and political, social and economic arrangements have a point, a *telos*, for the sake of which they are pursued. There is some kind of teleological order within which everything has its own natural goal. At the same time, this teleological order has a limiting function. Everything has its place in the natural and fixed order of things. The value of this view is that it makes us understand of infinity as the loss of reference to a certain goal that is other than itself. That means, a goal that is other than infinite movement, a goal that functions as a natural terminus. Endless continuation becomes a goal in itself, and therefore becomes, in a Borgesian sense, meaningless movement, without rest, a horrible endless spiral. Aristotle's main concern about this possibility of infiniteness is about its invasion in ethical and political life, consequently installing chaos and disorder. Ethical theory and political arrangements would then, no longer be concerned with their true goal, which has to do with concentration on *eudaimonia* and on securing the necessary conditions for the good human life.

2.4.4.2 Reassessing Scarcity in Health Care

Taking the above analysis on the relationship between meaning and finitude into account, it should not surprise us that the virtue of temperance counts as an important virtue for the Greeks.[173] This virtue keeps us from loosing oneself in infiniteness. Contemporary life, however, is generally not so much determined by the classical virtue of temperance, but on the other hand, by an optimism, filled with various kinds of expectations, projects, hopes for success, possibilities, all within the typically modern framework of the self-made human condition, shaped by the idea of personal control and freedom from all fixed orders.

This has led Hans Achterhuis to rethink the phenomenon of scarcity. He argues that contrary to the modern belief that increasing knowledge and technological progress would overcome ontological scarcity, scarcity was not overcome but socially and anthropologically *constructed* in the sixteenth and seventeenth centuries.[174] This is a very interesting thought for it provides a context to ask the question how one could in its rightful mind say that our contemporary Western societies suffer from scarcity (and in this case, from scarcity of health-care resources).

[172] *Ibid.*, 1257b 28ff.

[173] Among the ancients there were four cardinal virtues, the preeminent virtues of temperance, fortitude, prudence and justice. Cf. Plato in Section 1.3.1.1 *Suum Cuique Tribuere*. Three theological virtues were added in scholasticism: faith, hope and charity, which bring us to the well-known seven cardinal virtues that express a standard of moral excellence and upright conduct.

[174] H. Achterhuis, *Het rijk van de schaarste*, Baarn: Ambo, 1988.

From a general and abstract perspective it is indeed true that contemporary Western societies do not suffer from scarcity but rather from abundance. One can barely say that Western countries lack resources in the way that Third World countries do. How come then, that we nevertheless experience scarcity; that it is a hot topic in contemporary academic, social and political debates? The idea of scarcity, Achterhuis argues, is a paradox caused by anthropological and social mechanisms that come into being in the first place, because of the modern loss of a *fixed order* of things, both the fixed natural order, as well as the order of social relations.

With regard to the *natural* order, it is reasonable to hold that since Francis Bacon, nature counts as something that needs to be controlled, and subjected to the wants of men. With regard to the *social* order, it is so that since Modernity the old class society with a fixed order of class, rank and status, slowly erodes as a result of the eighteenth century. commercial revolution and the increasing wealth and influence of the middle classes. The idea of vertical mobility of status and of equality of opportunity starts to make its way in the general minds of people before being (very imperfectly) realised in political democracy and in the implementation of human rights. They become essential ideas in modern society.[175] As a result of the fact that individual identity becomes part of the self-made human condition, the road is open for individual desires to become mimetic.[176] This desire, however, is infinite. It has no fixed object. As such, it counts as the main catalyst of the rising spiral.[177]

What do these reflections add to the discussion of what we can and should reasonably expect from just health care? They add three things. Firstly, they make us more aware of the *relative meaning of scarcity* of health-care resources, as was already suggested by the analysis on the internal dynamics of scarcity in health care which shows a tendency towards infinity, and which was supported later on by the analysis on the relation between needs and preferences, showing that needs can be misused in an opportunistic way. It focuses our attention on

---

[175] Cf. N. Xenos, *Scarcity and Modernity*, London: Routledge, 1989.

[176] Against the common sense idea that we fix our desire for an object in a fully autonomous, independent, and linear (subject desires object) way, René Girard reveals a different mechanism for the human desire by analysing the novelist masterpieces (Cervantes, Stendhal, Proust and Dostoïevski). According to Girard, the human desire does not fix itself in an autonomous way according to a linear path between the subject and an object, but by imitation of the desire of another person, according to a triangular plan: subject – model – object. Consequently, the object of my desire is not fixed, but moves along with the desire of the other (the model). This triangular structure is better able to explain envy or jealousy. In short, the analysis of the mimetic desire refers to the fact that I desire – by way of *mimesis* – what the other desires. Cf. R. Girard, *Mensonge romantique et vérité romanesque*, Paris: Grasset, 1961; *Id.*, *Des choses cachées depuis la fondation du monde*, Paris: Grasset, 1978.

[177] It also counts as the main reason of the importance of our above analysis on the difference between needs and preferences and the identification of the main characteristics of objective non-volitional needs. Cf. Section 2.3.2 *Needs and Preferences in Health Care*.

the fact that instead of being able to meet individual's health-care needs more adequately, contemporary Western societies face a population whose health-care needs increasingly exceed the medical possibilities society is able to afford. One of the basic enigma's of modern medicine is established in the fact that the growth of medical possibilities increasingly falls short of the health-care needs it happens to produce itself. Many health-care needs did not exist until the means of meeting them came into existence. As such, Achterhuis is right in arguing that scarcity results from social and anthropological mechanisms.

However, I do not agree with Achterhuis's argument that *all* needs are socially and anthropologically constructed. There remains, as I showed in the above analysis on needs and preferences, a category of universal non-volitional needs. These needs are independent from time and space. They entail the things that are necessary for the fundamental projects of *every* person. Meeting these needs has significance for the surviving of persons. As such, they differ from particular needs that originate from specific wants and desires, i.e. adventitious needs, in the fact that having a non-volitional course-of-life need does not depend on the person's free will, and failure to meet the need causes fundamental and crucial harm. Basic health-care needs are such universal needs. They are not limited to the Western welfare states, which have developed organised health-care systems. The fact that many Third World countries do not have a similar system that meets the health-care needs of their citizens does not mean that these persons do not *have* these needs. Having the need does not depend on the existence of the system. Being in good health is of fundamental value for every person and eliminating or reducing barriers that undermine this value – such as disease, illness or injury – is a positive obligation of every just society.[178] In this line of reasoning, I follow Martha Nussbaum in arguing that to give up the notion that some human needs and functional capacities are universal, is 'to turn things over to the free play of forces in a world situation in which the social forces affecting the lives of women, minorities and the poor are rarely benign'.[179]

So secondly, it is very important to be aware of both the aspect of social and anthropological construction of needs as well as the universal and generic character of some needs before we start talking about scarcity in health care. With regard to health care, it implies that as long as *basic and custodial care for those who truly need* it, that is, the dependent, the poor, the disabled and the elderly is not adequately fulfilled, all sophisticated and expensive high-tech medicine that only realises a limited improvement of life quality for a very small number of patients, is problematic. Against this horizon, we must honestly and critically ask ourselves whether every prolongation of life, every application of medical technology is truly

---

[178] Cf. Y. Denier; T. Meulenbergs, 'Health Care Needs and Distributive Justice. Philosophical Remarks on the Organisation of Health Care Systems', in R.K. Lie; P.T. Schotsmans (eds.), *Healthy Thoughts. European Perspectives on Health Care Ethics*, Leuven: Peeters, 2002, pp. 265–297, 272.

[179] M.C. Nussbaum, 'Human Functioning and Social Justice: In Defense of Aristotelian Essentialism', in *Political Theory* 20(1992), pp. 202–246, 212.

that important as it may seem to be at first sight. To what extent are we willing to give up certain goods and services for the use of others who need them more?[180] These critical questions can only be adequately answered against the horizon of a positively refined attitude towards dependence and human finitude.

Finally, a positively refined valuation of human life, its limits and its finitude may set the stage for reflections on equal social valuation of respectful, decent and attentive care for citizens's *quality* of life. It would entail breaking up the sole focus on *quantity* of life and on the quantity of independent healthy normal functioning. As such, the normative valuation of health becomes less strictly one-sided. It may avoid the risk of therapeutic obstinacy and install the possibility of becoming better able to deal respectfully with dependency and mortality as essential characteristics of the human condition, even though we find it tragic.

2.4.4.3 Equal Respect for the Dignity of All Persons

Dealing respectfully with dependency and mortality as essential characteristics of the human condition entails furthermore that we are willing to ask ourselves whether our social institutions regarding health care truly take the equal dignity of *all* persons into account.[181] It raises critical questions concerning the so-called hierarchical relation between the health care pillars of prevention, cure and care. Finitude equally applies to all human beings, and dependence is a generic characteristic of human existence. Just health care must take this into account. Concretely, this implies that it must attribute equal recognition to care for the least-advantaged, i.e. for those who need it the most. Equal respect for the dignity of all persons entails that the sick, the disabled, the old and the severely handicapped are considered to be equally worthy as the healthy, wealthy and productive. For at bottom, the goal of health care remains taking care for the persons who need it, i.e. taking care for the dependent, no matter whether this dependency is only temporary, lifelong or irreversible and terminal. In principle, this entails that the discussion regarding setting limits on health care in the first place should entail a discussion on setting limits on goods and services, never on people.

2.4.4.4 The Paradox of Scarcity and Abundance in Care

Having said all this, it may be a bit clearer now to the reader what I said at the beginning of this first part, namely that the viewpoint of interpreting scarcity as an economic translation of *finitude* would be more apt to incorporate *care* in the discussion. In order to grasp a truly adequate conception of just health care, we need efficiency, justice and care to function as a trinity. This might happen more easily by a positive valuation of finitude.

---

[180] This is a Dworkinian question. I will come back to this in detail in Section 5.4.2 *The Prudent Insurance Principle*.

[181] This will be one of the main concerns of the philosophical analyses of Eva Feder-Kittay, Martha Nussbaum, and Ronald Dworkin. Cf. Section 4.1 *A Second Extension to Long-Term Care?* and Section 5.1 *Two Intuitions on Limits*.

I also said that we would meet a strange but interesting paradox: it seems that the logic of *abundance* that is implied by the concept of *care* enables us better to deal with the logic of ontological *scarcity* because it motivates us to cope with *finitude* in a fully different way. Three reflections indicate somewhat more precisely what I mean by this paradox.

Firstly, it is clear by now that the criticism of the finitude approach is mainly directed against the dynamics of the endlessly upward spiral movement, the tendency towards infinity that characterises contemporary medicine and health care to a large degree. In short, it comes down to a revaluation of the Aristotelian aversion to *pleonexia* and *kapēlikē* or *chrematistike*, to insatiability and endless accumulation.[182]

Secondly, organising health care against the horizon of consciousness of human finitude and the limits of human existence, presupposes a kind of abundance. This abundance is nothing of the kind that is enclosed in the one-sided glorification of medicine's continuous relocation of natural limits, draining away all resources for the glory of technological progress, in the endlessly upward spiral movement. On the contrary, it presupposes the abundance of authentic care, of equally respectful, good, decent, attentive and qualitative care for all against the horizon of finitude. This means, an equal social respect for the care for all those persons whose healthy, normal functioning can be restored (whose care is 'cost-effective', so to say), as well as for the care for those persons who cannot or can no longer be brought to the level of normal functioning, and who need care throughout the length of their lives. It presupposes attention for the quality of life, rather than its quantity.[183] Against the horizon of finitude, we might be better able to deal meaningfully with finitude and the quality of human life, even though we find it hard and often tragic. For remember Borges: a sole focus on quantity of life most probably sacrifices the quality of it.

Finally then, what can be considered as the rightful task of just health care from a contemporary Socratic perspective: what *can* and *should* we reasonably expect from it? As the gap between what is technically possible and what should be equitably accessible increases, our moral responsibility and the answer to the question what we owe to each other at the bar of just health care has less and less to do with doing all we can in the name of health as the *summum bonum*, and ever more with providing *equal access* to the *abundance* of respectful, good, decent, attentive and qualitative *basic care* for all. The logic of scarcity and its consequent narrow focus on efficiency and cost-effectiveness of care, bringing back the sick to the level of healthy functioning as soon as possible, may not colonise the discourse in health care, thereby marginalising authentic dependency care. The just society should provide the framework that fulfils the necessary conditions for health care that shows equal respect for the three pillars of prevention, cure, and care.

---

[182] See also O. Höffe, *Medizin ohne Ethik?* Frankfurt: Suhrkamp, 2002, esp. pp. 174–196.

[183] Cf. H. De Dijn, *De herontdekking van de ziel. Voor een volwaardige kwaliteitszorg*, Nijmegen: Valkhof Pers, 1999.

CONCLUSION OF PART I

In this first part, *Just Health Care: Presuppositions and Objectives*, the centre of attention has been the question: how best to square the proverbial welfare circle of efficiency, justice and care? How can health-care resources be matched to needs, or needs to resources in socially acceptable and economically feasible ways?

Accordingly, I have investigated the various elements that constitute the characteristic problem of contemporary health care. The essence of the argument is that the complexity of the debate on cost-containment and setting limits on health care is due to an internal inconsistency of three parameters: finding an acceptable balance between economic *efficiency*, distributive *justice* and comprehensive *care* seems to be a quasi-impossible task. Nevertheless, it is generally held to be a central goal of just health care to provide decent-quality care for all on the basis of need in a way that is socially efficient, thus keeping the system from breaking down.

The problem of scarcity complicates realisation of an acceptable balance. This has to do with the multiple structure of the concept of scarcity (analysed in Section 2.1). The first and most common interpretation defines scarcity as the given fact of limited resources, as a natural condition of life. Put in general terms, it comes down to acknowledging the fact that we are not in the Garden of Eden. An alternative interpretation of scarcity entails, however, that there is an important social and anthropological dynamics linked to it. The fact that we are not in the Paradise is to a great extent intensified by social and anthropological mechanisms. This second interpretation is able to explain the fact that scarcity of health-care resources is experienced even (or maybe even more strongly) in the wealthy industrialised Western countries that have disposal of an extensive framework of health-care institutions and services. By way of conclusion, it is interesting to show the complexity and difficulty of reconciling the goals of efficiency, justice and care in the following threefold way.

EFFICIENCY, JUSTICE AND THE LANGUAGE OF SCARCITY

Firstly, the goals of efficiency and justice seem to match quite easily with the language of scarcity in its interpretation as a natural condition of limited resources. In trying to balance efficiency, effectiveness and equity in health care, we have seen (in Sections 2.2 and 2.3) that the arguments of providing equal access for equal need – thereby guaranteeing fulfilment of the small objective truncated scale of basic course-of-life needs that have a non-volitional structure – justified by the importance of preserving the normal opportunity range and guaranteeing fair equality of opportunity for all, together provide a well-founded framework of arguments that can serve as a basis for agreement on what we owe to each other at the bar of just health care. As such, it can serve as a guiding framework for social policy within which we can tackle the question what *should* we and what *can* we reasonably expect from a just health-care system? This framework furthermore allows us to achieve a clear understanding of the role and task of the health-care pillars of *prevention* and *cure* within the framework of justice. Preventing disease and handicap, maintaining and restoring health when possible, is an important aspect of guaranteeing everyone a fair chance in life.

JUSTICE, CARE AND THE LANGUAGE OF ABUNDANCE

However, care and support for the long-term dependent persons who cannot or can no longer be brought to the level of normal functioning, does not match with the picture of securing efficiency and effectiveness in preserving the normal opportunity range and guaranteeing fair equality of opportunity for all. Nevertheless, care and social support for the elderly people, the long-term disabled, the chronically and terminally ill is an important aspect of health care as well. When cure is not or no longer possible, caring for the patients, supporting them and easing their suffering as much as possible, is a very important task of medicine and health care.

Against the dominant hierarchical relation between the pillars of prevention, cure and care, I have argued for equal social acknowledgement of the importance of long-term dependency care (cf. Section 2.4). The difficulty is that, whereas prevention and cure easily lend themselves to be arranged by the goals of efficiency, effectiveness and equity within a framework that is dominated by the language of scarcity, it seems that the essence of authentic care escapes the language of scarcity, efficiency and effectiveness in guaranteeing equal opportunity. When is care for the long-term dependent efficient? When is it cost-effective?

If the general goal of just health care is solely determined according to its contribution to normal healthy functioning, and through this, to the normal opportunity range, the framework of health care threatens to intensify the hierarchical relation between prevention, cure and care, and threatens to colonise the domain of care by the language of scarcity. This is highly problematic since the essence of truly authentic long-term care is not a matter of prevention and cure in a cost-efficient way, but of caring for the comfort and quality of life of the dependent. It is about respectful patience and cautious, attentive care for decent quality support. As such, it takes place within a framework that is characterised by the language of abundance. Not so much abundance of material resources, but rather of time, attention and support. If it is the goal of just health care to express equal respect for the dignity of *all* human beings, the macro-level decisions that shape the framework within which micro decisions arise, must equally secure provision of a framework within which authentic care for the elderly and for the lifelong disabled can take place. As such, a contemporary Socratic perspective must be included in the field of just health care.

JUSTICE AS A MEAN BETWEEN EFFICIENCY AND CARE

Finally, it is important to note that within the triadic relation between the goals of efficiency, justice and care, both efficiency and care fall *within* the framework of justice. How should we understand this? The best way to do this, I believe, is by recalling attention to the Aristotelian characterisation of justice as the harmonising, ordering and balancing virtue, installing a mean between too much and too little. Within the structure of this work on *Efficiency, Justice and Care*, justice serves a similarly harmonising goal. It functions as an umbrella-concept. This means that just health care should include a well-balanced concern for economic efficiency as well as for authentic care, in a way that both efficiency and care express the virtue of being a mean between too much and too little. Just health care combines neither too much nor too little focus on efficiency with neither too much nor too little concentration on authentic care. As such, justice avoids endless concentration on efficiency, as well as infinite responsibility in the field of care.

Having all this in hand, it is time now to move to Part II, *Distributive Justice and Health Care*, where I will assess whether and to what extent we can *incorporate* these various issues of just health care, its presuppositions and objectives into a theory of justice.

PART II

DISTRIBUTIVE JUSTICE AND HEALTH CARE

INTRODUCTION TO PART II

To what extent can we *incorporate* the various issues of just health care, its presuppositions and objectives as analysed in the previous chapters into a theory of justice? This will be the central question in the chapters below. Part II provides a detailed analysis of the function and significance of just health care within three contemporary theories of justice: the Rawlsian theory of Norman Daniels, the capabilities approach of Martha Nussbaum, and the resource-egalitarian proposal of Ronald Dworkin.

Chapter 3, *Justice as Fairness: a Rawlsian View on Just Health Care*, starts with a section in which I first highlight the basic assumptions of Rawls's theory and stress those aspects, which especially intersect my problem about the scope and design of just health care. I will be as brief as possible since by now, the theory is no doubt reasonably familiar to the reader. In working to an answer to the implications of Rawls's theory for the matter of health care, I will stress some features of his theory and leave others undiscussed. In Section 3.2, I will raise a range of questions concerning Rawls's view on health care, followed by a detailed analysis of his dealings with this issue. Section 3.3 provides reflections on the way in which the Rawlsian theory has been expanded by Ronald Green and Norman Daniels to cover implications for health-care policy. I will end this section by examining the theory of Daniels in view of the problem of long-term care.

Chapter 4, *Nussbaum's Capabilities Approach: a Non-Contractarian Account of Care*, starts with a detailed analysis of the fierce critiques of Martha Nussbaum and Eva Kittay, which provide a real challenge to Rawlsian-inspired theories on health care. Subsequently, I investigate the possibilities of Nussbaum's capabilities approach with regard to health care in Section 4.2 In Section 4.3, I come back to the problem of integrating long-term care into a theory of justice. Finally, in Section 4.4, I enquire into the qualities and limits of Nussbaum's capabilities approach with regard to my problem of just health care.

Chapter 5, *Setting Limits: Dworkin's Proposal*, has a similar structure. Sections 5.1 and 5.2 contain a dialogue both with the central and with the problematic issues that have turned up in the previous two chapters; issues to which Dworkin's theory aims to provide an alternative answer. Subsequently, I present a brief analysis of the main ideas on which his theory is built in Section 5.3 In Section 5.4, I provide a view on the implications of his theory for just health care. Section 5.5 contains a concluding analysis of the qualities and drawbacks of the Dworkinian approach.

It is interesting to point out in advance that the complexity of finding a balance between efficiency, justice and care will intersect the following chapters. Put generally, we will see that the Rawlsian approach of Norman Daniels incorporates the idea of *efficiency* in his conception of just health care but cannot, however, take comprehensive *care* into account. Its contractualist basis excludes care for the irreversible and long-term dependent from the domain of justice. In reaction to this, the capabilities approach of Martha Nussbaum reconciles both *justice* and long-term *care*. It is a rich account that concentrates on the terms that determine human flourishing. However, references to economic *efficiency* are mostly absent. Finally, Ronald Dworkin's proposal seems to offer an interesting solution to our problem of combining efficiency, justice and care, by giving personal responsibility a more central role. Recent literature, however, has shown that overemphasising the role of personal responsibility not only meets difficulties in practical applicability, but also clashes with a consistent understanding of what we owe to each other at the bar of justice. Consequently, the problem returns: how can we set limits on health care while realising an acceptable balance between efficiency, justice and care? Within this analysis, it is my philosophical task to obtain a clear understanding of the various arguments used, and to assess, explain and justify or modify the relevant categories and distinctions and their interrelations. Let us start the discussion.

CHAPTER 3

JUSTICE AS FAIRNESS: JOHN RAWLS

John Rawls's *Theory of Justice*, first published in 1971, has had an enormous impact not only in academic cloisters but in many areas of public and human service where questions of fairness arise.[1] As Robert Nozick later observed 'political philosophers now must either work with Rawls's theory or explain why not'.[2] As for Rawls himself, all his later works are reflections on the sense and scope of his theory as presented in 1971.

3.1 SOME BASIC FEATURES OF RAWLS'S THEORY

In general, the Rawlsian theory is best understood as a threefold response to utilitarianism, perfectionism and intuitionism. His *anti-utilitarian* concern appears in his proposition that maximising the greatest good for the greatest number – in which 'good' is welfaristically understood as utility, happiness, pleasure or preference satisfaction – cannot be the subject of justice. This is based on the Kantian idea that persons are ends in themselves and therefore possess an inviolability that may not be sacrificed for the benefit of others or for the greater good of society.[3] In the just society, the equal liberties and rights of all are taken as settled.

Furthermore, Rawls argues that the categories of happiness, pleasure and preference satisfaction are too subjective to serve as a ground for compensation at the bar of justice. There is no possibility of interpersonal comparison of subjective welfare and individual preference satisfaction. Instead of leaning on tastes and preferences, one should use objective judgements about the importance of goods for people. Thus, Rawls proposes a form of resource egalitarianism. There are certain primary goods, basic resources, that every person needs, and it is possible to compare these. Answering the underpinning Kantian conviction that persons are ends in themselves, his proposal holds that all human beings are entitled to equally valuable shares of these basic resources.

But, however objectively important the possession of these basic resources is, they neither entail nor establish a standard of excellence.[4] In his *anti-perfectionist* study, Rawls looks for a basis of agreement that is independent from any

---

[1] J. Rawls, *A Theory of Justice*, Cambridge, MA: Harvard University Press, 1971. In 1999, a revised edition has been published. The references in the text below include both editions.

[2] R. Nozick, *Anarchy, State and Utopia*, New York: Basic Books, 1974, p. 183.

[3] J. Rawls, *A Theory of Justice*, 1971, pp. 3–4, 22–27; 1999, pp. 3–4, 19–24.

[4] J. Rawls, *Political Liberalism*, expanded paperback edition, New York: Colombia University Press, 1996, p. 188.

comprehensive idea of the basic values of the good human life. For a modern democratic society is no longer dominated by one shared conception of the good life.[5] On the contrary, it is characterised by a plurality of many comprehensive religious, philosophical and moral doctrines, of which none is shared by all citizens. This pluralism, says Rawls, is 'the normal result of the exercise of human reason within the framework of free institutions of a constitutional regime'.[6] Therefore, the basic resources have to form an objective basis that is important to every person, no matter what his particular conception of the good is.

Finally, Rawls also responds to *intuitionism*, the doctrine that holds that there is a plurality of first principles that have to be weighed against each other by asking ourselves which balance, in our considered judgement is the most just.[7] According to this position, there are no explicit methods, no priority rules for weighing these principles against one another. We are simply to strike a balance by intuition, by what seems to us most nearly right.[8] Rawls's practical aim, on the other hand, is to reach a reasonably reliable agreement in judgement, which provides us with a common conception of justice. In view of this, he formulates a plurality of principles in lexical order that are to serve as a reasonable and generally acceptable proposal for bringing about the desired agreement in judgements.[9]

3.1.1 Primary Goods and the Basic Structure of Society

Rawls begins by noting that many different kinds of things are said to be just or unjust, including big things such as laws, institutions and social systems and small things such as persons themselves, or their attitudes, judgements and actions. His particular concern, however, is with what he calls 'social justice', taking the basic structure of society as primary subject of justice.[10] The basic structure of society concerns the unified system of the major political, social and economic institutions that distribute fundamental rights and duties, and determine the appropriate distribution of the benefits and burdens of social cooperation according to a set of principles, the so-called principles of justice. Compendious, however, through this focus, it is highly important because the basic structure of society profoundly affects people's life prospects and chances, 'what they can expect to be and how well they can hope to do'.[11]

---

[5] *Ibid.*, pp. 179–180.

[6] *Ibid.*, p. xviii.

[7] J. Rawls, *A Theory of Justice*, 1971, pp. 34–45; 1999, pp. 30–40.

[8] An example of classic intuitionism is the prima facie theory as presented by W.D. Ross, *The Right and the Good*, Oxford: Clarendon Press, 1939, esp. Chapters I and II, and his *The Foundations of Ethics*, Oxford: Clarendon Press, 1939. Another representative work of intuitionism is G.E. Moore, *Principia Ethica*, Cambridge: University Press, 1903, esp. Chapters 1 and 6.

[9] A serial or lexical order is 'an order which requires us to satisfy the first principle in the ordering before we can move on to the second, the second before we can consider the third. A principle does not come into play until those previous to it are either fully met or do not apply. A serial ordering avoids, then, having to balance principles at all; those earlier in the ordering have an absolute weight, so to speak, with respect to later ones, and hold without exception'. See: *A Theory of Justice*, 1971, pp. 42–43; 1999, pp. 37–38.

[10] *A Theory of Justice*, 1971, p. 7; 1999, pp. 6–10.

[11] *Ibid.*, 1971, p. 7; 1999, pp. 6–7.

A just social structure, Rawls argues, would allocate a set of what he calls 'primary social goods'. In *A Theory of Justice*, he lists the primary social goods in broad categories, as rights and liberties, opportunities and powers, income and wealth, and the social bases of self-respect.[12] Later, he lists them more specifically in five categories:[13]

(a) A set of basic rights and liberties, including freedom of thought and association, freedom defined by the integrity of the person, and so on. These are the background institutions necessary for the development and exercise of the capacity to decide upon, revise, and rationally to pursue, a conception of the good. Similarly, these liberties allow for the development and exercise of the sense of right and justice under political and social conditions that are free.
(b) Freedom of movement and free choice of occupation against a background of diverse opportunities. These are required for the pursuit of final ends as well as to give effect to a decision to revise and change them, if one so desires.
(c) Powers and prerogatives of offices as well as positions of responsibility in the political and economic institutions of the basic structure. They are needed to give scope to various self-governing and social capacities of the self.
(d) Income and wealth, understood broadly as all-purpose means for achieving directly or indirectly a wide range of ends, whatever they happen to be.
(e) The social bases of self-respect. These are those aspects of basic institutions that are normally essential if citizens are to have a lively sense of their own worth as moral persons and to be able to realise their highest-order interests and advance their ends with self-confidence.[14]

These goods are *primary* goods because they are things that persons need in their status as free and equal citizens, and as normal and fully cooperating members of society over a complete life.[15] They are *social* primary goods in view of their

[12] *Ibid.*, 1971, pp. 62, 440; 1999, pp. 54, 386.

[13] See: J. Rawls, 'Social Unity and Primary Goods', 1999/1982, pp. 362–363, 366; *Id.*, *Political Liberalism*, 1996, p. 181.

[14] An example of such institutions that is *not* covered by the primary goods under (a–d) is provided by Rawls: 'Society as employer of last resort through general or local government, or other social and economic policies. Lacking a sense of long-term security and the opportunity for meaningful work and occupation is not only destructive of citizens' self-respect but of their sense that they are members of society and not simply caught in it. This leads to self-hatred, bitterness, and resentment.' In: *Political Liberalism*, 1996, p. lix.

[15] *A Theory of Justice*, 1999, p. xiii. It is important to note that this account of primary goods replaces the famous original account according to which primary goods were said to be 'things that every rational person is presumed to want, whatever he may hope or plan to get out of life' (*A Theory of Justice*, 1971, p. 62; 1999, p. 54). The weakness of the original account was that it left ambiguous whether something's being a primary good depends solely on the natural facts of human psychology or whether it also depends on a moral conception of the person that embodies a certain ideal. With the revised account, this ambiguity is resolved in favour of the latter, including a specific political conception of the person. This account is given since Rawls's essay 'Social Unity and Primary Goods', published in 1982.

connection with the basic structure: liberties and opportunities are defined by the rules of major institutions and the distribution of income and wealth is also regulated by them. Together they form the social bases of self-respect.[16] Rawls acknowledges the presence of other kinds of primary goods, including 'natural goods' such as health and vigour, intelligence and imagination; but the distribution of these, he argues, is not so directly influenced by the basic structure.[17] As such, they are morally arbitrary: 'The natural distribution is neither just nor unjust .... These are simply natural facts. What is just and unjust is the way that institutions deal with these facts.'[18] A just society would be one in which its basic structures were so arranged that everyone had a fair share of the primary social goods, thereby compensating for the arbitrariness of fortune.

3.1.2 The Principles of Justice

Rawls argues that two principles of justice, lexically ordered according to two priority rules, if fully met, would qualify a society as just. The first principle is the *principle of greatest equal liberty* and states that:

> 1. Each person has an equal right to a fully adequate scheme of equal basic liberties which is compatible with a similar scheme of liberties for all.[19]

This first principle is an important element in his search for an alternative theory to utilitarianism for it states that the equalisation of the primary goods under (a) and (b) must be complete and guaranteed for everyone. No extension for some is allowed if the cost of doing so is their exclusion for others.

Whereas the principle of equal liberty is concerned with the primary goods under (a) and (b), the second principle is concerned with the distribution of the other primary goods, such as social (c) and (e) and economic goods (d). The second principle is subdivided into two: the *principle of fair equality of opportunity*, followed by the *difference principle*:

> 2. Social and economic inequalities are to satisfy two conditions. First, they must be attached to offices and positions open to all under conditions of fair equality of opportunity; and second, they must be to the greatest benefit of the least advantaged members of society.[20]

To the extent that inequalities exist in the distribution of social and economic goods throughout a society, they should be capable of being overcome through institutions to which everyone has an equal and fair chance of access. The principle of fair equality of opportunity would require, for instance, that people have an equal chance of getting a job relative to their capacity to do the job. It would require that an applicant with the relevant qualifications should not be overlooked because of gender, sexual orientation, religion or other morally arbitrary characteristics.

---

[16] *Ibid.*, 1971, p. 92; 1999, p. 79.
[17] *Ibid.*, 1971, p. 62; 1999, p. 54.
[18] *Ibid.*, 1971, p. 102; 1999, p. 87.
[19] *Political Liberalism*, 1996, p. 291. See also: *A Theory of Justice*, 1971, pp. 60–65; 1999, pp. 52–56.
[20] *Ibid.*, 1996, p. 291. See also: *A Theory of Justice*, 1971, pp. 60–90; 1999, pp. 52–78.

The least advantaged people referred to in the difference principle are those with the lowest prospects of enjoying the primary social goods. Importantly, the principle allows the possibility that an unequal share-out of social goods may be compatible with a just distribution – but only if it works to the advantage of the least advantaged members of the society.

The principles of justice are to be ranked in lexical order and two priority rules determine the ranking.[21] The first rule determines the *priority of liberty*, 'the basic liberties can be restricted only for the sake of liberty'. The second priority rule determines the *priority of justice over efficiency and welfare*, 'The second principle is lexically prior to the principle of efficiency and to that of maximising the sum of advantages; and fair opportunity is prior to the difference principle.' This means that Rawls advocates first choosing a politico–economic system that maximises the degree to which the primary goods under (a) and (b) can be provided to all citizens on an equal basis, and then choosing the distribution of economic resources and the forms of economic and political institutions that will maximise an index of the remaining three categories of primary goods going to the least well-off group.

3.1.3 Reflective Equilibrium

Why are these principles and rules postulated as the defining hallmarks of justice? Why is Rawls's system claimed to be superior to utilitarian approaches? The answer lies in what he calls 'the reflective equilibrium'. This is a dynamic process, seeking coherence between the principles of justice, people's considered judgements – or, what we *actually* think – about particular cases and questions, and the theoretical apparatus that generates the principles of justice, i.e. in this case, the original position – or, what we *would* think – about these cases and questions.[22] This equilibrium is not necessarily stable for circumstances can change and particular cases may lead us to revise our considered judgements.[23]

3.1.3.1 Considered Judgements

Rawls argues that his principles and rules can be held to be valid if they match up to the considered judgements that people *actually* make about what is just or unjust in society.[24] Ultimately, his system rests on an appeal to the well-considered judgements of people. A general example of this would flow from his concern, in the difference principle, not only to maximise the overall amount of wealth and income in society but also to protect the position of those at the bottom of the social pile. Widening inequalities in income and wealth might

---

[21] *A Theory of Justice*, 1971, pp. 302–303; 1999, pp. 266–267.

[22] *Ibid.*, 1971, pp. 17–22; 1999, pp. 15–19.

[23] *Ibid.*, 1971, pp. 17–22, 48–49, 579; 1999, pp. 15–19, 42–43, 507. See also: N. Daniels, *Justice and Justification. Reflective Equilibrium in Theory and Practice*, Cambridge: Cambridge University Press, 1996.

[24] *A Theory of Justice*, 1971, pp. 19–22, 48, 579; 1999, pp. 17–19, 42, 507.

well be achieved at the expense of worsening the position of those who are very poor; but this would be identified as an unjust act by the difference principle. In contrast, classical utilitarianism has no such concern with the least advantaged: its only mission is to maximise the total amount of benefit in a society, whatever the cost to those with the least enjoyment of it. It is the former approach, Rawls argues, that better corresponds with people's considered views about what is just. On mature reflection, people would not consider a structure that ignored the plight of the least advantaged to be a just one.

3.1.3.2 The Original Position

Rawls has a fall-back position to support this argument, which comes down to 'persuasion by philosophical reflection'.[25] The validity of his principles, he argues, can also be established by their correspondence with those judgements that free and rational persons *would* make in a hypothetical initial situation. Here, Rawls is drawing on the traditional idea of a social contract into which people voluntarily enter to regulate the major relationships between them; but he takes this idea into new territory by trying to imagine a society in which everyone is equally in the dark and then asking what, in such an imaginary society, people would contractually accept as a fair way of organising themselves. Four conditions characterise this 'original position' as he names it.[26]

THE CIRCUMSTANCES OF JUSTICE. Rawls takes up this idea from David Hume, thus resuming his observations.[27] The circumstances of justice are the normal conditions under which a society as a cooperative venture for mutual advantage, typically marked by a conflict as well as an identity of interests is both possible and necessary. These conditions may be divided into two kinds. Firstly, there are the *objective circumstances*: many individuals coexisting together at the same time on a definite geographical territory, being roughly similar in physical and mental powers, being all vulnerable to attack, and all subject to having their plans blocked by the united force of others. The context is that of moderate scarcity: natural and other resources are not so abundant that schemes of cooperation become superfluous, nor are conditions so harsh that fruitful ventures must inevitably break down. The *subjective circumstances* refer to the relevant aspects of the persons working together. They are roughly similar in needs and interests, but nevertheless have their own life plans, which lead them to different ends and beliefs and to making conflicting claims on the natural and social resources available.

---

[25] *Ibid.*, 1971, p. 21; 1999, p. 19.

[26] *Ibid.*, 1971, pp. 118–150; 1999, pp. 102–130.

[27] See: D. Hume, *A Treatise of Human Nature*, 2nd edn., ed. L.A. Selby-Bigge, rev. P.H. Nidditch, Oxford: Clarendon Press, 1978, p. 495: "t is only from the selfishness and confin'd generosity of men, along with the scanty provision nature has made for his wants, that justice derives its origin'. The idea of the circumstances of justice and the relation between Hume and Rawls is clearly summarised and commented on by Brian Barry in his *Theories of Justice*, Berkeley, University of California Press, 1989, pp. 152–163, 179–183.

THE CHARACTERISTICS OF THE PARTIES. The parties in the original position are free and equal in that they all have the same rights in the procedure for freely choosing the principles. The purpose of these conditions is to represent equality between human beings as moral persons, as creatures having two moral powers. Firstly, they are to have the capacity to understand, to apply, and to act upon the public conception of justice, that is, upon the principles that are adopted, and secondly, they are to have the capacity to form, to revise and rationally to pursue a conception of the good.[28]

Rawls appeals to Kant's notion of the autonomous agent, motivated by rational principles rather than personal or particular desires. The parties to the contract are assumed to be rational people pursuing their life plans in a rational way. This means that they attempt to win for themselves the highest index of primary social goods since this enables them to promote their conception of the good most effectively whatever it turns out to be.[29] In attempting to do this, they are mutually disinterested, that is, they are not bound by prior moral ties to each other; they are not willing to have their interests sacrificed to the others; they are not moved by affection or rancour. On the contrary, they feel entitled to press their rights on each other.[30]

In later works, Rawls stresses that justice as fairness is badly misunderstood if the deliberations of the parties, and the motives we attribute to them, are mistaken for an account of the moral psychology, either of actual persons or of citizens in the well-ordered society.[31] To avoid this error, Rawls specifically distinguishes in his later works three points of view: that of the *parties* in the original position, that of *citizens* in a well-ordered society, and finally that of *ourselves* – of you and me who are elaborating justice as fairness and examining it as a political conception of justice.[32] The first two points of view belong to the conception of justice as fairness and are specified by reference to its fundamental ideas. However, whereas the conceptions of a well-ordered society and of citizens as free and equal might conceivably be realised in our social world, the parties as rational representatives who specify the fair terms of social cooperation by agreeing to principles of justice are simply parts of the original position, set up by you and me in working out justice as fairness. They are the artificial creatures inhabiting the device of representation. The third point of view – that of you and

---

[28] See also: *A Theory of Justice*, 1971, p. 19; 1999, p. 17; J. Rawls, 'Justice as Fairness: Political not Metaphysical', 1999/1985, p. 398.

[29] *A Theory of Justice*, 1971, p. 144; 1999, p. 125.

[30] *Ibid.*, 1971, pp. 129–130; 1999, pp. 111–112.

[31] Examples of such an account are the well-known communitarian critique made by Michael Sandel in his work *Liberalism and the Limits of Justice*, Cambridge: Cambridge University Press, 1982. Related criticisms have been made in feminist thought by Anette Baier, in 'The Need for More than Justice', in V. Held (ed.), *Justice and Care: Essential Readings in Feminist Ethics*, Bolder: Westview, 1995, pp. 47–60; and by Alison Jaggar, in her work *Feminist Politics and Human Nature*, Totowa: Rowman and Littlefield, 1983.

[32] See: J. Rawls, 'Fairness to Goodness', 1999/1975, pp. 273–276; *Id.*, *Political Liberalism*, pp. 22–35.

me – is that from which the political conception of justice is to be assessed. Here the test is that of reflective equilibrium: how well the view as a whole articulates our more firmly considered convictions of political justice.

THE VEIL OF IGNORANCE. In order to guarantee that the procedure is fair and the principles agreed to will be just, the effects of specific contingencies, which put men at odds and tempt them to exploit social and natural circumstances to their own advantage, have to be nullified. Therefore, the parties of the imagined original situation are deprived of any information about certain kinds of particular facts:

> [N]o one knows his fortune in the distribution of natural assets and abilities, his intelligence and strength, and the like. Nor ... does anyone know his conception of the good, the particulars of his rational plan of life, or even the special features of his psychology such as his aversion to risk or liability to optimism or pessimism ... the parties do not know the particular circumstances of their society [and] have no information as to which generation they belong.[33]

Behind the veil of ignorance, the parties are obliged to evaluate principles solely on the basis of general considerations. They will not try to engineer a set of social relationships that systematically benefit some at the expense of others, for they do not know whether they themselves will be within the favoured groups.

THE FORMAL CONSTRAINTS OF THE RIGHT. The chosen principles of justice must have certain features that enable them to work effectively as principles.[34] They must be general principles that are universal in application. They must serve as a public conception of justice, that is, they must be publicly acknowledged to be effective moral constitutions of social life. And finally, they must be capable of acting as a final court of appeal for settling disputes about the justice of any particular act.

3.1.4 Maximin

With all these conditions met, Rawls argues that people would indeed choose his principles of social justice in preference to any others that are on offer. But why? Rawls puts forth two arguments that are later categorised by Brian Barry respectively as 'justice as impartiality' and 'justice as mutual advantage'.[35] The first is a direct attempt to show that only an allocation of resources that maximises the index of primary goods going to the worst-off group is fair. The second is an indirect attempt to consolidate this view by invoking the original position.

Fundamental is the case for the prima facie justice of equality. On Rawls's conception of the morally arbitrary, all differences in achievement are based on morally arbitrary factors. Perhaps a plausible presentation would be to talk of three lotteries: there is the natural lottery, which distributes genetic endowments;

---

[33] *A Theory of Justice*, 1971, pp. 136–137; 1999, p. 118.

[34] *Ibid.*, 1971, pp. 130–136, 175–178; 1999, pp. 112–118, 153–155.

[35] B. Barry, *Theories of Justice*, Berkeley: University of California Press, 1989, pp. 3–9, 213–234. See also: J. Roemer, *Theories of Distributive Justice*, Cambridge, MA: Harvard University Press, 1996, pp. 172–183.

there is the social lottery, which distributes more or less favourable home and school environments; and there is the element of good luck: the lottery that distributes illnesses, accidents and the chance of being on the right place at the right time. Since all inequalities are morally arbitrary, equal distribution counts as a benchmark. Nobody can be asked to accept a less preferred outcome to the one defined by this benchmark, however much others might gain from such a move. However, if it is possible to make *everyone* better off, then it is legitimate to back away from straight equality. Now, it is Rawls's argument that everyone gains from some degree of inequality, because total production will increase (presumably due to incentive effects) and the increased product can be distributed in a way that improves everyone's prospects. But the limits of permissible inequality are reached when the prospects of the worst-off reach their highest point. Any greater inequality than that does not meet the requirement that everyone should gain. The worst-off are to be made as well-off as possible.

According to the *impartiality argument* – which holds that people should not look at things from their own point of view alone but seek to find a basis of agreement that is acceptable from *all* points of view – inequality can be justified only on the grounds that it is rational for representatives of all social positions to accept a certain amount of it. What can be said is simply this: the worst-off gain as much as they possibly can from inequality, so they have no reasonable complaint; and the rest will still be doing better than the worst-off, so they have no reasonable complaint. Thus, all groups gain as much as they can reasonably demand. From an impartial viewpoint the two principles of justice form a reasonable agreement for *all* social positions.

While the first argument attempts to establish the principles of justice as an instance of justice as impartiality, the second tries to establish it as an instance of justice as *mutual advantage*. When faced with choices from within the veil of ignorance, rational people who are intent on advancing their *own* interests will choose the option in which the most disastrous result is the least damaging *for themselves*. As Rawls puts it: 'The maximin rule tells us to rank alternatives by the worst possible outcomes: we are to adopt the alternative the worst possible outcome of which is superior to the worst outcomes of the others.'[36] People in the original position will choose his principles in preference to utilitarianism or perfectionism because they will know that the worst that could happen *to themselves* under a Rawlsian contract would be better than the worst that could happen to themselves in a utilitarian or perfectionist situation. Utilitarianism, for example, allows the liberties of some to be severely restricted if the result is greater overall liberty for many. Perfectionism would allow the same restrictions in the name of a single religious or moral doctrine that everyone in society is obliged to endorse.

[36] *A Theory of Justice*, 1971, pp. 152–153; 1999, p. 133.

3.1.5 Four Important Qualities of the Rawlsian Theory

Since Rawls excludes health care from his catalogue of primary social goods, there can be no simple or straightforward application of his ideas to our problem in hand. Nevertheless, the principles upon which his theory is constructed find sympathetic echoes in contemporary debates about the allocation of health care and these may profit from a careful application in this field. Authors who have attempted the task are Ronald Green and Norman Daniels. But before analysing in detail the implications of the Rawlsian theory in the field of health care, I first want to stress four general qualities of the Rawlsian theory with regard to our thinking about human well-being, which lead to the conclusion that the Rawlsian approach overcomes most problems of utilitarianism, perfectionism, intuitionism and libertarianism.

3.1.5.1 The Inviolability of the Person

By determining the fair shares of primary social goods, thereby guaranteeing equal liberty for all, securing fair equality of opportunity, and not allowing the worst-off to be victims of social and economic inequalities, the two principles of justice express the way in which the Rawlsian approach is deeply and deliberately antithetical to the classical utilitarian view that the aims of justice are served by securing the greatest happiness of the greatest number, however much unhappiness there may be among those who are passed over in the process. The first priority rule, determining the absolute priority of liberty, consolidates this position. We may plainly confirm that this approach acts as a barrier against making victims for the sake of the common good. The Rawlsian theory successfully expresses the Kantian viewpoint that 'each person possesses an inviolability founded on justice that even welfare of society cannot override. For this reason, justice denies that the loss of freedom for some is made right by a greater good shared by others.'[37]

3.1.5.2 Objective Goods Theory

A second important advantage is the fact that social primary goods serve as a small-scale objective measure of well-being. Thereby, this approach successfully answers to a special case where our moral judgements would incline us not to use a subjective satisfaction scale as a measure of well-being, namely the case of social hijacking by persons with expensive tastes.[38] Supposing we judge how well-off a person is by reference to the full range of individual preferences in a satisfaction scale. Let us further suppose that moderate people adjust their tastes and preferences so that they have a reasonable chance of being satisfied with their

[37] *Ibid.*, 1971, p. 3; 1999, p. 3.
[38] I borrow the term 'social hijacking' from N. Daniels, *Just Health Care*, Cambridge: Cambridge University Press, 1985; also cf. Section 2.3.2.2 *Needs in Philosophical Disrepute?*

share of social goods. Extravagant people, however, form exotic and expensive tastes and they are desperately unhappy when their preferences are not satisfied. Since the extravagants are so unhappy compared to the moderates, should we increase their shares? They seem, after all, to be much less well-off. Yet, it seems unjust to deny the moderates equal claims to further distributions just because they have been modest in forming their tastes. Rather, it seems reasonable to hold people responsible for their unhappiness if it results from extravagant preferences, which could have been otherwise.

This hijacking case suggests a more general division of responsibility. John Rawls urges that we hold society responsible for guaranteeing the individual a fair share of social primary goods as basic liberties, opportunities and all-purpose means like income and wealth, needed for pursuing individual conceptions of the good. But we should hold individuals responsible for choosing their ends in such a way that they have a reasonable chance of satisfying them under just arrangements. Consequently, the special features of an individual's conception of the good – here his extravagant tastes and resulting dissatisfaction – do not give rise to any special claims by justice to special resources: 'The immediate object of justice is not, then, happiness or the satisfaction of desires, though just institutions provide individuals with an acceptable framework within which they may pursue happiness. But in this pursuit, individuals remain responsible for the choice of their ends, so there is no injustice in not providing them with sufficient means to reach extravagant ends.'[39]

Thomas Scanlon raises the point that this division of responsibility relies on the *capacity* of persons to assume responsibility for their ends and to moderate the claims they make on their social institutions accordingly.[40] This is indeed the case for Rawls. According to his theory, it is an essential characteristic of citizens as free and equal persons possessing the two moral powers that they are at liberty to take charge of their lives and they can be expected to adapt their conception of the good to their expected fair share of primary goods.[41]

In Rawls's case, the full argument involves the claim that adopting a satisfaction scale commits us to an unacceptable view of persons as mere 'containers' for satisfaction, one that departs significantly from our moral practice. Satisfaction scales leave us no basis for *not* wanting to *be* whatever person, construed as a set of preferences. To borrow the term from Bernard Williams they leave us with

---

[39] In: N. Daniels, *Just Health Care*, 1985, p. 38. See also: J. Rawls, 'Fairness to Goodness', 1999/1975, pp. 284–285; 'Social Unity and Primary Goods', 1999/1982, pp. 369–374; 'Justice as Fairness: Political not Metaphysical', 1999/1985, p. 398; *Political Liberalism*, 1996, pp. 189–190.

[40] T.M. Scanlon, 'Preference and Urgency', in *Journal of Philosophy* 72(1975)19, pp. 665–666; *Id.*, 'The Significance of Choice', in S. McMurrin (ed.), *The Tanner Lectures on Human Values*, Vol. 8, Salt Lake City: University of Utah Press, 1988, pp. 149–216; *Id.*, *What We Owe to Each Other*, Cambridge: The Belknap Press of Harvard University Press, 1998, pp. 248–294.

[41] J. Rawls, 'Social Unity and Primary Goods', 1999/1982, pp. 370; 'Justice as Fairness: Political not Metaphysical', 1999/1985, pp. 407–408; *Political Liberalism*, pp. 189–190.

no basis for insisting on the *integrity* of persons.[42] Thus, the satisfaction scale is incompatible with a view of the nature of persons, which in turn underlies much of our moral practice, such as the free and reasonable person, capable of mastering and adjusting his wants and desires.

3.1.5.3 Strategic Role of the Primary Social Goods

A third advantage is that the narrow objective measure of well-being, provided by the index of primary goods, provides a common ground that we can all agree to no matter what each person's particular life plan is. Rawls writes: 'The thought behind the introduction of primary goods is to find a practicable public basis of interpersonal comparisons based on objective features of citizen's social circumstances open to view, all this given the background of reasonable pluralism.'[43] The focus on primary goods helps us to achieve the sort of overlapping consensus that is required for an enduring social cooperation among members of society with different conceptions of the good.

Thus, the primary goods approach provides a universal account of the goods that are needed to live a good human life. Using the words of Derek Parfit, Rawls provides an 'objective list theory', according to which the assessment of a person's well-being involves a substantive judgement about what goods, conditions and opportunities make life better.[44] This, however, does not mean that primary goods are substitutes for comprehensive doctrines.[45] On the contrary, the moral importance of the primary goods depends on their strategic role in the pursuit of diverse individual aims. The primary social goods function as instruments, all-purpose means that serve to enable all citizens to pursue their conception of the good. As such, the approach avoids the ethical danger of perfectionist paternalism, which is insensitive to different conceptions of the good that are of worth and value in the lives of people. On the other hand, the Rawlsian conception is not unconditionally tolerant towards all conceptions of the good but only to 'permissible' conceptions, that is, those that comply with the list of primary goods.[46]

---

[42] See: B. Williams, 'A Critique of Utilitarianism', in J. Smart; B. Williams, *Utilitarianism: For and Against*, Cambridge: Cambridge University Press, 1973, pp. 77–150; B. Williams, 'Utilitarianism and Moral Self-Indulgence', in H. Lewis (ed.) *Contemporary British Philosophy*, Series 4, London: Allen and Unwin, pp. 306–321, repr. in B. Williams, *Moral Luck Philosophical Papers 1973–1980*, Cambridge: Cambridge University Press, 1981, pp. 40–53; B. Williams, 'Persons, Character, and Morality', in A. Rorty (ed.), *The Identities of Persons*, Berkeley: University of California Press, 1976, pp. 197–216; repr. in B. Williams, *Moral Luck*, 1981, pp. 1–19; A. Sen; B. Williams (eds.), *Utilitarianism and Beyond*, Cambridge: Cambridge University Press, 1982, pp. 1–21. This view on the nature of persons is also pursued by Derek Parfit in his 'Later Selves and Moral Principles', in A. Montefiore, *Philosophy and Personal Relations*, London: Routledge, 1973, pp. 137–169; J. Rawls, 'Independence of Moral Theory', 1999/1975, pp. 286–302; *Id.*, 'Social Unity and Primary Goods', 1999/1982, pp. 359–387; and in N. Daniels 'Moral Theory and the Plasticity of Persons', in *Monist* 62(1979)3, pp. 265–287.

[43] *Political Liberalism*, p. 181.

[44] D. Parfit, *Reasons and Persons*, Oxford: Oxford University Press, 1984, pp. 493–502.

[45] *Political Liberalism*, pp. 187–189.

[46] *Ibid.*, pp. 190–200.

3.1.5.4 Real Equality of Opportunity

Finally, the two principles of justice provide a challenge to theories like libertarianism – which Rawls calls 'an impoverished form of liberalism'[47] – that view formal equality of opportunity as necessary and sufficient for distributive justice. By formal equality of opportunity, I mean the condition that there is no legal barrier against access to education, to all advantaged social positions, etc. Libertarian theory holds that only a minimal state limited to the narrow functions of protection against force, theft, fraud, enforcement of contracts, and so on, is justified; and that any state with more comprehensive powers violates the rights of individuals. This is based on the starting point that an absolute respect for the right to private property and for negative freedom constitutes the basis and guideline for the legitimate role of the state and for the basic principles of individual conduct. However, since there is no effort to preserve an equality of social conditions, the initial distribution of assets is strongly influenced by natural – talents and abilities – and social circumstances and such contingencies as accident and good fortune. Rawls strongly criticizes this libertarianism or 'system of natural liberty' as he calls it because it permits distributive shares to be improperly influenced by morally arbitrary factors and because it allows excessive social and economic inequalities, approved by the formal criterion of individual property rights.[48] From the beginning, the Rawlsian set of primary goods intends to do more, by providing a theory of resource egalitarianism that is explicitly based on an elimination of the morally arbitrary factors, and is only prepared to accept inequality when it adds to the benefit of the least advantaged.

Whereas *formal* equality of opportunity involves only respect for negative freedom, the Rawlsian viewpoint on *real* equality of opportunity is based on the triadic structure of freedom, which is explained by reference to three items: 'the agents who are free, the restrictions or limitations, which they are free from, and what it is that they are free to do or not to do'.[49] Rawls holds that freedom is always, first of all, the freedom of the agent who is free or not, and includes then, both the negative aspect of freedom *from* something and the positive aspect of the freedom *to do* something, or to become something.

In stressing the inviolability of the person, Rawls focuses on the so-called negative freedom not to be harmed, not to be wronged intentionally in some

---

[47] *Ibid.*, p. lviii.

[48] *A Theory of Justice*, 1971, pp. 65–67, 72–73, 106–108; 1999, pp. 57–58, 62–63, 91–93; *Political Liberalism*, 1996, pp. 262–269, 324–331. His refusal to let the opportunities and expectations of those with the same abilities and aspirations be affected by their social class has been importantly inspired, Rawls says, by Bernard Williams's argumentation in B. Williams, 'The Idea of Equality', 1973/1962.

[49] *A Theory of Justice*, 1971, p. 202; 1999, p. 177. In this, Rawls follows the classic article on freedom as a triadic relation: G. MacCallum, 'Negative and Positive Freedom', in *Philosophical Review* 76(1967), pp. 312–334. The triadic relation view challenges the idea that we may usefully distinguish between two distinct kinds of political and social freedom, that is, between negative freedom on the one hand, and positive freedom on the other, while at the same time defending that one of these is the one and only true version of freedom. For such a dualist distinctive view, see: I. Berlin, *Four Essays on Liberty*, Oxford: Oxford University Press, 1969.

specific way. However, in the absence of just background institutions which guide resource allocation, there would be nothing to guarantee the integrity of citizens in the positive sense as free, intelligent agents, capable of deliberating over personal, political, civil, and socio-economic matters in a reasoned way. This brings us to a class of positive freedoms Isaiah Berlin did not discuss in his essay on liberty. It does not per definition involve the dangerous perversion of political authority telling people what is in their 'real' interest, thus justifying policies of oppression, a perversion that Berlin rightly feared. In his response to perfectionism, Rawls expresses the same rejection to oppressive comprehensive doctrines. On the other hand, he talks of positive freedom as the extent to which a person has the freedom to shape *his own* life, in absence of coercion, and he concentrates on the necessary background institutions – primary goods – as the means by which persons can reasonably pursue their own reasonable conception of the good.

3.2 RAWLSIAN REFLECTIONS ON JUST HEALTH CARE

In view of the enormous impact of *A Theory of Justice* in the field of social and political philosophy, it is worth asking what the implications of Rawls's view for health-care policy are. The aforementioned qualities – respect for the person as person; the objective list and the ensuing division of responsibility; related to this, the focus on the strategic role of the primary goods as a result of respect for the different conceptions of the good; and finally, the focus on real equality of opportunity – together give the impression that Rawlsian theory provides good grounds for reflections on just health care. However, although questions concerning the supply and distribution of health care are central to contemporary discussions of social justice – and it is social justice that is the object of Rawls's inquiry – little space is devoted to health or health care in *A Theory of Justice*. The index of primary goods itself contains no reference to health, sickness, medicine, or medical care. Although it is true that there are some very brief references to these topics throughout the text, these are but mere side remarks or afterthoughts to other discussions.[50] Why is this major work virtually silent on a matter that many persons believe to be at the forefront of questions of justice today? Several explanations might be offered.

---

[50] These remarks include the definition of health as a 'natural primary good' (*A Theory of Justice*, 1971, p. 62; 1999, p. 54); the assessment that reasonable regulations and efficient measures for public health and safety promote the common interest (*Ibid.*, 1971, p. 97; 1999, p. 83); the designation of inoculation procedures and health services as 'public goods' in the special economic sense of that term (*Ibid.*, 1971, pp. 268, 270; 1999, pp. 237, 239); and the brief, unexpanded suggestion that the social minimum of a just society, that is, those payments made to the least-favoured groups, will include 'special payments for sickness' in the form of a negative income tax (*Ibid.*, 1971, p. 275; 1999, p. 243).

3.2.1 Health as a Natural Primary Good

Firstly, there is the fact that Rawls is primarily interested in developing a theory for the distribution of those basic goods that are created by social cooperation and that are distributed by the social system. He calls these the 'social primary goods', and they include various civil rights and liberties, socially bestowed powers and opportunities, the material goods of wealth and income, and the important social good of self-respect.[51] As we have seen, Rawls distinguishes these from the 'natural primary goods' of intelligence, vigour, imagination, and good health whose distribution, he maintains, is only indirectly affected by the social structure, and he places the natural primary goods outside the scope of his concern.[52]

On closer view, however, this move on his part is not really valid. For a natural good like intelligence, the distinction between what is directly and indirectly mediated by the social structure is artificial. In view of all we know about the effects of family circumstances on learning, for example, it does not seem right to say that intelligence is not in large measure distributed when income shares are decided. If this point has force with respect to intelligence, it would seem to be even more pertinent where health is concerned. Modern medical technology, with its enormous preventive and therapeutic powers, renders almost archaic the notion that health depends solely on natural contingencies. Social decisions concerning medical care have a vital impact on everyone's health, even when health is construed only in the narrowest sense as freedom from physical disease.[53] Health is class-related.

3.2.2 Ideal Theory

Secondly, in Rawls's ideal theory the fundamental idea is to find the most appropriate conception of justice for specifying the fair terms of social cooperation between free and equal citizens who are *normal and fully cooperating* members of society over a complete life. He presupposes that everyone has physical needs and psychological capacities *within the normal range* and that these basic needs are met. Within the political conception of justice, worked out for the basic structure of society, the citizens have two moral powers to a *requisite minimum degree*: the capacity for a sense of justice and the capacity to form, to revise, and rationally to pursue a conception of one's rational advantage or good.[54] Such idealisation allows Rawls to first construct a theory of justice for the simpler, idealised case,

---

[51] *Ibid.*, 1971, p. 92; 1999, p. 79.

[52] *Ibid.*, 1971, p. 62, 102; 1999, p. 54, 87.

[53] See: N. Daniels, B. Kennedy; I. Kawachi, 'Justice is Good for Our Health', in N. Daniels; B. Kennedy; I. Kawachi, *Is Inequality Bad For Our Health?* Boston, 2000.

[54] Though this implication of his ideal theory is only implicitly present in 1971, Rawls, answering his critics, emphasises this explicitly in his later works and in the revised edition of *A Theory of Justice*. See his 'A Kantian Conception of Equality', 1999/1975, p. 259; 'Social Unity and Primary Goods',1999/1982, p. 368; *A Theory of Justice*, 1971, pp. 505–512; 1999, pp. xiii, 83–84, 442–449; *Political Liberalism*, 1996, pp. 3–9, 18–21, 183. About his 'ideal theory', see: *A Theory of Justice*, 1971, pp. 7–10, 245–247, 351–352; 1999, pp. 7–9, 215–217, 308–309.

and then to worry about extensions of the theory to contexts in which conditions are more realistic and people are not all normal. This implies that in his theory the problems of special health care and of how to treat the mentally and congenitally defective do not arise. To use the words of Norman Daniels: 'there is no distributive theory for health care because no one is sick!'[55]

3.2.3 *Income and Wealth*

A third reason why Rawls has neglected the issue of health care may have to do with his extensive concern with the problem of income distribution. Income and wealth are used as approximations to the whole index. The two principles of justice Rawls defends require the basic structure to be arranged 'so that the lifetime expectations of the least advantaged, *estimated by their income and wealth*, are as great as possible given fixed background institutions that secure the equal basic liberties and fair equality of opportunity'.[56] With regard to the fair opportunity principle, the strong emphasis on fair income distribution leads to a focus on fair competition in the quest for jobs and careers. Also the idea of special payments for sickness in the form of a negative income tax seems to illustrate income and wealth as the overall measure of well-being.[57]

One could ask then, whether Rawls believes that the matter of medical care can be made a function of just income distribution.[58] Indeed, this view may underlie his definition of health as a natural primary good only indirectly mediated by the social structure. Thus, he might believe that once society has been set up so that everyone receives a just share of income, medical care can be arranged for privately out of that share, with special provisions made for those at the bottom of the income ladder.[59] This view, however, would run against the general tenor

[55] N. Daniels, *Just Health Care*, 1985, p. 43.

[56] See: 'Social Unity and Primary Goods', 1999/1982, pp. 363, 368 (my italics, YD). See also: *A Theory of Justice*, 1971, p. 98; 1999, p. 84, where he points at the different possible criteria with which to identify the least advantaged (the criterion of the social position or that of relative income and wealth).

[57] Negative income tax means that the tax rate is negative in the case of income below a certain minimum. Thus, instead of paying a certain tax, the taxpayer is attributed a certain transfer.

[58] A comparable analysis can be found in Jon Elster's critique on Michael Walzer's analysis of 'Spheres of Justice' and 'blocked exchange' of money: the necessity of blocked exchange becomes important only in the case of great income inequalities among citizens. The principle held within the sphere of medical care, 'that care should be proportionate to illness and not to wealth' – i.e. that there should be a blocked exchange of money – is based, according to Elster, to a large extent on the fact that people are 'mainly opposed to *great inequalities of income*'. This implies that when income inequality would be reduced or mitigated, the idea of different spheres of justice and of blocked exchange would become less important. See: M. Walzer, *Spheres of Justice: A Defense of Pluralism and Equality*, New York: Basic Books, 1983, pp. 86, 95–128; J. Elster, *Local Justice. How Institutions Allocate Scarce Goods and Necessary Burdens*, Cambridge: Cambridge University Press, 1992, esp. pp. 11–14, 146.

[59] Note, however, that this statement only holds in the case of 'normal', curable diseases but not in the case of noncurable and uninsurable chronic handicaps and long-term disabilities. And even for 'normal' diseases it presupposes an ideal, well-functioning insurance market for health care. A classic article on the subject is K. Arrow, 'Uncertainty and the Welfare Economics of Medical Care', in *American Economic Review* 53(1963), pp. 941–973.

of Rawls's position. It is very doubtful that the rational agents, whose choices define social principles in this theory, would leave the distribution of medical care to be determined by one's income share. I will repeatedly consider various aspects of this issue in detail later.

Now, suffice to say that the three elements, i.e. the designation of health as a natural primary good, the presupposition of active and fully cooperative citizens and the assessment of income and wealth as approximations to the index, seem to put Rawls's theory at odds with my previous defence of an objective truncated scale of well-being in the case of health care. Indeed, the Rawlsian index of primary goods seems to be too truncated a scale.

3.2.4 *The Inflexibility Critique*

3.2.4.1 *Capabilities*

That Rawls's theory is too truncated is a concern that has been raised before by Kenneth Arrow and Amartya Sen and has been taken up and reconsidered with regard to health care by philosophers like Ronald Green, Norman Daniels, Eva Kittay, Martha Nussbaum and Ronald Dworkin.[60] This critique is known as the *inflexibility critique* and holds that the primary goods approach is insensitive to the many significant variations among persons in their capacities – moral, intellectual

---

[60] See: K.J. Arrow, 'Some Ordinalist-Utilitarian Notes on Rawls's *Theory of Justice*', in *Journal of Philosophy* 70(1973)9, pp. 245–263; esp. pp. 253–254; A.K. Sen, 'Equality of What?' in S. McMurrin (ed.), *Tanner Lectures on Human Values*, Vol. I, Cambridge: Cambridge University Press, 1980; *Id.*, 'Well-Being, Agency and Freedom', in *Journal of Philosophy* 82(1985)4, pp. 169–221, esp. pp. 195–202; *Id.*, 'Justice: Means versus Freedoms', in *Philosophy and Public Affairs* 19(1990), pp. 111–121; *Id.*, *Inequality Reexamined*, Cambridge, MA: Harvard University Press, 1992, esp. Chapter 5. See also: J. Rawls, 'Social Unity and Primary Goods', 1999/1982, pp. 368–369; and *Political Liberalism*, 1996, pp. 182–183; R. Green, 'Health Care and Justice in Contract Theory Perspective', in R.M. Veatch, R. Branson (eds.), *Ethics and Health Policy*, Cambridge, MA: Ballinger, 1976, pp. 111–126; N. Daniels, 'Health-Care Needs and Distributive Justice', in *Philosophy and Public Affairs* 10(1981)2, pp. 146–179, repr. in *Id.*, *Justice and Justification: Reflective Equilibrium in Theory and Practice*, Cambridge: Cambridge University Press, 1996, pp. 179–207; *Id.*, *Just Health Care*, Cambridge: Cambridge University Press, 1985; E. Feder Kittay, *Love's Labor: Essays on Women, Equality, and Dependency*, New York: Routledge, 1999; M.C. Nussbaum, 'The Future of Feminist Liberalism', in *Proceedings and Addresses of the American Philosophical Association* (2000)74, pp. 47–79; *Id.*, 'Disabled Lives: Who Cares?' in *The New York Review of Books* 48(2001)1, pp. 34–37; *Id.*, 'Long-Term Care and Social Justice: A Challenge to Conventional Ideas of the Social Contract', in World Health Organization, *Ethical Choices in Long-Term Care. What Does Justice Require?*, Geneva: World Health Organization, 2002, pp. 31–65; *Id.*, 'Capabilities as Fundamental Entitlements: Sen and Social Justice', in *Feminist Economics* 9(2003)2–3, pp. 33–59; *Id.*, 'Rawls and Feminism', in S. Freeman (ed.), *The Cambridge Companion to Rawls*, Cambridge: Cambridge University Press, 2003, pp. 488–520, esp. pp. 511–514; *Id.*, 'Beyond the Social Contract: Toward Global Justice', in G.B. Peterson (ed.), *The Tanner Lectures on Human Values*, Vol. 24, Salt Lake City: University of Utah Press, 2004, pp. 413–507; B. Barry, *The Liberal Theory of Justice*, Oxford: Oxford Unversity Press, 1973, pp. 55–57; R. Dworkin, 'Justice and the High Cost of Health', 1994, repr. in *Id.*, *Sovereign Virtue: The Theory and Practice of Equality*, Cambridge, MA: Harvard University Press, 2000, pp. 307–319; *Id.*, 'Justice in the Distribution of Health Care', 1993, repr. in M. Clayton, A. Williams (eds.), *The Ideal of Equality*, Basingstoke: Palgrave Macmillan, 2002, pp. 203–222.

and physical – and in their determinate conceptions of the good, as well as in their preferences and tastes. These variations are sometimes so great that it is hardly fair to secure the same index of primary goods for everyone to cover their needs as citizens and let it further be as it is. Arrow refers to variations in people's needs for medical care and how expensive it is for them to satisfy their needs. Thus, some people are inefficient converters of primary goods into relative advantage or well-being because they are ill or handicapped. Sen has stressed the importance of variations among people in their basic capabilities and therefore in their ability to use primary goods to attain their aims. The core of the problem is that once we allow people to differ in needs, we will see that people with equal resources will not be equally well-off. As a result, if we take the primary goods as the appropriate measure of well-being for purposes of justice, we may treat people unfairly.

There are really two related claims here. Firstly, the variability among persons implies that the primary goods are an inflexible measure of well-being, ignoring variations that matter. Secondly, this inflexibility should count as strong evidence that the primary goods are the wrong working space: we are not ultimately concerned with goods – primary or not – but with what people, given their variability, can do and be with these goods. Hence, Sen's famous charge that there is an element of fetishism in Rawls's use of primary goods.[61] What is of ultimate importance here, Sen argues, does not concern primary goods, but capabilities, which are the result of 'a *relationship* between persons and goods'.[62] Sen's own account of well-being concentrates on the concept of 'functioning'. How well-off we are, depends on what we can do and be, i.e. on how we function. For example our 'doings' and 'beings' include 'activities (such as eating, reading or seeing), or states of existence or being, e.g. being well-nourished, being free from malaria, not being ashamed of the poverty of one's clothing or shoes'.[63] Sen suggests that we represent 'the focal features of a person's living' by an n-tuple of different types of functionings; each component of the n-tuple reflects the extent of the achievement of a particular functioning. The n-tuple is thus not just an array of kinds of functionings, but it includes a measure of the level of achievement of each functioning in the array. Sen holds that a person's capability can be represented by the set of n-tuples of functionings from which the person can choose any one n-tuple. In this way, the capability set stands for 'the actual freedom of choice a person has over alternative lives that he or she can lead'.[64] Thus, given the same index of primary goods, a handicapped or ill individual may not enjoy the same capability set or freedom of choice as someone who is 'normal'.

3.2.4.2 Actual Choice

Another line of inflexibility criticism has been expressed by Richard Arneson and Gerald Cohen and entails that the primary goods approach fails to capture

[61] A.K. Sen, 'Equality of What?' 1980, p. 363.
[62] *Ibid.*, p. 366.
[63] *Id.*, 'Well-Being, Agency and Freedom', 1985, p. 197.
[64] *Id.*, 'Justice: Means versus Freedom', 1990, pp. 113–114.

a fundamental moral intuition that underlies our concerns about permissible inequalities.[65] The intuition is that whenever we are made worse off through no fault of our own, or as a result of something beyond our control we have a legitimate initial claim on others for assistance or compensation for our misfortune. Here too the primary goods are being accused of being insensitive to a type of variability among people, in particular, the variability among the preferences people have. For example, some people may have more expensive tastes than others and so be less efficient converters of primary goods into satisfaction or welfare. For the same assignment of primary goods, such people will be less well-off than people with more modest tastes. This line of objection makes the issue of choice of preferences or control over preferences central to egalitarian concerns. The suggestion that emerges is this: our egalitarian concerns do not require that we be compensated for being worse off than others when it is our *chosen* preferences that make us worse off. For his argument, Cohen makes use of the distinction between two kinds of luck, made by Ronald Dworkin. If I am made worse off because gambles I have made turned out badly, i.e. because I have had poor *option luck*, then egalitarian concerns are not triggered. If I fare worse than others because of matters outside my control, then I am a victim of poor *brute luck*, and egalitarian concerns appropriately come to the foreground.[66] If this claim is right, then Rawls's suggestion that 'we should hold people responsible' seems in need of qualification, for now people did not choose their preferences and so should not really be held responsible for them. Arneson suggests we restate Rawls's claim and make actual choice central.[67] This implies also a modification of the difference principle, since some people (e.g., the penniless artist who had the talents and education to be a prosperous business man; or the person with serious health injuries due to a very unhealthy lifestyle) might end up among the worst-off with regard to various expectations and opportunities open to them. However, this has been the result of their own choices. If actual choice is 'foregrounded', as Cohen puts it, in the way proposed here, then it may not be fair that the difference principle makes the worst-off as well-off as possible. It should only help those who are poor through no fault of their own.[68]

Rawls replies to these criticisms, first, by claiming that one should not fail to understand the basic aim of his theory, which is to specify within the political conception of justice the fair terms of social cooperation among citizens with an

[65] R.J. Arneson, 'Equality and Equal Opportunity for Welfare', in *Philosophical Studies* 54(1988), pp. 79–95; G.A. Cohen, 'On the Currency of Egalitarian Justice', in *Ethics* 99(1989), pp. 906–944.

[66] See: G.A. Cohen, *op. cit.*, p. 908. The distinction between these kinds of luck is made in R. Dworkin, 'What is Equality? Part 2: Equality of Resources', in *Philosophy and Public Affairs* 10(1981)4, pp. 283–345, esp. pp. 293–298; repr in: R. Dworkin, *Sovereign Virtue*, Cambridge, MA: Harvard University Press, 2000, pp. 65–119, esp. pp. 73–77.

[67] R.J. Arneson, 'Primary Goods reconsidered', in *Nous* 24(1990), pp. 429–454.

[68] I will consider this discussion in detail in Section 5.5.3 *On Personal Responsibility and the Right to Health Care*.

essential minimum of moral, intellectual, and physical capacities.[69] Rawls admits that, provided due precautions are taken, we can, if need be, *expand* the list of primary goods to other goods like leisure time, and even to certain mental states such as freedom from physical pain. However, he does not pursue these matters in his theory. Instead, he stresses that one should always recognise

> first, [to] stay within the limits of justice as fairness as a political conception of justice that can serve as the focus of an overlapping consensus; and second [to] respect the constraints of simplicity and availability of information to which any practicable political conception is subject.[70]

In working out a political conception of justice 'I have assumed throughout', Rawls repeats,

> and shall continue to assume, that while citizens do not have equal capacities, they do have, at least to the essential minimum degree, the moral, intellectual, and physical capacities that enable them to be fully cooperating members of society over a complete life.[71]

However, Rawls admits that Arrow and Sen are right that in some of these cases the same index for everyone would be unfair. He agrees with Sen 'that basic capabilities are of first importance and that the use of primary goods is always to be assessed in the light of assumptions about those capabilities', i.e. in the light of what people can 'do' with these goods.[72] At the basis of Rawls's reliance on primary goods is the recognition that these goods are essential all-purpose *means* to realise the higher-order interests connected with citizens's particular conceptions of the good and the opportunities to realise their life plans. Thomas Scanlon adds that as long as it is acknowledged that the moral importance of the primary goods depends on their 'strategic role in the pursuit of diverse individual aims', 'the value that we can agree to assign to these resources need not be "fetishistic" in the sense criticised by Sen'.[73] Important for our problem in hand is the fact that Rawls considers Sen's theory as a meaningful *complement* to his own theory:

> When we attempt to deal with the problem of special medical and health needs, a different or a more comprehensive notion than that of primary goods ... will, I believe, be necessary; for example, Sen's notion of an index which focuses on person's basic capabilities may prove fruitful for this problem and serve as an essential complement to the use of primary goods.[74]

Rawls rejects, on the contrary, the critique raised by Arneson and Cohen. Rawls has claimed that individuals be held responsible for their ends, whereas society is responsible for providing the just framework of all-purpose means within which

---

[69] Rawls replies to these criticisms in his 'Social Unity and Primary Goods', 1999/1982, pp. 368–371; and in *Political Liberalism*, 1996, pp. 181–187.

[70] J. Rawls, 'The Priority of Right and Ideas of the Good', 1999/1988, pp. 454–455; *Political Liberalism*, p. 182.

[71] J. Rawls, 'Social Unity and Primary Goods', 1999/1982, p. 368; *Political Liberalism*, p. 183.

[72] *Political Liberalism*, p. 183.

[73] T. Scanlon, 'Value, Desire, and Quality of Life', in M.C. Nussbaum; A. Sen, *The Quality of Life*, Oxford: Clarendon Press, 1993, pp. 185–200, 197.

[74] J. Rawls, 'Social Unity and Primary Goods', 1999/1982, p. 369, fn. 8. I will take up this discussion in detail in Section 4.2.2 *Complementary Theory?*

individuals can pursue their conceptions of the good. However, in this argument, Rawls never assumes an intuitive notion of merit, related to the question whether people *actually chose* to have the preferences they have. For that would require too much information and would thus run afoul of the constraints of simplicity and availability of information.[75] Holding people responsible for their ends, in the Rawlsian sense, means that we are acting as if they can exercise their underlying moral power to form and revise their conceptions of the good. We want to back away from holding people responsible only if we have reason to think the underlying capacity is compromised, not if we think certain actual choices have or have not been made. An illuminating example to support the argument is provided by Norman Daniels:

> If we think it is a concern of justice to intervene in each person's life to rectify unequal opportunities for welfare or advantage wherever preferences were not actually chosen, then we must think it is a task of justice to restrict in many quite intrusive and coercive ways the autonomy we grant people to pursue their plans of life, including their autonomy in childrearing.[76]

For if we think parents have a responsibility to teach their children to be virtuous and to convey to them what they think is valuable, then we can expect children to grow up with some unchosen preferences. From the Rawlsian perspective, it is not reasonable to make actual choice central in the manner proposed by Cohen and Arneson, although we are interested in protecting the underlying *capacity* for choice, the capacity to form, and revise one's conception of the good in a normal fashion. Thus, a person's complaints about having some unchosen preferences only qualifies for compensation if these preferences *disable* him in his *capacity* to form, pursue, and eventually to revise his conception of the good in a normal fashion.

By way of systematic reply to both critiques, Rawls distinguishes, in *Political Liberalism*, four main kinds of variations in transforming the primary goods into different functionings and proposes then to ask whether a variation places people above or below the line, that is, whether it leaves them with more or less than the minimum essential capacities required to be a normal cooperating member of society.[77] These variations are:

(a) variations in moral and intellectual capacities and skills;
(b) variations in physical capacities and skills, including the effects of illness and accident on natural abilities;
(c) variations in citizen's conceptions of the good (the fact of reasonable pluralism); and
(d) variations in tastes and preferences.

Rawls admits to Sen and Arrow that his account of primary goods is adequate for all cases, except for case (b), which covers instances of illness and accident

---

[75] *Id.*, *Political Liberalism*, p. 182, fn. 11.
[76] N. Daniels, 'Equality of What: Welfare, Resources, or Capabilities?' 1996/1990, p. 223.
[77] J. Rawls, *Political Liberalism*, pp. 184–185.

placing people below the line. The variations under (a) are dealt with by the social practices of qualifying for positions against the background of fair equality of opportunity and by the difference principle. As for case (c), justice as fairness is fair to permissible conceptions of the good. For case (d), variations in preferences and tastes are seen as our own responsibility. This is part of what free citizens with realised moral powers may expect of one another. In case (b), the variations that put some citizens below the line as a result of illness and accident can only be dealt with, according to Rawls, at the legislative stage when the prevalence and kinds of these misfortunes are known and the costs of treating them can be ascertained and balanced along with total government expenditure.

Therefore, what we arrive at are the limits of the Rawlsian political conception of justice, as the question of health care seems unanswerable from within the political conception. Consequently, Rawls treats normal health care as one of the four 'problems of extension'. The other three concern extension of justice to cover our duties to future generations, the extension to cover the principles that apply to international law, and the extension to our relations to animals and the order of nature.[78] Although Rawls himself has not pursued the implications of his theory for health policy, others have. In Section 3.3, I will analyse what I would like to call the *first* extension to health care, encompassing the theories of Green and Daniels that try to extend the Rawlsian theory to normal health care. Chapter 4 contains the *second* extension and concerns the problem of long-term care for the people with special and/or long-term medical needs.

3.3 EXTENDING RAWLS TO NORMAL HEALTH CARE

Undoubtedly a simple, but fundamental question must have turned up regularly in the mind of the reader: why not situate health behind the veil of ignorance and work out a conception of just health care from there? For indeed, health is one of those natural assets, like strength and intelligence; information on which the parties in the original position are deprived of. A person's health status is the result of the natural lottery, in combination with the social lottery and the element of good luck. As such, health is to a great extent morally arbitrary and *not* a matter of justice. However, the way in which social institutions deal with such issues, *is* a matter of justice, according to Rawls.[79] Taking the two arguments within the original position that lead to the maximin route – justice as impartiality and justice as mutual advantage – into account, it would be unreasonable to think that the parties would leave health care outside the realm of just social institutions. When we would eventually find ourselves in a situation of ill health, we would want to be taken care of. If we were to be a victim of some physical handicap, we would not want to be deprived of the possibility to lead a 'life of our own'. And when it would turn out that we are mentally disabled, we would

[78] *Ibid.*, pp, 20–22, 244–245, 272.

[79] Cf. Section 3.1.1 *Primary Goods and the Basic Structure of* Society.

want to be taken care of and be treated with respect for our human dignity, even if we would not or no longer be aware of it.

This answer, however, does not solve our problem. On the contrary, it is but the starting point. Our philosophical task is to integrate health care within the theory, i.e. to work out a clear understanding of the relationship between health care and the index of primary goods, distributed according to the two principles of justice.

3.3.1 Health Care: An Additional Primary Social Good?

A simple solution to the problem seems to be within reach. The Rawlsian list of primary goods is too truncated. Could we not simply add health care to the index and treat it as a distinct primary social good? Would it not be a good idea to extend the principles of justice so that they also encompass the just distribution of health care?

This approach has been defended by Ronald Green.[80] Given that certain social determinants have a vital impact on health, Green argues, it can no longer be said to be *solely* a natural good. Therefore, health care ought to be considered a primary social good in Rawlsian terms. Even more, Green holds that access to health care is not only a primary social good, but possibly one of the most important of those goods. Health care stands near to the basic civil liberties, since mental and physical well-being is rationally just as important as the primary social goods distributed by the first principle.[81] As such, it is more important than income and wealth.[82] As a result, it seems improbable that the rational agents in the original position whose choices define the principles of justice would neglect to establish separate principles for medical care or would substantially leave its distribution to be determined by one's income share. On the contrary, Green argues that we could expect the parties finally to opt for a principle of equal access to health care. This could be understood as a third principle of justice, being lexicographically prior to the fair opportunity principle and to the difference principle: 'each member of society, whatever his position or background, would be guaranteed an equal right to the most extensive health services the society allows'.[83] Health care is a basic right of all persons regardless of income, and universal basic health care is a desirable goal of the just society.

However, Kenneth Arrow points at two problems that follow from merely adding health care to the list of primary goods. First is the problem of the bottomless pit, and second is the problem of interpersonal comparisons. Firstly, he argues that the difference principle, which requires inequalities to work to the benefit of the least advantaged, could easily lead to the choice of medical procedures that

---

[80] R. Green, 'Health Care and Justice in Contract Theory Perspective', in R.M. Veatch; R. Branson (eds.), *Ethics and Health Policy*, Cambridge, MA: Ballinger, 1976, pp. 111–126.

[81] *Ibid.*, *p.* 112, 117.

[82] *Ibid.*, p. 119.

[83] *Ibid.*, p. 117.

serve the needs of persons with special health and medical needs that are so costly as to reduce society to a poverty level.[84] Since each member of the original position seeks to protect his most vital interests, why should he not insist upon all possible medical expenditures for making the worst-off life even a little more satisfying and fruitful, as soon as the resources become available? Given the sophisticated development of medical technology, there are virtually no limits to what can be spent to preserve and improve individual life and health.[85] Arrow's critique can be applied analogously to Green's proposal of adding a third principle. Holding that health care is probably the *most important* social primary good, and thereby defending a principle that is *lexicographically prior* to the other two principles, and that intends to guarantee an equal right to *the most extensive* health services that society allows, simply intensifies the problem of the bottomless pit to which Arrow referred regarding the difference principle.

Secondly, adding health care to the list of primary goods would force us into a trade-off with income and wealth. After all, when taking different health-care needs into account, equal income no longer means equality. According to Arrow this forces Rawls into the interpersonal comparisons of utility he had hoped to avoid by taking income and wealth as approximation to the whole index. As long as there is more than one primary good, there is a problem of commensurating the different goods.[86] In health care, the matter becomes even more complicated by the fact that, apart from objective differences in health, some persons experience a greater subjective need for health care than others. Rawls's contract theory bypasses this problem both because it is not concerned with maximising some social total of well-being and because its focus is not on case to case allocations but on the basic structure of society. It is his intention to formulate a limited amount of very general principles of justice and primary social goods with which we can identify the relevant social positions. If one seeks the solution in merely adding principles and primary goods, one risks generating a long list of such goods, one to meet each important need. In adding principles that should guarantee all important basic needs – such as food, clothing, health care and shelter – the identification of the least advantaged becomes impossible – Only the poor? Or also the hungry, the sick, the homeless and the illiterate?

These difficulties should not surprise us, given how little people in the original position *know* about their own ends or about their society. *In se*, the Rawlsian theory of justice as fairness chosen in the original position is really only a schema for a theory of justice. We have to treat the index, which is intended to give specificity and content to the distributive theory, as a variable whose values cannot be filled in without more specific knowledge of the society to which the theory is to be applied. The degree of abstractness of the structure of the index and

---

[84] K. Arrow, 'Some Ordinalist-Utilitarian Notes on Rawls's *Theory of Justice*', in *Journal of Philosophy* 70(1973)9, pp. 245–263, 251.
[85] Cf. Section 2.1.1.1 *The Twofold Dynamics of Scarcity*.
[86] *Ibid.*, p. 254.

the principles of justice determined within the original position is very high, only settling certain constraints on the weighting as illustrated by the priority rules.[87] We can turn them into a more substantive determination only in the legislative stages of designing a just society, for in these the veil of ignorance is partially lifted.[88] This means that simply adding health care to the index would ignore the limits of justice as fairness, as a political conception of justice. For if we treat health-care services as a distinct primary social good, we abandon the constraints of simplicity and availability of information to which the political conception is subject. To put it simply: Rawls does not hold that his list contains *all* important goods and adding health care to the list of primary goods and formulating a third principle simply because the rational agents would consider it very important, would go against his basic intention.

3.3.2 Fair Equality of Opportunity: Norman Daniels

An alternative and more refined solution to the extension problem is offered by Norman Daniels, a neo-Rawlsian philosopher, who has tried over many years to construct an approach to the fair allocation of health care that takes account of established theories of justice. His account extends Rawls's theory to health and health care by means of answering three central questions. Firstly, is health care morally important in ways that justify the fact that many societies distribute health care more equally than many other social goods?[89] Secondly, how much equality should there be? That is to say, which inequalities in health care are morally acceptable and which are unjust?[90] Thirdly, how can we meet competing health-care needs fairly under reasonable resource constraints? Is there a fair process for making rationing decisions?[91] I will consider these three questions in detail below.

---

[87] J. Rawls, 'A Kantian Conception of Equality', 1999/1975, p. 261.

[88] *Id.*, *A Theory of Justice*, 1971, 195–201; 1999, 171–176; 'the weights for the index of primary goods need not be established in the original position once and for all, and in detail, for every well-ordered society. What is to be established initially is *the general form* of the index and such *constraints* on the weights as that expressed by the priority of basic liberties. Further details necessary for practice can be filled in progressively in the stages sketched in TJ, Section 3.1, as more specific information is available', in: *id.*, 'Social Unity and Primary Goods', 1999/1982, p. 369, fn. 8 (my italics, YD).

[89] N. Daniels, 'Health-care Needs and Distributive Justice', in *Philosophy and Public Affairs* 10(1981)2, pp. 146–179, repr. in *Id.*, *Justice and Justification: Reflective Equilibrium in Theory and Practice*, Cambridge: Cambridge University Press, 1996, pp. 179–207; *Id.*, *Just Health Care*, Cambridge: Cambridge University Press, 1985; *Id.*, 'Justice, Health, and Healthcare', 2001, p. 2.

[90] N. Daniels; B. Kennedy; I. Kawachi; A. Sen, *Is Inequality Bad for Our Health?* Boston: Beacon Press, 2000; N. Daniels, 'Justice, Health, and Healthcare', in *American Journal of Bioethics* 1(2001)2, pp. 2–16.

[91] *Id.*, 'Four Unsolved Rationing Problems: a Challenge', in *Hastings Center Report* 24(1994)4, pp. 27–29; *Id.*, 'Rationing Fairly: Programmatic Considerations', 1996/1993, pp. 317–326; N. Daniels, D.W. Light; R.L. Caplan, *Benchmarks of Fairness for Health Care Reform*, New York: Oxford University Press, 1996; N. Daniels, J.E. Sabin, 'Limits to Health Care: Fair Procedures, Democratic Deliberation, and the Legitimacy Problem for Insurers', in *Philosophy and Public Affairs*, 26(1997), pp. 303–350; N. Daniels, 'Justice, Health, and Healthcare', in *American Journal of Bioethics* 1(2001)2, pp. 2–16.

3.3.2.1 The Special Moral Importance of Health Care?

Daniels begins by posing the question why a concern with distributive justice should extend to health care at all.[92] What is it about health care that makes it morally more important than many other things, which improve the quality of people's lives? His initial answer appeals to public sentiment in moral practice: people seem to feel that health care is sufficiently important to personal well-being that its availability should be determined by nothing other than medical need.[93] Non-medical features, particularly those that have to do with social or economic fortune, should not determine people's access to care. But why is this so? As expounded in Chapter 2, fulfilment of the relevant medical needs seems to be necessary for maintaining normal species functioning.[94] Do such needs have the two objective properties noted earlier?[95] Clearly, they are *objectively ascribable*. The line between normal functioning and disease and disability can be drawn in the relatively objective and non-evaluative context provided by the biomedical sciences. Are these needs *objectively important* in the appropriate way? In a broad range of contexts we do treat them as such. However, it is of interest to see what gives health-care needs their special importance.

Daniels offers three alternative routes.[96] A first answer to the question might be: their strategic function as a necessary condition to realise our life projects, whatever they may be. Health has an important instrumental value. Whatever our specific chosen goals or tasks, we need our health, our normal species functioning in order to achieve them.[97] In the case of health care we might say that whatever our chosen goals, we need our health, and so appropriate health care. However, as Daniels argues, strictly speaking, this is not completely true. For a person's *particular goals* are not *necessarily* undermined by failing health or disability. Moreover, people often cope with it and adjust their goals and projects to fit better with their dysfunction or disability. This does not necessarily diminish happiness or satisfaction in life to a level below that is achievable with normal functioning. It is questionable, however, to what extent coping reflects an example of adaptive preferences.[98]

A second answer might be that though normal functioning is not by definition a necessary condition for satisfaction or happiness in life, meeting health-care

[92] N. Daniels, 'Health-care Needs and Distributive Justice', 1996/1981, pp. 179–207; *Id., Just Health Care*, Cambridge: Cambridge University Press, 1985, pp. 1–18.

[93] *Ibid.*, 1996/1981, pp. 179–188; *Ibid.*, 1985, pp. 19–35. See also: R. Veatch; R. Branson (eds.), *Ethics and Health Policy*, Cambridge, MA: Ballinger, 1976; B. Williams, 'The Idea of Equality', in P. Laslett; G. Runciman (eds.), *Politics, Philosophy and Society II*, Oxford: Blackwell, 1962, repr. in B. Williams, *Problems of the Self. Philosophical Papers 1956–1972*, Cambridge: Cambridge University Press, pp. 230–249; M. Walzer, *Spheres of Justice*, New York: Basic Books, 1983, p. 86.

[94] *Ibid.*, 1996/1981, p. 184; *Ibid.*, 1985, pp. 26–27.

[95] Cf. Section 2.3.2.5 *Objective Truncated Scale of Well-Being*.

[96] *Ibid.*, 1996/1981, pp. 184–185; *Ibid.*, 1985, pp. 27–28.

[97] See also: D. Brock, 'The Separability of Health and Well-Being', in C. Murray et al. (eds.), *Summary Measures of Population Health. Concepts, Ethics, Measurement and Applications*, Geneva: World Health Organization, 2002, pp. 115–120.

[98] See Chapter 4 for further discussion.

needs may have a definite *tendency* to promote *happiness*. This idea may be all the utilitarian needs to guide public policy. It can be interpreted as a generalised version of the view that health care is special because it reduces pain and suffering, which is no doubt one of its more obvious merits. However, the utilitarian must weigh this reduction of pain and suffering against the satisfaction of all other kinds of preferences that promote happiness. If it is the empirical estimate of the strength of the tendency that makes health care special; the special character disappears when the estimate changes.

Daniels defends a third account and holds that the central moral importance, for purposes of justice, of preventing and treating disease and disability with effective health-care services, derives from the way in which protecting normal functioning contributes to protecting the *normal opportunity range*.[99] The normal opportunity range for a given society is the array of life plans that a person in this society could reasonably hope to pursue, given his talents or skills. Daniels acknowledges that the normal range is socially and individually relative. Various facts and key features of society – like its historical development, its level of material wealth and technological development, important cultural facts and facts about social organisation, including the conception of justice regulating its basic institutions – determine what the normal range is and how it is distributed in society. Furthermore, the share of the normal range open to an individual is also determined by his particular talents and skills. As such, the normal opportunity range abstracts from important individual differences in *effective* opportunity. Individuals generally choose to develop only some of their talents and skills, effectively narrowing their range of opportunities. However, Daniels argues that maintaining normal functioning preserves their broader, fair share of the *normal* opportunity range, giving them the chance to revise their plans of life over time. Taking the social and individual aspect of the normal range into account, he concludes that we should use the normal opportunity range open to an individual in general as a benchmark for assessing the relative importance of health-care needs at the macro level.[100]

Thus, it is the idea of *enablement* that gives Daniels the key to the question. Contrary to the first alternative, Daniels holds that health-care needs are indeed strategically important, but within the framework of the range of opportunities open to individuals, not as a *necessary* means to *any* goal. Responding to the utilitarian alternative, he maintains that the normal opportunity range functions

---

[99] *Id.*, 'Health-Care Needs and Distributive Justice', 1996/1981, pp. 185–188; *Id.*, *Just Health Care*, 1985, pp. 27–35; *Id.*, 'Justice, Health, and Healthcare', 2001, p. 3.

[100] 'Of course, impairment of normal species functioning … can diminish individual happiness or satisfaction, which depends on the individual's conception of the "good". Such effects are important at the micro level – for example to individual decisions about utilizing health care. At this micro level, an individual is deciding which services to use from among those society is obliged to provide – if the society is just. Here individual choices about happiness will be the final determinant of what is done. But this appeal to individual happiness will not solve [our problem in hand], which concerns what society is obliged to provide. [We are here seeking] the appropriate framework within which to apply principles of justice to health care.' In: N. Daniels, *Just Health Care*, 1985, p. 35.

at the macro level. With reference to Rawls, Daniels writes that the purpose of justice is not to ensure happiness or the satisfaction of desires but to provide a fair and acceptable framework within which people can be enabled to pursue their own life plan, according to their own expectations of happiness within it.[101] Specifically, by keeping people close to normal functioning, health-care preserves for people the ability to participate in the political, social and economic life of their society. It sustains them as fully participating citizens in all spheres of social life. It is this effect on opportunity that allows us to explain why people treat these medical needs as special and important: people have a fundamental interest in protecting their share of the normal range of opportunities.

However, this description and explanation of our moral practice falls short of giving us a normative account. Is there a social obligation to protect our share of the normal opportunity range? For simply having an interest in such protection does not mean that society *ought* to provide it. The relationship between health care and the protection of opportunity, Daniels argues, suggests that the appropriate principle of distributive justice for regulating the design of a health-care system is a principle protecting *fair equality of opportunity*.[102] Any theory of justice that supports fair equality of opportunity could thus in principle be extended to health care. Daniels holds that Rawls's ideas can be extended to health care by the simple expedient of including health-care organisations among the basic arrangements in society that help to promote fair equality of opportunity. Because health-care needs, like educational needs are, firstly, more unequally distributed by the natural lottery than other basic needs like food, clothing and shelter (some people need considerably more health care than others, while on the other hand, people's need for food and clothing is generally the same); secondly, because they can be highly unpredictable due to the element of simple bad luck; thirdly, because they are strategically important within the normal opportunity range; and fourthly, because they can be catastrophically expensive; they are appropriately seen as the object of institutions that provide for fair equality of opportunity. Whereas it might be reasonably expected that people can adequately provide for food, clothing and shelter from their fair shares of income and wealth, this does not apply to goods like health care and education that are therefore an appropriate object of social insurance schemes or subsidy schemes to buy private insurance. It would be unreasonable to expect that individuals generally should be able to gain sufficient access to health care, by relying solely on their own private resources.

Nevertheless, by subsuming the health-care system under Rawls's principle of fair equality of opportunity, Daniels at the same distances himself from the

---

[101] *Id.*, 'Health-Care Needs and Distributive Justice', 1996/1981, p. 190; *Id.*, *Just Health Care*, 1985, p. 38.

[102] *Ibid.*, 1996/1981, pp. 188–194; *Ibid.* 1985, pp. 39–48; *Id.*, 'Justice, Health, and Healthcare', 2001, pp. 3–4.

Rawlsian idea that health is solely a natural as opposed to a social primary good because its possession is less influenced by basic institutions.[103] However, in this dissociation, Daniels takes a route different from Ronald Green's by explicitly stressing that 'opportunity, not health care or education is the primary social good here'.[104] The primary social goods themselves remain general and abstract properties of social arrangements – basic liberties, opportunities, and income and wealth as all-purpose means.

This account has several implications.[105] Firstly, the principle of fair equality of opportunity not only prohibits discriminatory barriers to access, but requires positive social measures that correct for the negative effects on opportunity. Our social obligation is to provide institutions that protect opportunity, such as health-care institutions and the provision of public education. With regard to the design of health-care institutions and related issues of resource allocation, the account supports the provision of universal access to appropriate health care – including traditional public health and preventive measures – through public or mixed insurance schemes. Health care aimed at protecting fair equality of opportunity should not be distributed according to the ability to pay, and the burden of payment should not fall disproportionately on the ill. Against the background of the relationship between the concepts of normal species functioning, the normal opportunity range and the fair equality of opportunity principle, Daniels proposes a hierarchy of four institutional levels in health care.[106] At the first level, preventive health-care institutions act to minimise the likelihood of departures from the normality assumption. Included here are 'institutions which provide for public health, environmental cleanliness, preventive personal medical services, occupational health and safety, food and drug protection, nutrition education, and educational and incentive measures to promote individual responsibility for healthy life styles'. However, not all departures from normal functioning can be prevented. The second layer consists of institutions that deliver 'personal medical and rehabilitative services that restore normal functioning'. Similarly, not all treatments are cures and some institutions and services are needed to maintain persons in a way that is as close as possible to normal functioning. This third layer of institutions is involved with 'more extended medical and social support services for the (moderately) chronically ill and disabled and the frail elderly'. Finally, a fourth layer involves health care and related social services for those who can in no way be brought closer to the level of normal functioning: 'Terminal care and care for the seriously mentally and physically disabled fit here'.

---

[103] See also: N. Daniels; B. Kennedy; I. Kawachi; A. Sen, *Is Inequality Bad for Our Health?* Boston: Beacon Press, 2000.

[104] N. Daniels, *Just Health Care*, 1985, p. 45, fn. 3.

[105] *Id.*, 'Health-Care Needs and Distributive Justice', 1996/1981, pp. 192–194; *Id.*, *op. cit.*, 1985, pp. 45–48; *Id.*, 'Justice, Health, and Healthcare', 2001, p. 4.

[106] *Id.*, *op. cit.*, 1996/1981, p. 194; *Id.*, *op. cit.*, 1985, pp. 47–48.

Secondly, though the fair equality of opportunity account asks for universal access to basic institutions, this does not require opportunity to be equal for all persons. It requires only that it be equal for persons with similar skills and talents. It does not imply *levelling* all individual differences. With regard to health, this would mean that health inequalities are not by definition unjust.

3.3.2.2 Which Health Inequalities Are Unjust?

Although Daniels construed health care broadly to include public health as well as individual preventive, acute and chronic care, the focus on just health care alone added to the misconception that our vastly improved health in the last century is primarily the result of health care. Inspired by the enormous amount of literature that has emerged in the last two decades exploring the social determinants of health, he concentrates in later works on the correlation between general socio-economic factors and health.[107] He writes:

> [U]niversal access to health care does not necessarily break the link between social status and health. Our health is affected not simply by the ease with which we can see a doctor – though that surely matters – but also by our social position and the underlying inequality of our society ... while the exact processes are not fully understood, evidence suggests that there are *social determinants of health*.[108]

Daniels investigates five plausible and identifiable correlations, i.e. pathways through which social inequalities appear to produce health inequalities.[109] In doing this, he emphasises caution for correlations do not necessarily imply causation.

LEVEL OF ECONOMIC DEVELOPMENT. To begin, we have long known that a country's prosperity is related to its health, as measured, for example, by life expectancy: in richer countries, people tend to live longer. This well-established finding suggests a natural ordering of societies along some fixed path of economic development: as a country or region develops economically – providing access to the basic material conditions of health, such as clean water, adequate nutrition and housing, and general sanitary living conditions – average health improves.

---

[107] N. Daniels; B. Kennedy; I. Kawachi; A. Sen, *Is Inequality Bad for Our Health?* Boston: Beacon Press, 2000; N. Daniels, 'Justice, Health and Health Care', in *American Journal of Bioethics* 1(2001)2, pp. 2–16; N. Daniels; J. Bryant; R.A. Castano; O.G. Dantes; K.S. Khan; S. Pannarunothai, 'Benchmarks of Fairness for Health Care Reform: A Policy Tool for Developing Countries', in *Bulletin of the World Health Organization* 78(2000)6, pp. 740–750. See also: A. Sen, 'Health in Development', in *Bulletin of the World Health Organization* 77(1999)8, pp. 619–623.

[108] N. Daniels et al., *Is Inequality Bad for Our Health?* Boston: Beacon Press, 2000, pp. 3–4; N. Daniels et al., 'Benchmarks of Fairness for Health Care Reform: A Policy Tool for Developing Countries', in *Bulletin of the World Health Organization* 78(2000)6, pp. 740, 742; N. Daniels, 'Justice, Health, and Healthcare', 2001, pp. 6–9. See also: R.G. Wilkinson, *Unequal Societies: the Afflictions of Inequality*, London: Routledge, 1996; I. Kawachi; B. Kennedy; R.G. Wilkinson, *Income Inequality and Health: a Reader*, New York: New Press, 1999; M. Marmot et al., 'Contribution of Psychosocial Factors to Socioeconomic Differences in Health', in *Milbank Quarterly* 76(1998), pp. 403–408; I. Kawachi et al., 'Social Capital, Income Inequality, and Mortality', in *American Journal of Public Health* 87(1997), pp. 1491–1498; G. Dahlgren; M. Whitehead, *Policies and Strategies to Promote Social Equality in Health*, Stockholm: Institute of Future Studies, 1991.

[109] N. Daniels et al., *Is Inequality Bad for Our Health?* 2000, pp. 9–14; *Id.*, *op. cit.*, 2001, pp. 6–8.

RELATIVE INCOME. Furthermore, even within this relationship, variations in factors other than wealth and level of economic development may explain many of the differences in health outcomes among nations. Culture, social organisation, and government policies also help determine population health. One especially important factor in explaining the health of a society, Daniels argues, is the distribution of income: 'the health of a population depends not just on the size of the economic pie, but also on how the pie is shared'.[110] Daniels defends the relative-income hypothesis, which holds that differences in health outcomes among developed nations correspond to income distribution. Moreover, research shows that developed countries with more equal income distributions, such as Sweden, have higher life expectancies than countries with less equal income distribution but higher gross domestic product (GDP) per capita, like the USA. Therefore, the more unequal a society is in economic terms, the more unequal it is in health terms.

INDIVIDUAL SOCIO-ECONOMIC STATUS. When we move from comparing societies to comparing individual members, we find, once more, that inequality is important. At each step along the socio-economic ladder, we see improved health outcomes. This suggests that differences in health outcomes are not confined to the extremes of rich and poor, but are observed across *all* levels of socio-economic status. Research shows that this does not appear to be explicable by differences in access to health care, but rather by the level of income inequality.

EDUCATIONAL STATUS. Furthermore, Daniels points at another correlation through which social inequalities appear to produce health inequalities, such as the correlation between income inequality, educational inequality, and health inequality. There is ample evidence of the relation between unequal income distribution and educational spending. In the USA, the states with the most unequal income distributions invest less in public education, have larger uninsured populations and spend less on social safety nets. Differential investment in human capital – in particular, education, going from adult literacy, over gender disparities in access to basic education to general educational opportunities for children – is a strong predictor of health.

POLITICAL PARTICIPATION. Daniels concludes by arguing that these social mechanisms are tightly linked to the political processes that influence government policy. Income inequality appears to affect health by undermining civil society. For it erodes social cohesion, as measured by the higher levels of social mistrust and reduces participation in civic organisations; lack of social cohesion leads to lower participation in political activity (such as voting, serving in local government, volunteering for political campaigns); and lower participation, in turn, undermines the responsiveness of government institutions in addressing the needs of the worst-off. States with the highest income inequality, and thus lowest levels of social capital and political participation, are less likely to invest in human capital and provide less generous safety nets.

[110] *Ibid.*, p. 9. See also: *Id.*, 'Justice, Health, and Healthcare', 2001, p. 7.

These social determinants show that health is produced not merely by having access to medical prevention and treatment – in that sense, access to health care makes a *distinct*, but *limited* contribution to the protection of equal opportunity – but also – to a measurably greater extent – by the cumulative experience of social and educational conditions over the course of one's life.

Within this framework of correlations, the question 'when are health inequalities between groups of persons or individuals inequitable?' becomes particularly significant. At first sight, the answer is simple: when they are avoidable, unnecessary and unfair.[111] This implies that when they are rooted in biological differences that we do not know how to overcome they are unavoidable, and therefore not an inequity. Thus, health inequalities due to determinants as unequal access to clean water, sanitation, adequate shelter, basic education, vaccinations, and prenatal and maternal care are unjust for we believe that these inequalities are avoidable by a just and responsible social policy that supplies these missing determinants.

However, matters become more complicated when we ask ourselves whether the poorer health status of groups whose members smoke and drink heavily is unfair.[112] We may be inclined to say that it is not unfair, provided that the participation in such risky behaviour is truly *voluntary*. But if many people in a cultural group or class behave similarly, then the behaviour might acquire the qualities of a social norm – in which case we might wonder just how voluntary the behaviour is.

Furthermore, we cannot simply eliminate health inequalities by eliminating poverty, for instance. Health inequalities will continue to persist, even in societies that provide the poor with access to all standard public health and medical services, as well as basic income and education, and they persist as a gradient of health throughout the socio-economic hierarchy. What then, are we to think of the health inequalities that would persist? Should we eliminate all socio-economic inequalities? We might believe them all unjust, or at least those we did not freely choose. However, very few people embrace such a radical egalitarian view. To answer this question, Daniels argues that some degree of socio-economic inequality is unavoidable, or even necessary, and therefore not unjust. For this, he turns to Rawls's theory.

To the extent that social policies are responsible for the social and economic inequalities that produce these health effects, we are forced to look upstream from the point of medical delivery and ask about the fairness of the distribution of these social and economic goods. Rawls's theory of justice as fairness, Daniels argues, contains principles that give a plausible account of the fair distribution

---

[111] *Ibid.*, pp. 14–16.

[112] Cf. Section 5.5.3 *On Personal Responsibility and the Right to Health Care*. See also: Y. Denier, 'On Personal Responsibility and the Human Right to Health Care', in *Cambridge Quarterly of Healthcare Ethics* 14(2005)2, pp. 224–234.

of those determinants, thus providing an answer to our problem in hand.[113] Let us start by considering what a just society would require with regard to the distribution of the social determinants of health.

(a) *Equal Basic Liberties*. The first principle says that in an ideal society everyone is guaranteed equal basic liberties, including the right to participate in politics. In addition, there are safeguards aimed at assuring for all, whether rich or poor, the worth or value of those rights. Since there is evidence that political participation is a social determinant of health, the Rawlsian ideal assures institutional protections that counter the usual effects of socio-economic inequalities on political participation – and thus on health.
(b) *Fair Equality of Opportunity*. This principle condemns discriminatory barriers and requires robust measures aimed at mitigating the effects of socio-economic inequalities and other contingencies on opportunity. In addition to high-quality public education at all levels, such measures would include the provision of early childhood interventions intended to promote the development of capabilities independently of the advantages of family background. The equal opportunity principle also requires provision of universal access to comprehensive health care, including extensive public health, primary health care, and medical and social support services aimed at promoting normal functioning for all.
(c) *The Difference Principle*. Finally, a just society restricts inequalities in income and wealth to those that benefit the least advantaged. Rawls's difference principle permits inequalities in income only if the inequalities work to make the worst-off as well-off as possible. Daniels agrees with Cohen that this principle is not a simple principle which tolerates any inequality so long as there is some benefit that flows down the socio-economic ladder but rather one which requires a maximal flow downward. It would therefore flatten socio-economic inequalities in a robust way, assuring far more than a 'decent minimum'.[114] In addition, the assurances of the value of political participation and fair equality of opportunity would further constrain allowable income inequalities.

In short, Rawlsian justice – though not devised for the case of health – regulates the distribution of the key social determinants of health. There is nothing about the theory that should make us focus narrowly on medical services. Properly understood, justice as fairness tells us what justice requires in the distribution of all *socially controllable* determinants of health. The implication is that we should view health inequalities that derive from social determinants as unjust unless the determinants are distributed in conformity with these robust principles.

However, a theoretical question remains. Even if a just distribution of the social determinants of health flattens health gradients further than what we observe in the most egalitarian developed countries, we must still expect a residue of health inequalities: people who are less well-off in economic terms will

---

113 N. Daniels, *Is Inequality Bad for Our Health?*, 2000, pp. 17–23.

114 G.A. Cohen, 'On the Currency of Egalitarian Justice', in *Ethics* 99(1989), pp. 906–944.

continue to be less healthy. Should we aim to reduce further those otherwise justifiable economic inequalities because of the inequalities in health status they create? Furthermore, we know that in real life people routinely trade health risks for other benefits. They do so when they commute longer distances for a better job, practice certain sports, go on a skiing holiday or on an exotic hiking tour. When is it an appropriate exercise of autonomy? Some such trades are unfair; others will be restricted only by paternalists. Rawls gave priority to the principle of protecting equal basic liberties because he believed that once people achieve some threshold level of material well-being, they will not trade fundamental liberties for other goods. Making such a trade might deny them the liberty to pursue their most cherished ideals, whatever they turn out to be. Could we make the same argument about trading health for other goods?

There is some plausibility to the claim that rational people should refrain from trading their health for other goods. Loss of health may preclude us from pursuing what we most value in life. We do, after all, see people willing to trade almost anything to regain health once they lose it. In that sense, health is certainly considered to be very important *Bonum* in life. Nevertheless, there is also a strong reason to think this priority is not clear-cut, especially when the trade is between a risk to health and other goods that people value highly. Refusing *ex ante* to allow *any* trades of health risks for other goods, even when the background conditions on choice are otherwise fair, may seem unjustifiably paternalistic, perhaps in a way that a refusal to allow trades of basic liberties is not.[115]

Daniels concludes with a pragmatic proposal. Fair equality of opportunity is only approximated even in an ideally just system, because we can only *mitigate*, not eliminate, the effects of family and other social contingencies. Although we give general priority to equal opportunity over the difference principle, we cannot achieve complete equality in health any more than we can achieve complete equal opportunity. Daniels says, 'Justice is always rough around the edges.'[116] Decisions about whether to further reduce residual inequalities are matters for the democratic process. Justice itself does not command their reduction.

3.3.2.3 When Are the Limits to Health-Care Fair?

Stressing the social determinants of health does not suggest, however, that our society should immediately reallocate resources away from medicine to schools, for example, in the expectation that a better-educated population will be healthier. Rather, the arguments make clear that *some* reallocation of resources to improve the social determinants is justifiable. More importantly, it makes clear that there are limits to what can be reasonably expected from the health-care system. Earlier, the internal and external aspects of scarcity have shown that rationing in health care is unavoidable.[117] Firstly, society simply cannot afford to treat

---

[115] I will discuss the issue of voluntary health risks in detail below. Cf. I will discuss the issue of voluntary health risks in detail below. Cf. Section 5.5.3 *On Personal Responsibility and the Right to Health Care*

[116] N. Daniels et al., *Is Inequality Bad for Our Health?* Boston: Beacon Press, 2000, p. 22.

[117] Cf. Section 2.1 *On Scarcity of Health Care Resources.*

people in all the ways their needs require or the accepted distributive principles seem to demand. And secondly, health care is not the only important good. All societies must decide which needs should be given priority and in which case resources are better spent elsewhere.

Priority setting and rationing, Daniels holds, raises questions of fairness in a two-fold way. Firstly, justice requires efficiency in management and allocation, since inefficient use of constrained resources means that some needs that could have been met, will not be met.[118] Furthermore, even efficient rationing concerns situations in which losers as well as winners have plausible claims to have their needs met. When society knowingly and deliberately refrains from meeting some legitimate needs, it had better have acceptable justification for the distributive choices it makes.[119]

Rationing decisions both at the micro and macro levels, share three key features.[120] Firstly, unlike money, the goods to be provided are not sufficiently divisible without loss of benefit. We thus cannot avoid unequal or 'lumpy' distributions. Meeting the needs of some will mean that the requirements of others will go unsatisfied. Secondly, when we ration, we deny benefits to individuals who can plausibly claim they are owed to them in principle. Thirdly, the general distributive principles (for instance, fair equality of opportunity) appealed to by claimants and rationers, alike do not by themselves provide adequate reasons for choosing among claimants. They are too schematic and thus fail to yield specific solutions to these rationing problems. Solving these problems should bridge the gap between principles of distributive justice and problems of institutional design.[121]

How should fair decisions about limits to health care be made? Under what conditions should we view such decisions as a legitimate exercise of moral authority? Daniels stresses that his answers to the first two questions – the question of the special moral importance of health care and the question of which health inequalities are inequitable – fail to give guidance to the matter of rationing.[122] We need to supplement the principled account of justice for health and health care with an account of fair process for setting limits or rationing of care. Daniels's account is provided by four conditions that comprise 'accountability for reasonableness'. The argument consists of three steps: reasonable pluralism in health care, fair process, and accountability for reasonableness.

Reasonable Pluralism in Health Care. Firstly, answering the question when limits to health care are fair, would be simple if people could agree on principles of distributive justice that would determine how to set fair limits on health care.

---

[118] N. Daniels, 'Benchmarks of Fairness for Health Care Reform', 2000, p. 740.

[119] *Id.*, 'Rationing Fairly: Programmatic Considerations', 1996/1993, p. 317.

[120] *Id.*, 'Four Unsolved Rationing Problems', 1994, p. 27; *Id.*, 'Rationing Fairly: Programmatic Considerations', 1996/1993, p. 317–328.

[121] For a clear and detailed view on the difficulty of bridging the gap between principles and institutions, and the complexity of the arguments held by the different actors in the field, see: J. Elster, *Local Justice. How Institutions Allocate Scarce Goods and Necessary Burdens*, Cambridge: Cambridge University Press, 1992.

[122] *Id.*, 'Four Unsolved Rationing Problems: a Challenge', 1994, p. 27; *Id.*, 'Rationing Fairly: Programmatic Considerations', 1996/1993, pp. 317–318; *Id.*, 'Justice, Health, and Healthcare', 2001, pp. 2–3.

Unfortunately, this is not the case. Distributive issues in rationing matters remain highly contested for they meet the fact of reasonable pluralism.[123] Reasonable people with diverse moral and religious views disagree about what constitutes a fair allocation of resources to meet competing health-care needs, even when they agree on other aspects of just health-care systems, such as the importance of universal access to whatever services are provided. We should expect and respect such diversity in views about rationing in health care. Nevertheless, we must reach acceptable social policies despite our disagreements. This moral controversy raises problems of legitimacy. Under what conditions should we accept the moral authority of those who make rationing decisions as legitimate?

To support the first step of the argument, Daniels sketches a problem that has been labelled the *Priorities Problem*.[124] How much priority should we give to treating the sickest or most disabled patients? Daniels begins his argument with two extreme positions. The maximin position – give complete priority to the worst-off patients; and the maximise position – give priority to whatever treatment produces the greatest net benefit regardless of which patient:[125]

> Suppose [we could invest] in Technologies A or B, but the resources are 'lumpy' (we cannot introduce some A and some B) …. The Maximin position would settle the matter by determining whether patients treated by A are worse off before treatment than patients treated by B. If so, we introduce A; if patients treated by B are worse off, we introduce B. If the two sets of patients are equally badly off, we can break the tie by considering to whom we can provide the most benefit. The Maximize position chooses between A and B solely by reference to which produces greatest net benefit. In practice, most people are likely to reject both extreme positions. If the benefits A and B produce are nearly equal, but patients needing A start off much worse than patients needing B, most people seem to believe we should introduce A. They prefer to provide A even if they know we could produce somewhat more net health benefit by introducing B. But if the net benefit produced by A is very small, or if B produces significantly more net benefit, then most people will overcome their concern to give priority to the worst-off and will prefer to introduce B to A. Some people who would give priority to patients needing A temper their preference if those patients end up faring much better than patients needing B. Disagreements persists.[126]

---

[123] N. Daniels, 'Justice, Health, and Healthcare', 2001, p. 9.

[124] N. Daniels, 'Four Unsolved Rationing Problems', 1994, p. 28; *Id.*, 'Rationing Fairly: Programmatic Considerations', 1996/1993, pp. 320–321; *Id.*, 'Limits to Health Care: Fair Procedures, Democratic Deliberation, and the Legitimacy Problem for Insurers', 1997, pp. 319–320; *Id.*, 'Justice, Health, and Healthcare', pp. 9–10. See also: F. Kamm, *Morality, Mortality. Volume I: Death and Whom to Save from It*, New York: Oxford University Press, 1993.

[125] At the same time, Daniels takes the opportunity to put the idea that his fair equality account implies an immediate choice for the maximin position in rationing into perspective: 'though I believe my account is only committed to giving some priority to the worst off, placing it in a broad family of views that leave the degree of priority unspecified.' In: N. Daniels 'Justice, Health, and Healthcare', 2001, p. 9.

[126] N. Daniels, 'Justice, Health, and Healthcare', 2001, pp. 9–10. See also: *Id.*, 'Limits to Health Care', 1997, pp. 319–320. For these insights, Daniels appeals to the research of Erik Nord, 'The Relevance of Health State After Treatment in Prioritising Between Different Patients', in *Journal of Medical Ethics* 19(1993), pp. 37–42; *Id.*, 'The Person-Tradeoff Approach to Valuing Healthcare Programs', in *Medical Decision Making* 15(1995), pp. 201–208; and *Id.*, *Cost-Value Analysis in Healthcare: Making Sense out of QALY's*, Cambridge: Cambridge University Press, 1999. For cross-national evidence that people are not straight maximizers, see: E. Nord; J. Richardson; A. Street; H. Kuhse; P. Singer, 'Maximizing Health Benefits vs Egalitarianism: An Australian Survey of Health Issues', in *Social Science and Medicine* 41(1995)10, pp. 1429–1437.

It turns out that a definite but very small minority are inclined to be straight maximiners and a definite but very small minority are inclined to be straight and consistent maximisers. Most people fall in between and they vary considerably in how much benefit they are willing to sacrifice in order to give priority to the worst-off patients. The same goes for the *Fair Chances/Best Outcomes Problem* and for the *Aggregation Problem*.[127] The first asks how much we should favour producing the best outcome with our limited resources above granting everyone who might benefit a fair chance (through lottery, for instance). The second problem asks when lesser benefits to many, outweigh greater benefits to a few. These rationing problems also suggest that most people are not straight health maximisers, and that a variety of factors and their interrelations (such as the amount of benefit; the importance of fairness and rights; how badly off the patient is; how many people could benefit; elements of personal responsibility; and the character of actors involved) always come into play. People think very differently about the same things. However, we lack principled characterisations of acceptable middle-course solutions.[128]

Daniels takes up two possible strategies: the philosophical approach of Frances Kamm and the empirical approach of Erik Nord. Frances Kamm proposes a hypothetical approach that consists of subtle examination of varied hypothetical cases, seeking to reveal agreement on a complex set of underlying principles that can account for the judgements the philosophical inquirer makes about these cases.[129] Kamm insists on hypothetical cases rather than real ones, attempting to isolate more clearly in these cases the relevant features that motivate our judgements. She believes that the method will uncover the underlying moral structure in our beliefs. Erik Nord also explores hypothetical cases by asking groups of people 'person-trade-off' questions.[130] These questions are a variation on a standard economic approach seeking 'indifference points' or curves reflecting when an individual finds two benefits or outcomes equally valuable. For example, if we can invest only in treatments A and B, and A is used for people more seriously ill than B, we might ask how many treatments with B someone would trade for some number of treatments with A. Nord hopes this approach can uncover the structure of moral concerns in a population of people. Daniels holds that the insights from both approaches are important information inputs for a fair,

---

[127] For a discussion of these two problems, see: N. Daniels, 'Four Unsolved Rationing Problems: A Challenge', 1994, pp. 27–28; *Id.*, 'Rationing Fairly: Programmatic Considerations', 1996/1993, pp. 318–322.

[128] N. Daniels, 'Four Unsolved Rationing Problems: A Challenge', 1994, pp. 28–29; *Id.*, 'Justice, Health, and Healthcare', 2001, p. 10.

[129] F. Kamm, *Morality, Mortality. Volume I: Death and Whom to Save From It*, Oxford: Oxford University Press, 1993.

[130] E. Nord, 'The Person-Tradeoff Approach to Valuing Healthcare Programs', in *Medical Decision Making* 15(1995), pp. 201–208; *Id.*, *Cost-Value Analysis in Healthcare: Making Sense out of Qaly's*, Cambridge: Cambridge University Press.

deliberative process of decision-making, but they are not a substitute for such a fair process.[131]

FAIR PROCESS. The second step of the argument is a Rawlsian-inspired retreat to procedural justice. Since we lack prior consensus on the relevant distributive principles as solutions to a variety of morally controversial rationing problems, and since general principles of justice for health and health care (as Daniels formulated in his answers to the first two questions) fail to give guidance on how to solve them, we should rely on a *fair process* to establish legitimacy for complex and difficult resource allocation decisions.

ACCOUNTABILITY FOR REASONABLENESS. Thirdly, a fair process will have to meet several constraints that Daniels calls 'accountability for reasonableness'.[132] Daniels holds that we would take a giant step towards solving the problems of legitimacy and fairness that face public agencies and private health plans making limit-setting decisions if the following four conditions were satisfied.

1. The Publicity Condition: limit-setting decisions and their rationales must be publicly accessible to all affected by them. The benefits of publicity are both internal and external to the decision-making institution. Firstly, the quality of decision-making improves if reasons must be articulated. Secondly, fairness improves over time, both formally – since like cases are treated similarly – and substantively – since there is systematic evaluation of reasons.
2. The Relevance Condition imposes two important conditions on the rationales that are made publicly accessible. The rationales for coverage and limit-setting decisions should aim to provide a reasonable construal of how the organisation or society should provide 'value for money' in meeting the varied health needs of a defined population under reasonable resource constraints. Specifically, a construal will be 'reasonable' if it appeals to reasons and principles that are accepted as relevant by fair-minded people who seek the terms of cooperation that are mutually justifiable.
3. The Appeals Condition: there is a mechanism for challenging and disputing resolutions regarding limit-setting decisions, including the opportunity for revising decisions in the light of further evidence or arguments.
4. The Enforcement Condition: there is either voluntary or public regulation of the process to ensure that conditions 1–3 are met.

The guiding idea behind these conditions is to convert limit-setting decisions into part of a larger public deliberation about how to use limited resources to

---

[131] N. Daniels, 'Kamm's Moral Methods', in *Philosophy and Phenomenological Research* 54(1998)4, pp. 947–954; *Id.*, 'Distributive Justice and the Use of Summary Measures of Population Health Status', in *Summarizing Population Health: Directions for the Development and Application of Population Metrics*, Institute of Medicine, Washington: Academy Press, 1998, pp. 54–71.

[132] N. Daniels; J. Sabin, 'Limits to Health Care', 1997, pp. 322–343; N. Daniels; J. Sabin, 'The Ethics of Accountablility and the Reform of Managed Care Organisations', in *Health Affairs* 17(1998)5, pp. 50–69; N. Daniels, 'Enabling Democratic Deliberation: How Managed Care Organisations Ought to Make Decisions About Coverage for New Technologies', in S. Macedo (ed.), *Deliberative Politics: Essays on Democracy and Disagreement*, New York: Oxford University Press, 1999; *Id.*, 'Justice, Health, and Healthcare', 2001, pp. 11–13.

fairly protect the health of a population with varied needs. Daniels stresses that the broader public deliberation envisioned here is not necessarily an organised democratic procedure, though it could include the deliberation underlying public regulation of the health-care system. Rather, it may take place in various forms in an array of institutions, spilling over into legislative politics only under some circumstances. Meeting these conditions also serves an educational purpose. The public is made familiar with the need for limits and appropriate ways to reason about them.

Together, these conditions hold institutions – whether public or private – and decision-makers in them 'accountable for the reasonableness' of the limits they set. All must engage in the process of establishing their credentials for fair decision-making about such fundamental matters every time they make such a decision. Whether in public or in mixed systems, establishing the accountability of decision-makers to those affected by their decisions is the only way to show, over time, that arguably fair decisions are being made and that those making them have established a procedure we should view as legitimate. This is not to say that public participation is an essential ingredient of the process in either public or mixed systems, but the accountability to the public in both cases is necessary to facilitate broader democratic processes that regulate the system. Only through such accountability and the way in which it facilitates or enables a broader social deliberation there will be a wider perception that rationing decisions are fair and are made through an exercise of legitimate authority. As an answer to the question of setting limits fairly, of limiting access *To Whom?* and *To What?*, Daniels holds that wherever we have no prior consensus on fair outcomes, establishing a fair process is the best we can do.

Finally, it is clear that there are four ways in which Rawls's theory also provides support for Daniels's approach to the third question. Firstly, his recourse to procedural justice as an answer to the fact of reasonable pluralism is at the heart of the Rawlsian theory. Secondly, Rawls places great emphasis on the importance of publicity as a constraint on theories of justice. Principles of justice and the grounds for them must be publicly acknowledged. This constraint is central to the conditions that establish accountability for reasonableness. Thirdly, Rawls develops the view that public reason must constrain the content of public deliberations and decisions about fundamental matters of justice, avoiding special considerations that might be elements of the comprehensive moral views that people hold.[133] Accountability for reasonableness pushes decision-makers towards finding reasons of which all can agree that they are relevant to the goals of cooperative health delivery schemes. And fourthly, put all together, this involves a form of institutional reflective equilibrium, a commitment to both transparency and coherence in the giving of reasons.[134]

---

[133] See also: J. Rawls, 'The Idea of Public Reason Revisited', 1999/1997, pp. 573–615.

[134] See: N. Daniels, *Justice and Justification. Reflective Equilibrium in Theory and Practice*, Cambridge: Cambridge University Press, 1996.

3.3.3 *Qualities of Daniels's Theory*

What are the merits and limitations of Daniels's extension of Rawls to health care? Does his analysis do justice to Rawls's objective, non-welfarist account of primary social goods? Does it provide a useful answer to the pressing problems in contemporary health care – like the issue of cost control, the role of personal responsibility, of high-tech medicine, etc.? Regarding the merits of the approach, two qualities should be highlighted.

3.3.3.1 The Objective and Small-Scale Moral Function

Among contemporary theories of justice in health care, Daniels's approach is generally acknowledged as the most elaborated one available.[135] His theory provides an elaborate legitimisation of the moral function of a health-care system from the perspective of justice: health care belongs to the basic social institutions that help guarantee fair equality of opportunity. After due consideration, his theory on just health care would appear to meet the general qualities of the Rawlsian primary goods approach – i.e. the objective list and the ensuing division of responsibility; the strategic role of the primary goods; respect for the person as person; and finally, real equality of opportunity.[136]

In his enablement approach, Daniels connects the fair equality of opportunity principle with a narrow and objective biomedical concept of health and disease and with a needs concept that is oriented towards the normal opportunity range.[137] By doing so, he takes an *objective course*. The moral importance of meeting health-care needs is thus located in the more objective impact on opportunity rather than in the more subjective impact on happiness. His theory on just health care is an objective goods theory within which not *all* health-care needs are moral bedrock, but only those that are *strategically* important within the fair opportunity framework. As such, Daniels's approach provides a liberal-egalitarian challenge to utilitarian and libertarian theories of health care.[138] A *person* is considered

---

[135] Other approaches are the libertarian approach, based on Robert Nozick's views, of H. Tristam Engelhardt; the social-utilitarian approach of Daniel Callahan; the communitarian perspective, based on Michael Walzer's position, of Margo Trappenburg; and the liberal-communitarian approach of Ezekiel Emmanuel.

[136] In later works, Rawls appears to endorse Daniels's extension of his theory to health care. See: J. Rawls, *Political Liberalism*, 1996, pp. 184, fn. 14, 185, fn. 15; *Id., Justice as Fairness: A Restatement*, ed. E. Kelly, Cambridge, MA: The Belknap Press of Harvard University Press, 2001, p. 175.

[137] Cf. Section 2.3.4.1 *Health Care Needs and Normal Functioning.*

[138] A utilitarian inspired approach of health care is provided by John Harris in his article 'The Survival Lottery', in *Philosophy* 50(1975), pp. 81–87, discussed by P. Singer, 'Utility and the Survival Lottery', in *Philosophy*, 52(1977), pp. 218–222. A libertarian approach is provided by H.T. Engelhardt, *The Foundations of Bioethics*, New York: Oxford University Press, 1986 and in his article 'Health Care Allocation: Response to the Unjust, the Unfortunate and the Undesirable', in E.E. Shelp (ed.), *Justice and Health Care*, Dordrecht: Reidel, 1981, pp. 3–21. See also: Y. Denier; T. Meulenbergs, 'Health Care Needs and Distributive Justice. Philosophical Remarks on the Organisation of Health Care Systems', in R.K. Lie; P.T. Schotsmans, *Healthy Thoughts. European Perspectives on Health Care Ethics*, Leuven: Peeters, 2002, pp. 265–297.

to be anything but a drop in the ocean of overall utility or a mere container for preference satisfaction. For Daniels, all individuals, irrespective of ability to pay or utility status, are entitled to equal access to basic health care because proper appreciation of the worth of individuals as such and of the importance of health care to them, warrants this entitlement.[139] Contrary to libertarian theory, the fair opportunity perspective defends a positive social obligation to mitigate barriers that prevent fair equality of opportunity. A universal health-care system is one of the basic institutions of the just society, thus providing the grounds for *real* – rather than mere formal – access to health care. With this theory Daniels has developed an objective small scale basis for justice in health care, thus answering the Rawlsian aim to find a common ground that we can all reasonably agree to.

3.3.3.2 Answering the Inflexibility Critique

Furthermore, Daniels takes up the challenge of defending Rawls against both lines of the inflexibility critique.[140]

Firstly, he provides a better solution than Green to the problem posed by Sen and Arrow. By subsuming health under the fair equality of opportunity principle, instead of merely adding it to the list of primary social goods, he responds to the variability of health-care needs without abandoning the Rawlsian conditions of objectivity and simplicity. The primary goods list remains an abstract index. Opportunity, not health care is the primary good. However, health care is needed, among other things, to protect opportunity. According to Daniels, we can still simplify matters in using the index by looking solely at income and wealth – assuming a background of equal liberties and fair equality of opportunity.[141] Thus, he avoids the problem of generating endless additions of primary goods to the list, one to meet each important need, consequently requiring complex interpersonal comparisons to identify the least advantaged.

Secondly, he mitigates the bottomless pit problem by arguing that there are limits to what can be reasonably expected from a health-care system in a threefold way. (1) He stresses: 'Fairness also includes efficiency in management and allocations, since when resources are constrained their inefficient use means that some needs will not be met that could have been met.'[142] Meeting health-care needs should not and need not be a bottomless pit. Efficiency is a matter of justice. (2) He appeals to the Rawlsian distinction between fairness and goodness by posing that just health care is not about the fulfilment of unlimited desires in the name of happiness – thus countering the internal aspect of scarcity – but about guaranteeing a safety net or minimum floor below which no one is allowed to

---

139 On the difference between utilitarian and Kantian theories on health care, see: A. Buchanan, 'Philosophic Perspectives on Access to Health Care: Distributive Justice in Health Care', in *The Mount Sinai Journal of Medicine* 64(1997)2, pp. 90–95.

140 N. Daniels, 'Equality of What: Welfare, Resources, or Capabilities?' 1996/1990, pp. 208–231.

141 N. Daniels, *Just Health Care*, 1985, p. 45, fn. 3.

142 N. Daniels et al., 'Benchmarks of Fairness for Health Care Reform: A Policy Tool for Developing Countries', in *Bulletin of the World Health Organization* 78(2000)6, p. 740.

fall, thus securing fair equality of opportunity for all.[143] (3) Finally, by specifying the broad field of various social determinants of health he clarifies the way in which there are external limits to infinity matters in health care.[144]

Thirdly, regarding the responsibility issue raised by Arneson and Cohen, who argue that, because of the distinction between brute luck and option luck, actual choice of preferences should be the decisive matter in egalitarian concerns, Daniels argues in the line of Rawls that this would require such detailed information about each particular situation as to violate the conditions of simplicity.[145] In view of the social and psychological determinants of preference formation – as shown by the arguments of the contented slave and of the sour grapes – and in view of the related metaphysical question of how free the free will really is, the problem of justice and responsibility no longer turns out to be solved by the mere distinction between ambitions and endowments.[146] Although in Daniels's approach, responsibility plays an important role in specifying the scope of justice – 'assuming that people actually chose to have the preferences they have and that they can always, at least over time and with some cost and effort, revise them',[147] thus excluding expensive and offensive tastes from the field of rightful claims – Daniels ultimately avoids ending up in an infinite series of responsibility questions, by proposing to look solely at the moral importance of the situation *in se*. Unchosen preferences that make us worse off than others do not arouse egalitarian concerns of compensation *unless* they can be assimilated to cases of psychological and/or physical disability, i.e. to a departure from normal functioning, and thus form an example of a 'handicapping taste'.[148] Holding people responsible for their ends, Daniels argues, means that, in assuming the presence of just arrangements and institutions, we are acting as if they can exercise their underlying moral power to form and revise their conceptions of the good. If we think this underlying capacity is compromised, we back away from the responsibility cut and rely on compensation at the bar of justice.[149]

---

[143] N. Daniels, 'Health-Care Needs and Distributive Justice', 1996/1981, pp. 179–207; *Id.*, 'Justice, Health, and Healthcare', 2001, p. 4.

[144] N. Daniels et al., *Is Inequality Bad for Our Health?* 2000; *Id.*, 'Justice, Health, and Healthcare', 2001, pp. 6–9.

[145] J. Rawls, *Political Liberalism*, 1996, pp. 182, fn. 11, 185, fn. 15; N. Daniels, 'Equality of What', 1996/1990, pp. 208–231; *Id.*, 'Justice, Health, and Healthcare', 2001, p. 5.

[146] J. Elster, 'Sour Grapes – Utilitarianism and the Genesis of Wants', in A. Sen; B. Williams, *Utilitarianism and Beyond*, Cambridge: Cambridge University Press, 1982, pp. 219–238; R. Dworkin, 'What is Equality? Part 2: Equality of Resources', in *Philosophy and Public Affairs* 10(1981)4, pp. 283–345; repr. in R. Dworkin, *Sovereign Virtue*, Cambridge, MA: Harvard University Press, 2000, pp. 65–119. K. Devooght, *Essays on Responsibility-Sensitive Egalitarianism and the Measurement of Income Inequality* (nonpublished Ph.D., K.U. Leuven, Faculty of Economics and Applied Economics), Leuven, 2003, esp. Chapter 1, pp. 7–36. Also, cf. Section 5.5.3 *On Personal Responsibility and the Right to Health Care*.

[147] N. Daniels, 'Equality of What', 1996/1990, p. 218.

[148] Daniels borrows the term from R. Dworkin, 'Equality of Resources', 1981, pp. 302–304; 2000, pp. 82–83.

[149] This is closely related to the viewpoint of Harry Frankfurt, who places great moral weight into the harm principle. See: H. Frankfurt, 'Necessity and Desire', in *Id.*, *The Importance of What We Care About*, Cambridge, 1988, pp. 104–116.

3.3.4 Problems and Challenges

Allow me to classify this analysis into three categories. The first category refers to an issue that could be experienced as problematic, but need not necessarily be so. The second category refers to a variety of issues that were addressed by Daniels himself and are for the most part, dealt with in a satisfactory way. The third and most interesting category refers to the problem that, for my view, forms the real challenge for Daniels's theory. This problem will be discussed in Chapter 4.

3.3.4.1 A Thin Theory

In Rawls's theory of *Justice as Fairness* the thick veil of ignorance results in abstract, formal and procedurally characterised principles. His conception of justice thus functions as a metatheory in the light of which the principles have to be fleshed out and evaluated. Norman Daniels has carried out the process of fleshing out the fair equality of opportunity principle within the realm of health care. This, however, does not mean that suddenly all health-care matters become clear in a concrete and detailed way. His theory still holds a high level of abstraction and idealisation.[150] As became clear in the discussion on priority setting and rationing in health care, the fair opportunity principle remains indeterminate and a thinner veil is needed to make allocation decisions.[151] To weigh health-care expenditures against other spending in the public household, and to deliberate about conflicting needs within the field of health care we need detailed empirical knowledge about the particular society – like its level of wealth and development; its level of industrialisation; which particular institutions increase production and which guarantee increase of chances; what its social, moral, and economic traditions and 'shared understandings' about 'the good' are, etc. – and about the effects of alternative allocations.[152] We can answer these and similar questions only in the legislative stage for in these the veil of ignorance is partially lifted.[153]

Still, two important things, Daniels argues, can be elaborated at this abstract level of theory. That is, determining both the formal conditions and more or less substantive benchmarks within which the detailed discussion among the different actors will take place. All this happens in the light of the general Rawlsian principles of justice and on a high level of abstraction. The formal conditions of the discussion are comprised by 'accountability for reasonableness', establishing the accountability of decision-makers to those affected by these decisions, thus

---

[150] Daniels acknowledges this. With regard to the relation between social determinants of health, political participation and residual health inequalities, he writes: 'It may still be that the use of a democratic process in nonideal conditions is the fairest resolution we can practically achieve, but it still falls well short of what an ideally just democratic process involves.' See: N. Daniels, *Is Inequality Bad for Our Health?* 2000, p. 23. See also: *Id.*, *Just Health Care*, 1985, pp. 34–35.

[151] *Id.*, *Just Health Care*, 1985, p. 47.

[152] *Ibid.*, p. 54: 'Deciding which needs are to be met and what resources are to be devoted to doing so requires careful moral judgements and a wealth of empirical knowledge about the effects of alternative allocations. The various institutions which affect opportunity must be weighed against each other.'

[153] J. Rawls, *A Theory of Justice*, 1971, pp. 195–201; 1999, pp. 171–176.

showing their fair and legitimate character.[154] Secondly, it is possible to develop benchmarks of fairness on which a wide agreement exists, that help the integrated examination of objectives that often involve trade-offs with each other.[155]

However, abstract Daniels's extension of Rawls still is, this need not be a problem. For it is our philosophical task to assess, explain and justify or modify the relevant categories and distinctions and their interrelations. The high level of abstraction is only a problem when the expectations are very concrete.

3.3.4.2 Too Narrow – Too Vague – Too Strong

The second category of problems contains those that are addressed by Daniels himself and are, for my view, partially answered.[156]

Too Narrow. Firstly, it could be criticised that the fair equality of opportunity account is too narrow. When we reflect on the importance of meeting health-care needs, factors other than their effects on opportunity come into play. Some health care reduces pain and suffering and no fancy analysis of opportunity is needed to show why people value reducing them. Similarly, disease reminds us of the fragility of life and the limits of human existence and the solidarity we show with the ill by caring for them has come to have deep religious and moral significance in many cultures. Most of the time, people will want to know what is wrong with them and what will keep them well, and some medical care has this

---

[154] Cf. Section 3.3.2.3 *When are Limits to Health Care Fair?*

[155] These benchmarks involve health care in technologically advanced as well as developing countries. Daniels proposes nine benchmarks, each of which contains various criteria for evaluating specific aspects of the fairness of reform proposals. (1) *Intersectoral Public Health* – containing questions concerning basic nutrition, housing, environmental factors and public safety; (2) *Financial Barriers to Equitable Access* – containing issues regarding public and private coverage; (3) *Nonfinancial Barriers* – geographical maldistribution, gender, culture, discrimination; (4) *Comprehensiveness of Benefits and Tiering* – what tiering (i.e. inequalities in the coverage of quality and care) is fair? What is unfair? And what is unavoidable? (5) *Equitable Financing* – tax-based? Premium-based? Or Out-of-Pocket based? (6) *Efficacy, Efficiency, and Quality of Care* – primary health care focus, evidence-based practice and measures to improve quality (7) *Administrative Efficiency*; (8) *Democratic Accountability and Empowerment* – explicit, public procedures for evaluation of services and deliberation on resource allocation, fair grievance procedures, adequate privacy protection, strengthening civil society; (9) *Patient and Provider Autonomy* – degree of consumer and of practitioner autonomy. For benchmarking in reform of a technologically advanced but inefficient and inequitable system that lacks universal coverage, see: N. Daniels, D. Brock, 'Ethical Foundations of the Clinton Administration's Proposed Health Care System', in *Journal of the American Medical Association* 271(1994), pp. 1189–1196; N. Daniels, *Seeking Fair Treatment: From the AIDS Epidemic to National Health Care Reform*, New York: Oxford University Press, 1995; N. Daniels; D. Light; R. Kaplan, *Benchmarks of Fairness for Health Care Reform*, New York: Oxford University Press, 1996. For an adaptation of the benchmarks to use in health systems in countries at different levels of development, see: N. Daniels et al., 'Benchmarks of Fairness for Health Care Reform: A Policy Tool for Developing Countries', in *Bulletin of the World Health Organization* 78(2000)6, pp. 740–750.

[156] N. Daniels, *Just Health Care*, 1985, pp. 49–55; *Id.*, 'Justice, Health, and Healthcare', 2001, pp. 2–16.

function. Some care postpones death, sometimes it reduces pain and suffering, and often it merely improves the quality of life in other ways. Furthermore, his definition of health-care needs in terms of normal species functioning, based on the biomedical concept of disease, may make it difficult to acknowledge many forms of psychological counselling as health-care needs.

Daniels explicitly stresses that he does not want to deny that health care has these varied effects and functions, and emphasises the non-homogeneous character of it. Health care does many things for us. However, his account abstracts from these varied effects of health care a *central function* – the maintenance of species-typical functioning – and emphasises its *central effect* on opportunity. It is his contention that most uses of health-care services, which we find intuitively important, such as those in which we seek to reduce pain or psychological therapy in the case of serious mental dysfunction, will be encompassed by the central effect on opportunity. While Daniels's analysis is not exhaustive, it focuses on that general benefit, which is most relevant from the point of view of distributive justice.[157]

Too Vague. Secondly, as the notion of fair equality of opportunity in Rawls's theory is a narrow notion – focused on producing fairness in the competition for jobs and careers – Daniels makes a shift to a broad – and thus vaguer – notion of equal opportunity. On the broad construal, we are concerned with individual shares of the normal opportunity range, the array of life plans it is reasonable for persons to choose in a given society. Talents and skills determine an individual's share of the normal range, as does normal species functioning. The complaint is that we have a fairly good idea of what goods interfere with fair equality of opportunity in the quest for jobs and careers, but it is much more complicated to pick out what constitutes interference with a fair share of the normal opportunity range. The price of Daniels's modification of Rawls's opportunity principle is that we have a less clear idea of how to justify the modified principle. A related concern is that the broader principle risks making the principle expansive in a way the narrower does not. Do we have to include *everything* that enhances opportunity?[158] In reply, Daniels stresses that his approach rests on the idea that certain institutions meet needs, which quite generally have a central impact on individual shares of the normal opportunity range and should therefore be governed directly by the opportunity principle. Even though on the narrow account of opportunity, health care may turn out to be an important social good, Daniels argues that it does not capture the full importance we do and ought to attribute to health care, whereas the broader notion gives a more elaborated understanding of our problem in hand.

---

[157] At the end of this section and in Chapter 4, I will concentrate in detail on another critique, which holds that Daniels's approach is too narrow, i.e. that of Sen and Nussbaum.

[158] Daniels considers the following as an example of the way in which this expansiveness might infect the principle: 'Suppose that supplying a car to everyone who cannot afford one would do more to remove individual impairments of the normal opportunity range than supplying certain health care services to those who need them. Does the fair equality of opportunity principle commit us now to supply cars instead of or in addition to treatment?' See: N. Daniels, *Just Health Care*, 1985, p. 51.

An important advantage of the broad account is that it can avoid age-bias better than the narrow approach. Job and career opportunities are more important in early and middle stages of life than in later ones and young people have more opportunities than the elderly. At the same time, our health-care needs increase later in life. We can avoid this by not biasing our allocations in favour of one stage of life and instead considering the age-relative opportunity range. Still, treating people differently at different stages of life does not produce inequalities across persons in the way that differential treatment by race or gender does. We all age, though we do not change gender or race. Fairness between age groups in designing health care is appropriately modelled by the idea of prudent allocation over a lifespan.[159] Under some conditions of scarcity, this implies that pure rationing by age – i.e. where age is not a proxy for other traits – is permissible.[160]

Too Strong. Thirdly, it is possible to interpret the fair equality of opportunity account as 'too strong' an approach in a twofold way. The first refers to the requirement of 'levelling' all differences between persons in their shares of the normal opportunity range. Daniels stresses that his distancing from the purely *formal* equality of opportunity concept, does not entail that his account of *fair* shares of the normal opportunity range means *equal* shares.[161] Health-care institutions have the limited function of maintaining normal species functioning: they eliminate individual differences due only to disease or disability.[162]

The second complaint can be traced to another source, namely the fear that health-care needs themselves are so expansive and expensive, given the advance of technology, that they create a bottomless pit. Here, we meet a different kind

---

[159] N. Daniels, *Just Health Care*, 1985, pp. 86–113; *Id., Am I My Parent's Keeper? An Essay on Justice Between the Young and the Old*, New York: Oxford University Press, 1988.

[160] Daniels's account of age-based rationing is different in rationale from that advocated by Daniel Callahan, who thinks the old have a duty to step aside in favour of the young; see: D. Callahan, *Setting Limits: Medical Goals in an Aging Society*, New York: Simon & Schuster, 1987. It is also different from those who argue for a version of the 'fair innings' view, which gives priority to the young on the grounds that the old have already had their opportunity to acquire years; see: D. Brock, 'Justice, Healthcare, and the Elderly', in *Philosophy and Public Affairs* 18(1989)3, pp. 297–312; A. Williams, 'Intergenerational Equity: An Exploration of the "Fair Innings" argument', in *Health Economics* 6(1997), pp. 117–132. It is also different in rationale from Frances Kamm, who argues that the young would be worse off than the old and in that sense 'need' years more than the old; see: F. Kamm, *Morality, Mortality. Volume 1: Death and Whom to Save From It*, Oxford: Oxford University Press, 1993. The considerable disagreement about what justice permits, even among those who accept some form of rationing, agues for the importance of the type of fair process Daniels defends. Unfortunately, I am unable to go into the complex discussion on age-based rationing here.

[161] Cf. Section 3.3.2.1 *The Special Moral Importance of Health Care?*

[162] See: N. Daniels, *Just Health Care*, 1985, pp. 52–53, answering the critiques of Lawrence Stern, 'Opportunity and Health Care: Criticisms and Suggestions', in *Journal of Medicine and Philosophy* 8(1983)4, pp. 339–362; and of Allen Buchanan, 'The Right to a Decent Minimum of Health Care', in *Philosophy and Public Affairs* 13(1984)1, pp. 55–78.

of hijacking, namely, hijacking by needs rather than preferences.[163] Consequently, recognising individual right claims to the satisfaction of health-care needs would, in the age of expensive high-tech medicine, force society to forego realising other social goals. In the same line of reasoning, Ezekiel Emmanuel writes: 'Daniels has offered us a criterion for determining what medical care constitutes the basic services which justify providing "Presidential" medicine to all.'[164] Daniels offers two points in response. Firstly, the narrow model of health-care needs, based on the normal opportunity range, excludes using medical technology to enhance normal capacities or functions.[165] The latter would involve giving priority to *altering* the normal opportunity range, rather than to protecting individuals whose normal range is compromised. Arguments about the relative merits of such use of scarce resources may be advanced, but they would not rest on claims about basic health-care needs and thus may have a different justificatory force. However, technology does expand the ways and costs we have for meeting basic health-care needs. Therefore, Daniels's account of needs at best reduces, but does not eliminate this worry.[166]

Secondly, this worry is stirred up by the idea that if we posit a fundamental individual right to have needs satisfied, no other social goals will be able to override the right claims to all health-care needs. However, no such immediate, direct and blind rights to health care are posited in Daniels's account. Rather, the particular rights and entitlements of individuals are specified only *indirectly*, as a result of the basic health-care institutions acting in accordance with the general principle governing opportunity. The various institutions affecting opportunity must be weighed against each other. And similarly, the resources required to provide for fair equality of opportunity must be weighed against what is needed to provide for other important social institutions. This is true even though guaranteeing fair equality of opportunity has lexical priority over principles of justice promoting well-being in other ways. The point is that health-care institutions capable of protecting opportunity can be maintained only in societies whose productive capacities they do not undermine.[167] Rights to health care are thus

---

[163] C. Fried, *Right and Wrong*, Cambridge, MA: Harvard University Press, 1978; D. Braybrooke, 'Let Needs Diminish that Preferences May Prosper', in *Studies in Moral Philosophy, American Philosophical Quarterly Monograph Series*, No. 1, Oxford: Blackwell, 1968; E. Emanuel, *The Ends of Human Life: Medical Ethics in a Liberal Polity*, Cambridge, MA: Harvard University Press, 1991. Cf. Section 2.3.2.2 *Needs in Philosophical Disrepute?*

[164] E. Emanuel, *op. cit.*, p. 123. I have been arguing in the same line in Section *Care and the Boundaries of Human Life*.

[165] See: N. Daniels, 'Normal Functioning and the Treatment-Enhancement Distinction', in *Cambridge Quarterly of Healthcare Ethics* (2000)9, pp. 309–322.

[166] N. Daniels, *Just Health Care*, 1985, pp. 51–55; See also: A. Buchanan; D. Brock; N. Daniels; D. Wikler, *From Chance To Choice. Genetics and Justice*, Cambridge: Cambridge University Press, 2000.

[167] For instance, when the basic needs are met according to the opportunity principle, the so-called add-on technologies can be applied according to the second priority rule which makes room for the aspect of efficiency. See: N. Daniels; J. Sabin, *Setting Limits Fairly: Can We Learn To Share Medical Resources?* Oxford: University Press, 2002.

not some prior set of individual rights, but are always derivative, defined within a broad and various set of basic institutions governed by the fair equality of opportunity principle.

3.3.4.3 Inefficacious Health Care

The third category of critique refers to the role of normal functioning and reciprocity in Rawlsian contract theory. For the sake of the argument, let me return to a few fundamental ideas.[168] In Rawlsian social contract theory the parties are imagined as normal and fully cooperating members of society, roughly equal in ability, and possessing the two moral powers – capacity for a sense of justice and for a conception of the good. Furthermore, the social cooperation is imagined as one that is advantageous to all: by cooperating they each get more than they could get by not cooperating.

Previously, I have pointed out that Rawls himself did not devote much attention to health and medical care.[169] I have also set forth three reasons for this neglect. Firstly, Rawls considers health to be a natural good instead of a primary social good, being roughly independent of social structure. Secondly, he presupposes the parties of the contractual bargain to be normal, active and fully cooperative citizens over a complete lifespan. Thirdly, income and wealth are used as approximations to the whole index of primary social goods estimating the life expectations of the citizens.

In extending the Rawlsian theory to health care, it turned out that Norman Daniels first shifts emphasis from Rawls by characterising health as a primary social good, because our health status is much influenced by the social structures of society[170] With regard to the second presupposition, Daniels adheres as closely as he can to the idealising conditions of Rawls's imaginary original position, in which people are normal, fully functioning individuals with a complete lifespan. Health is understood in terms of normal species functioning and disease as impairment of normal functioning. As for Rawls's presupposition of income and wealth as approximations, Daniels avoids this discussion by subsuming health care under the principle of fair equality of opportunity. For Daniels, the primary goal of public health and medical care is

> to keep people as close to the idealisation of normal functioning as possible under reasonable resource constraints. Maintaining normal functioning, in turn, makes a limited but significant contribution to protecting the range of opportunities open to individuals.[171]

Given that in this theory the moral importance of health is situated, not in its effect on well-being in general but on opportunity, things could become

[168] Also analysed in Y. Denier, 'Public Health, Well-Being, and Reciprocity', in *Ethical Perspectives* 12(2005)1, pp. 41–66.
[169] Cf. Section 3.2 *Rawlsian Reflections on Just Health* Care.
[170] Cf. Section 3.3 *Extending Rawls to Normal Health* Care.
[171] N. Daniels et al., *Is Inequality Bad for our Health?*, 2000, p. 17.

less complex. Moreover, this is precisely what Daniels wants. The biomedical model of normal species functioning allows us to draw a fairly sharp line between uses of health-care services to prevent and treat diseases and uses which meet other social goals.[172] The effect of disease or disability on the normal opportunity range is easier to assess than their effect on a person's well-being. Differences in the impact of a particular impairment or disability will generally be greater on a person's well-being – given specific features of that person and her life plan – than on her access to a fair share of the opportunity range for her society. Daniels's reason for choosing the objective truncated scale of health-care needs is that he, in keeping with the basic Rawlsian view of finding agreement, wants to define a generally uncontroversial baseline that sets the stage on which we can all agree what principles of justice are relevant to distributing health-care services.

In Chapter 2, I have underlined the importance of health quantity. In that sense, health is important because it gives an important extent of independence, activity and mobility. Furthermore, in thinking about health and health care, it is most reasonable to take normal functioning as a benchmark. In prevention, cure and care we indeed try as much as possible to support people in reaching and maintaining the possibility to lead a normal life, i.e. a life 'of one's own'. However, as I have suggested in Section 2.4, taking the small contractual basis of the moral importance of health care as benchmark might imply that we have to pay a price that we do not want to pay. The importance attached to normal functioning is not unproblematic. In Chapter 4, I will show that Daniels's extension of Rawls, *only* accounts for normal health care that *restores* people so that once again they are fully cooperating members of society. With normal functioning as a benchmark, neither Rawls nor Daniels can deal with the problem of special health-care needs, that is, those needs for people who cannot possibly be brought to the level of normal species-typical functioning. Could this problem be solved by a second extension?

---

[172] Examples of this difference are treatment of dysfunctional noses (which are diseases, since noses have normal species functions and anatomy) and of ugly noses (which are simply deviations of individual or social conceptions of beauty). In the same line, infertility counts as a disease, even though some or many persons might prefer to be infertile and seek medical treatment to render themselves so. Similarly, unwanted pregnancy is not a disease. In such cases, uses of health technology are not a category of prevention and treatment of diseases but of other social goals. Daniels holds that they would not be justifiable as the fulfilment of health-care *needs*. Nontherapeutic abortions do not count as health-care needs, so if public health services have as its only function the meeting of health-care needs, then we cannot argue for funding the abortions just like any other procedure. But if public basic health care should serve other important goals – like ensuring that the poor as well as the well-off women can equally well control their bodies, then there is justification for funding abortions, although it is a different justification. See: N. Daniels, 'Health-Care Needs and Distributive Justice', 1981, pp. 156–158; *Id., Just Health Care*, 1985, pp. 31–32. See also: T.M. Scanlon, 'Preference and Urgency', 1975.

CHAPTER 4

NUSSBAUM'S APPROACH: A NON-CONTRACTARIAN ACCOUNT OF CARE

4.1 A SECOND EXTENSION TO LONG-TERM CARE?

4.1.1 Beyond the Social Contract: Martha Nussbaum's Critique

In recent publications, Martha Nussbaum develops a critical analysis of the role of normal functioning and reciprocity in Rawlsian contract theory.[1] In a first step, her critique concerns the Kantian conception of the person, which serves as a distinctive hallmark in Rawlsian theory. In a second step, the idea of contractarian productive reciprocity is brought under discussion. Finally, she concentrates on the implications of these two perspectives for the account of primary goods, especially on its dealing with the needs of people for long-term care in times of continuing asymmetrical dependency.

4.1.1.1 A Kantian Conception of the Person

John Rawls is explicit in tracing his conception of the person to Kant. His Kantian constructivism begins from a conception of the person and of practical reason – i.e. the ideal of *free* and *equal* moral persons who are both reasonable and rational and therefore are capable of taking part in social cooperation among persons so conceived – as central elements in the reasonable procedure of construction, the outcome of which determines the content of the principles of justice.[2]

Our nature as *free* rational moral agents consists in the two moral powers: the capacity for a sense of justice and the capacity for a conception of the good. The two moral powers are very important. They are the capacities that anyone needs if he is to occupy the role of citizen and engage in, benefit from and comply with the

---

[1] M.C. Nussbaum, 'The Future of Feminist Liberalism', in *Proceedings and Addresses of the American Philosophical Association* (2000)74, pp. 47–79; *Id.*, 'Disabled Lives: Who Cares?' in *The New York Review of Books* 48(2001)1, pp. 34–37; *Id.*, 'Long-Term Care and Social Justice: a Challenge to Conventional Ideas of the Social Contract'. in World Health Organization, *Ethical Choices in Long-Term Care: What Does Justice Require*? Geneva: World Health Organization, 2002, pp. 31–65; *Id.*, 'Capabilities as Fundamental Entitlements: Sen and Social Justice', in *Feminist Economics* 9(2003) 2–3, pp. 33–59; *Id.*, 'Rawls and Feminism', in S. Freeman (ed.), *The Cambridge Companion to Rawls*, Cambridge: Cambridge University Press, 2003, pp. 488–520, esp. pp. 511–514; M.C. Nussbaum, 'Beyond the Social Contract: Toward Global Justice', in G.B. Peterson (ed.), *The Tanner Lectures on Human Values*, Vol. 24, Salt Lake City: University of Utah Press, 2004, pp. 413–507.

[2] J. Rawls, *A Theory of Justice*, 1971, pp. 251–257; 1999, pp. 221–227; *Id.*, 'A Kantian Conception of Equality', 1999/1975, pp. 254–266; *Id.*, 'Kantian Constructivism in Moral Theory', 1999/1980, pp. 303–358; *Id.*, *Political Liberalism*, 1996, pp. 18–19, 48–54, 99–110, 115–116; *Id.*, *Justice as Fairness: a Restatement*, 2001, pp. 18–24.

demands of social cooperation in a democratic society. The capacity for a sense of justice – also called the reasonable – is the capacity to understand, apply and act according to the principles of justice. It entails our moral sensibility; the willingness to propose and abide by the fair terms of cooperation, given the assurance that others will likewise do so. The capacity for a conception of the good – also called the rational – is the capacity to form, revise and rationally pursue a rational plan of life. It entails our prudential rationality; the pursuit of our own conception of the good.[3] Both moral powers are, in effect, distinct but complementary capacities for practical reasoning as applied to justice. They are essential to our being free, self-governing agents who have a conception of our good and who can take responsibility for our actions and ends and participate in social life.

Having these two powers to the requisite minimum degree to be fully cooperating members of society makes persons *equal*. As such, all members of the well-ordered society have, and view themselves as having, a right to equal respect and consideration in determining the principles by which the basic arrangements of their society are to be regulated. The principles of equal liberty and fair opportunity are an expression of this equality. With regard to the other primary social goods of income and wealth, it is stated that everyone should have an equal share unless inequalities improve everyone's situation, including that of the least advantaged, provided these inequalities are consistent with equal liberty and fair opportunity. And thus we arrive at the difference principle.

In many ways the Kantian starting point is most appealing, as the central idea of Kantian ethics is that each human being possesses an inviolable dignity expressed in the prerequisite that each person is to be treated as an end in himself, and nobody as a mere means to the ends of others. However, Nussbaum argues, Kant's particular way of expressing his insights about human dignity is not unproblematic because for Kant human dignity and our moral capacity, which is the source of dignity, are radically separate from the natural world.[4]

---

[3] However, Rawls hastens to say that the reasonable is not the altruistic and the rational is not the egoistic, as if it would be respectively about saints and villains. Reasonable persons are not moved by the good as such but desire a social world in which they, as free and equal can cooperate with others on terms all can accept. They insist that reciprocity should hold within that world so that each benefits along with others. The rational applies to a single, unified agent with the powers of judgement and deliberation in seeking ends and interests peculiarly its own. However, rational agents are not solely self-interested in the sense that their interests are not always interests in benefits to themselves. Every interest is an interest of a self, but not every interest is in benefits to the self that has it. It is illuminating to understand both categories of reasonable and rational as respectively dialogic (or public) and monologic. The reasonable is public in a way the rational is not. By the reasonable we enter as equals the public world of others and stand ready to propose, or to accept, fair terms of cooperation with them. See: J. Rawls, 'Kantian Constructivism in Moral Theory', 1999/1980, pp. 303–358; *Id.*, 'Justice as Fairness: Political not Metaphysical', 1999/1985, pp. 388–414; *Id.*, *Political Liberalism*, 1996, pp. 48–54. See also: S. Freeman, 'Reason and Agreement in Social Contract Views', in *Philosophy and Public Affairs* 19(1990), pp. 141–147.

[4] M.C. Nussbaum, 'Future of Feminist Liberalism', 2000, p. 50; *Id.*, 'Disabled Lives: Who Cares?' 2001, pp. 35–36; *Id.*, 'Long-Term Care and Social Justice: a Challenge to Conventional Ideas of the Social Contract', 2002, pp. 44–48; *Id.*, 'Capabilities as Fundamental Entitlements: Sen and Social Justice', 2003, pp. 51–53; *Id.*, 'Beyond the Social Contract: Toward Global Justice', 2004, pp. 426–431.

Kant conceived of the human being as a fundamentally split being, being both a rational person in the realm of freedom and morality, as well as an animal being in the world of nature.[5] In so far as we exist merely in the realm of nature, i.e. through the animal side of human life, we are not ends in ourselves and do not have a dignity. Things in that realm simply have a price. Only in so far as we rise above the mere realm of nature and – through our capacity for moral rationality – exist in the realm of ends as well, we have dignity and transcend price. Although morality has the task of providing for human neediness, it is the person as a rational and moral being that serves as the goal of these provisions. The animal aspect of the human being itself is not an end. What is wrong with the Kantian split between personhood and animality? Nussbaum specifies four problematic aspects.

ANIMAL RATIONAL. Firstly, the split ignores that our human dignity is embedded in animality; that it is the dignity of a mortal and vulnerable being. It is a sort of dignity that could not be possessed by something that is not mortal and vulnerable, 'just as the beauty of a cherry tree in bloom could not be possessed by a diamond'.[6] Another way of putting the problem, Nussbaum argues, is to say that for Kant the most relevant genus under which we classify the human being is that of a rational being; 'our fellow genus members are the angels and any such further rational beings there may be. Within this genus, we are the animal species: the animal rational, … rather than the rational animal'.[7]

NO INDEPENDENT DIGNITY. Furthermore, the split denies that animality in itself possesses dignity; thus it leads us to disparage the natural aspect of our lives. It leads us to slight an important aspect of our lives that does have worth and dignity.

SELF-SUFFICIENCY. Thirdly, the Kantian split makes us think of the core of ourselves as self-sufficient, independent of the gifts of fate and fortune. In doing so, Nussbaum argues, we distort the nature of our own morality and rationality, which are embedded in material and animal selves.

A-TEMPORAL. Finally, it makes us think of ourselves as a-temporal beings. Thus, we may forget that the usual human life cycle brings with it periods of extreme dependency, in which our functioning is very similar to that of the mentally or physically handicapped.

It is important to note that Rawls does not endorse the metaphysical elements of Kant's position.[8] He rejects Kant's dualisms (between the necessary and the contingent, form and content, reason and desire, *noumena* and *phenomena*) and he does not presuppose the principles of justice to be a priori based in

---

[5] Cf. esp. I. Kant, *Grundlegung zur Metaphysik der Sitten*, Hamburg: Felix Meiner, 1999.

[6] M.C. Nussbaum, *op. cit.*, 2000, p. 50; *op. cit.*, 2001, p. 35; *op. cit.*, 2002, p. 45; *op. cit.*, 2003, p. 52; *op. cit.*, 2004, p. 427.

[7] *Ibid.*, 2004, p. 427, fn. 16.

[8] J. Rawls, *op. cit.*, 1971, pp. 251–257; 1999, pp. 221–227; *Id.*, 'A Kantian Conception of Equality', 1999/1975, p. 264; *Id.*, 'Kantian Constructivism', 1999/1980, pp. 304–307; *Id.*, *Political Liberalism*, 1996, pp. 99–101.

pure practical reason alone. On the contrary, he defends an empirical form of Kantianism by taking human nature and the fixed empirical conditions within which practical reason is exercised (the circumstances of justice) as relevant to discovering and justifying principles of justice. In Rawlsian theory, the contracting persons are understood from the beginning as being in need of material and other goods. Although through this the two elements of personality and animality are more thoroughly integrated than in Kant's theory, Rawls only departs from Kant to a degree. For the Rawlsian parties are repeatedly characterised as 'fully cooperating members of society over a complete life'.[9] They are imagined as competent, contracting adults, roughly similar in need, and capable of a normal level of social cooperation and productivity.

Nussbaum demonstrates that the general result of putting aside the dependency aspect of every human life is a sterile universe in which we are all considered to be active, socially cooperative, young and 'normal'. We learn to ignore the fact that disease, old age and accident can impede the moral and rational functions; and we are told to put aside the fact that the material and animal aspect of our being, i.e. our corporeality, is something to which we are always meaningfully and significantly related. It installs a dualism that goes wrong in two directions. It suggests, firstly, that our rationality is independent of our vulnerable animality; and secondly, that animality is brutish and stupid. Against this line of reasoning, Nussbaum focuses attention on the fact that reality is comprised of a complex continuum of types of rationality and of practical capacities expressed not only in the life cycle of 'normal' lives as people grow, become mature and decline, but also in the lives of mentally disabled people showing different types of rationality, communication and emotional expression.[10] This continuum expresses our *embedded* rationality. We cannot have a good understanding of ourselves without placing ourselves within that continuum of varieties.

4.1.1.2 Productive Reciprocity

In a second step, Nussbaum focuses on the contractarian presuppositions in Rawlsian theory that are based on the Humean idea that society is a cooperative venture for mutual advantage among rough equals within the circumstances of justice.[11] Although Rawls starts from a morally rich point of view, combining the rational (the prudential rationality of the parties in the original position, reflected in their pursuit of their own conception of the good) and the reasonable (which

[9] J. Rawls, 'Social Unity and Primary Goods', 1999/1982, p. 368; *Id.*, *Political Liberalism*, 1996, pp. 3, 9, 18, 20, 183; *Id.*, *A Theory of Justice*, 1999, p. xiii; *Justice as Fairness: a Restatement*, 2001, pp. 49, 60.

[10] M.C. Nussbaum, 'Future of Feminist Liberalism', 2000, p. 50; *Id.*, 'Disabled Lives: Who Cares?' 2001, pp. 34–37; *Id.*, 'Long-Term Care and Social Justice', 2002, pp. 31–32, 34, 46–48; *Id.*, 'Capabilities as Fundamental Entitlements', 2003, pp. 51–55; *Id.*, 'Beyond the Social Contract', 2004, pp. 419–422, 428, 440–443, 448–455. See also: A. MacIntyre, *Dependent Rational Animals: Why Human Beings Need the Virtues*, London: Duckworth, 1999.

[11] *Ibid.*, 2001, p. 35; *Ibid.*, 2002, pp. 41–47; *Ibid.*, 2003, pp. 51–52; *Ibid.*, 2004, pp. 423–426. See also: J. Rawls, *A Theory of Justice*, 1971, pp. 3–6, 126–130; 1999, pp. 3–6, 109–112.

includes the willingness to propose and abide by the fair terms of cooperation, modelled by the informational constraints behind the Veil of Ignorance), thus making it richer and less harsh than David Gauthier's theory – whose version of the social contract is a straightforward egoistic bargain for mutual advantage[12] – Rawls never gives up the contractarian starting point. This has lead Brian Barry to characterise justice as fairness as a theory that hovers uneasily between justice as impartiality and justice as mutual advantage.[13]

Rawls does not agree with this characterisation and argues in reaction that the fundamental idea of justice as fairness is that of a society as a fair system of cooperation in which the terms of cooperation that may reasonably be expected by each participant are specified by *a third position* that perches on reciprocity: all who are engaged in cooperation and who contribute their part are to benefit in an appropriate way.[14] He hastens to add that reciprocity does not equal mutual advantage: the idea of reciprocity is situated *between* the idea of impartiality, which is altruistic (being moved by the general good), and the idea of mutual advantage understood as everyone's being advantaged judged with respect to an appropriate benchmark of equality. The role of the difference principle is to define this ideal of reciprocity. All are made better off than they would be in a state of mere equality, and in a state where no one benefits at the expense of the worst-off.

Nevertheless, this nuance brought in by the idea of reciprocity does not solve our problem in hand for two reasons. Firstly, it should be emphasised that 'worst-off' is defined in terms of primary goods with special focus on income and wealth. From the viewpoint of justice, the worst-off are the poorest (not the unhappiest – welfarist view; or the most disabled – Sen's view). By identifying the least advantaged group largely on economic and agency terms, the difference principle does not include those with low productive capacities due to a physical or mental handicap and is hence insensitive to their relative unequal position in society.[15] In relation to this, questions arise regarding the fair opportunity principle and its focus on fair competition in the quest for jobs, careers and public offices. What about those who do not yet, those who never will and those who no longer compete, like children, chronically ill and handicapped persons, and elderly people?

---

[12] D. Gauthier, *Morals by Agreement*, New York: Oxford University Press, 1986.

[13] B. Barry, *Theories of Justice*, Berkeley: University of California Press, 1989.

[14] J. Rawls, *Political Liberalism*, pp. 16–18; *Id.*, *Justice as Fairness: a Restatement*, 2001, p. 49, fn. 14; pp. 60, 64, 76–77, 96, 122–124. See also: A. Gibbard, 'Constructing Justice', *Philosophy and Public Affairs* 20(1991), pp. 264–279.

[15] This critique has been uttered by many: see: A.K. Sen, 'Equality of What?' in S. McMurrin (ed.), *Tanner Lectures on Human Values*, Vol. I, Cambridge: Cambridge University Press, 1980; *Id.*, 'Well-Being, Agency and Freedom', in *Journal of Philosophy* 82(1985)4, pp. 169–221, esp. pp. 195–202; *Id.*, 'Justice: Means versus Freedoms', in *Philosophy and Public Affairs* 19(1990), pp. 111–121; *Id.*, *Inequality Reexamined*, Cambridge, MA: Harvard University Press, 1992, esp. Chapter 5; B. Barry, *Justice as Impartiality*, 1996, pp. 42, 59–60, 272, endn. 27 and 28; E. Feder Kittay, *Love's Labor: Essays on Women, Equality, and Dependency*, New York: Routledge, 1999, pp. 104–109; R. Dworkin, *Sovereign Virtue: the Theory and Practice of Equality*, Cambridge, MA: Harvard University Press, 2000, pp. 113–114; M.C. Nussbaum, *op. cit.*, 2001, p. 36; *op. cit.*, 2002, pp. 50, 52, 54–55, *op. cit.*, 2004, pp. 441–443.

Secondly, the same goes for the whole point of social cooperation: the purpose remains mutual advantage between rough equals, though within the constraints of fairness. According to Rawls, justice is about sharing out the gains from cooperation and his theory is replete with references to a society as a cooperative scheme for mutual advantage.[16] In contractarian theory, the idea of approximate equality plays a very important role in setting up the bargaining situation. Like Hume, he argues that justice arises only in situations of moderate scarcity and rough equality between the parties, such that none can dominate the others. Although Rawls explicitly assumes that the parties have various shortcomings of knowledge, thought and judgement, he nevertheless stipulates that these are all within *the normal range*, so as to retain the idea of rough equality. Thus, although once parties are inside the contracting situation the demands of moral impartiality affect them, the contractarian premise still affects those included at this initial stage, and prevents Rawls from taking a route denying them *all* knowledge of their mental and physical disabilities. Therefore, even though the bargain is going to be carried out from a morally rich point of view, and even though the parties are unable to bargain in the usual sense, the exercise remains a deal among rough equals, and it is supposed to give the parties something that they would not get by living on their own. Considerations of economic advantage and productive reciprocity play a large role in this.

4.1.1.3 Primary Goods

Influenced on the one hand by his Humean–contractarian starting point and on the other by his Kantian conception of the person, Rawls repeatedly characterises the parties as rough equals, all possessing a requisite normal degree of moral sensibility and prudential rationality. He considers the basic needs *fulfilled* and refers to citizens as fully cooperating members of society over a complete life.[17] The parties in the original position know that their endowments such as strength and intelligence all lie within the normal range. Rawls insists: '[T]he fundamental question of political philosophy is how to specify the fair terms of cooperation among persons so conceived.'[18]

In so conceiving of persons and circumstances, Nussbaum argues, Rawls explicitly omits from the situation of basic political choice the rather extreme forms of need and dependency that human beings may experience. According to Rawls, justice is not primarily about redressing inequalities imposed by nature or misfortune but about providing each person with *resources* that are sufficient to their moral powers of free, responsible and rational agency. Principles of justice are initially chosen for the ideal case of a well-ordered society, where it is assumed that all have the capacity for cooperation. Equally, his account of the

---

[16] See: B. Barry, *Theories of Justice*, 1989, pp. 241–254; *Id.*, *Justice as Impartiality*, 1996, pp. 59–60. See also for a broader perspective: *Id.*, 'Justice as Reciprocity', in *id.*, *Liberty and Justice*, Oxford: Clarendon Press, 1991, pp. 211–241.

[17] J. Rawls, *Political Liberalism*, 1996, pp. 7, 20, 183.

[18] *Ibid.*, p. 183. See also: pp. 15–22.

primary goods to be distributed by society, with its commitment to measuring relative social positions with reference to income and wealth, is framed as an account of the needs of citizens who are characterised by moral and prudential rationality and by the capacity to be fully cooperating.[19] Thus, it has no place for the needs of people who are not independent, i.e. for the needs for long-term care in times of continuing asymmetrical dependency.[20]

4.1.1.4 Implications

Nussbaum argues that these attachments to the social contract tradition create severe problems for Rawls in positioning people with both physical and mental disabilities, both temporary and permanent.[21] I agree with Nussbaum only partly, for I believe there is no problem in extending Rawls's theory to normal health care – i.e. in the case of temporary illness – as Daniels's extension has shown.[22] Normal health care can be integrated perfectly in contract thinking. Rawls's latest writings in which he explicitly confirms Daniels's extension offer additional support for this.

First of all, Rawls considers basic health care assured to all citizens as 'one of the essential *prerequisites* for a basic structure within which the ideal of public reason ... may protect the basic liberties and prevent social and economic inequalities from being excessive'. This is part of the requirement that citizen's basic needs are met 'at least insofar as their being met is necessary for citizens to understand and to be able to fruitfully exercise those rights and liberties'.[23] Without such provisions the basic liberties would be purely formal – like in

---

[19] M.C. Nussbaum, 'The Future of Feminist Liberalism', 2000, p. 51; *Id.*, *op. cit.*, 2001, p. 36; *Id.*, *op. cit.*, 2002, p. 47; *Id.*, 'Capabilities as Fundamental Entitlements', 2003, p. 52; *op. cit.*, 2004, p. 429. See also: E. Feder Kittay, *op. cit.*, pp. 75–113.

[20] Nussbaum's criticisms are motivated by a somewhat similar analysis, made by Eva Kittay, in which she argues that there are five presumptions in Rawlsian theory, which lead us to conclude that Rawls is unable to yield the egalitarian concerns that motivate the theory, because he is unable to confront facts of long-term asymmetrical neediness. Firstly, the Humean 'circumstances of justice' assume rough equality between the parties so that none can dominate the others. Secondly, the citizens in the well-ordered society are fully cooperating members of society over the course of a complete life and are situated within the normal range of functioning. Thirdly, the conception of free persons as those who think of themselves as self-originating sources of valid claims fails to make place for a freedom that might be enjoyed by someone who is not independent in that way. Fourthly, his account of the primary goods proposes an index of goods all persons require who possess the two moral powers: the ability to have a sense of justice and to form and revise one's conception of the good. And fifthly, the conception of social cooperation presupposes rough equality between the parties and has no place for relations of extreme dependency. See: E. Feder Kittay, *op. cit.*, 1999, pp. 75–113.

[21] M.C. Nussbaum, *op. cit.*, 2002, pp. 43–44, 53–54; *op. cit.*, 2003, p. 50; *op. cit.*, 2004, pp. 418, 431–446.

[22] In this context, it is important to mention the odd fact that Nussbaum in her own writings on the subject never takes Daniels's analysis into account. Eva Kittay on the other hand, does consider Daniels's theory by pointing at the problem of reciprocity in his theory. See: E. Feder Kittay, *op. cit.*, 1999, p. 208, endn. 171; *Id.*, 'Can Contractualism Justify State-Supported Long-Term Care Policies?' in World Health Organization, *Ethical Choices in Long-Term Care*, 2002, pp. 77–86.

[23] J. Rawls, *Political Liberalism*, 1996, pp. lix, 7; *Id.*, *Justice as Fairness: a Restatement*, 2001, p. 176 (my italics, YD).

libertarianism. In this case, Rawls urges, we cannot meet the necessary conditions of realism and stability for the right reasons. Specific institutions are necessary to support and encourage a reasonable constitutional regime. Basic health care assured to all citizens is one of them.[24]

Furthermore, Rawls hastens to stress that the assumption that persons as citizens have all the capacities that enable them to be normal and fully cooperating members of society does not imply that no one ever suffers from illness or accident: 'such misfortunes are to be expected in the ordinary course of human life; and provisions for these contingencies must be made'.[25] In *Justice as Fairness: a Restatement*, he formulates three features of the index of primary goods that in extension give the two principles of justice certain flexibility in adjusting to the differences in citizen's needs for medical care.[26] (a) In the original position, the general form and content of the basic rights and liberties are outlined and their priorities understood. The *further specification* of those rights and liberties is left to the constitutional, legislative and judicial stages as more information is made available and particular social conditions can be taken into account.[27] (b) The primary goods of *income and wealth* are not to be identified only with personal income and private wealth. For we have (partial) control over income and wealth not only as individuals but also as members of associations and groups: 'as citizens we are also the beneficiaries of the government's providing various personal goods and services to which we are entitled, as in the case of health care, or of its providing public goods, as in the case of measures ensuring public health (clean air, unpolluted water and the like). All of these items can ... be included in the index of primary goods.'[28] (c) Finally, Rawls stresses that the index of primary goods is an index of *expectations* of these goods over the course of a complete lifespan. Individual's expectations of primary goods can be the same *ex ante*,

---

[24] Other such institutions are: (a) public financing of elections and ways of assuring the availability of public information on matters of policy; (b) fair equality of opportunity in education and training. Without these, one cannot take part in the debates of public reason or contribute to social and economic policies; (c) a decent distribution of income and wealth, necessary to take intelligent and effective advantage of the basic freedoms. Otherwise, those with wealth and income tend to dominate those with less and tend to control political power in their favour; and (d) society as employer of last resort through general or local government, or other social and economic policies. Long-term security and the opportunity for meaningful work are important aspects of citizen's self-respect and of their sense of membership of society. See: J. Rawls, *Political Liberalism*, 1996, pp. lviii–lix, 6–8.

[25] *Id.*, 'Justice as Fairness: Political not Metaphysical', 1999/1985, p. 398.

[26] *Id.*, *Justice as Fairness: a Restatement*, 2001, pp. 172–173.

[27] See also: 'the variations that put some citizens below the line as a result of illness and accident ... can be dealt with ... at the legislative stage when the prevalence and kinds of these misfortunes are known and the costs of treating them can be ascertained and balanced along with total governmental expenditure.' J. Rawls, *Political Liberalism*, p. 184.

[28] J. Rawls, *Justice as Fairness: a Restatement*, 2001, p. 172. See also: *id.*, 'Fairness to Goodness', 1999/1975, pp. 271–273.

while the goods they actually receive are different *ex post*, depending on the various contingencies, in this case, on the illnesses and accidents that befall them.[29]

However, in all these qualifications and refinements, maintaining normal functioning is moral bedrock: 'The aim is to *restore* people by health care so that once again they are fully cooperating members of society.'[30] To attempt this extension Rawls interprets the assumption that citizens are normally cooperating members of society over a complete life 'to allow that they may be seriously ill or suffer from severe accidents *from time to time*'.[31] Rawls repeatedly refers to cases of medical needs as ones in which citizens fall *temporarily* below the minimum essential capacities for being normal and fully cooperating members of society. Everyone 'has physical needs and psychological capacities within the normal range' and 'no one suffers from *unusual* needs that are especially difficult to fulfil, for example, *unusual* and *costly* medical requirements'.[32]

I agree with Brian Barry that in such cases there is no doubt that any sort of insurance notion will cover medical care and earnings replacement for those who are normally contributing members of society.[33] However, and in this sense Nussbaum is right, the grim logic of reciprocity seems to exclude at least the congenitally and long-term disabled from the scope of justice.[34] The implicit result of the Kantian and contractarian presuppositions is that Rawls, as he acknowledges, cannot accommodate the problem of *special* long-term health care for people who can never be brought to the level of normal functioning.[35] In later works, Rawls is explicit: '[F]or our purposes here I leave aside *permanent* physical

[29] In the same line of reasoning, Norman Daniels sets the difference between the *normal* opportunity range and the *effective* opportunity range. In: N. Daniels, 'Health-Care Needs and Distributive Justice', 1996/1981, p. 188; *Id.*, *Just Health Care*, 1985, p. 34. Cf. Section 2.3.4.2 *Health, Disease and Opportunity* and Section 3.3.2.1 *The Special Moral Importance of Health Care?*

[30] J. Rawls, *Political Liberalism*, 1996, p. 184 (my italics, YD). See also: *ibid.*, pp. 21, 244–245.

[31] *Id.*, *Justice as Fairness: a Restatement*, 2001, p. 172 (my italics, YD).

[32] *Id.*, 'A Kantian Conception of Equality', 1999/1975, p. 259; *Id.*, 'Kantian Constructivism', 1999/1980, p. 332; *Id.*, *Political Liberalism*, 1996, p. 272, fn. 10 (my italics, YD).

[33] B. Barry, *Justice as Impartiality*, 1996, pp. 272–273, endn. 28. See also: M.C. Nussbaum, *op. cit.*, 2004, pp. 436, 438–440.

[34] There is, however, a possibility to include handicaps within a *hypothetical* insurance scheme. This is the proposal made by Ronald Dworkin. I will analyse this in detail in Chapter 5.

[35] *Id.*, 'Social Unity and Primary Goods', 1999/1982, pp. 368–369; *Id.*, *Political Liberalism*, 1996, p. 272; *Id.*, *Justice as Fairness: a Restatement*, 2001, pp. 168–176. In *A Theory of Justice*, Rawls already latently indicated the problem: 'I have said that the minimal requirements defining moral personality refer to a *capacity* and not to the *realization* of it. A being that has this capacity, whether or not it is yet developed, is to receive the full protection of the principles of justice. Since infants and children are thought to have basic rights (normally exercised on their behalf by parents and guardians), this interpretation of the requisite conditions seems necessary to match our considered judgments .... A full discussion would take up the various special cases of lack of capacity. That of children I have already commented upon briefly in connection with paternalism (§ 39). The problem of those who have lost their realized capacity *temporarily* through misfortune, accident or mental stress can be regarded in a similar way. But those more or less *permanently* deprived of moral personality may present a *difficulty*.' In *id.*, *A Theory of Justice*, 1971, pp. 509–510; 1999, pp. 445–446 (my italics, YD).

disabilities or mental disorders so *severe* as to *prevent* persons from being normal and fully cooperating members of society in the *usual* sense.'[36] The implicit admission is that Rawls cannot, any more than could Gauthier, accommodate the idea that justice demands support and care for the congenitally and permanently disabled. I agree with Nussbaum and Barry that even though Rawls does not want to put it in harsh terms – as Gauthier undeniably does[37] – the result is the same: the long-term disabled who can never be brought to the level of normal functioning seem to fall outside the realm of justice.[38]

What have Rawls and Daniels said about these problems? Rawls's first reaction is that if we can work out a viable theory for the normal range, we can worry about extensions of the theory to the hard cases, i.e. to contexts in which conditions are more realistic and people are not all normal, *later*.[39] In the case of special health care, this would mean *later still*. It would entail a *second* extension. But whereas we can integrate the first extension to normal health care within contract theory, this does not hold for the second extension for in these cases there is no reference whatsoever to the values of normal functioning and reciprocity. It is reasonable to think that Rawls is very much aware of this because he urges first, that 'we have a duty towards all human beings, however severely handicapped'. However, at the same time he holds that he is uncertain about how far justice as fairness can be successfully extended to cover the more extreme cases: 'At some point, then, we must see whether justice as fairness can be extended to provide guidelines for these cases; and if not, whether it must be rejected rather than supplemented by some other conception.'[40] Rawls puts it carefully: in such cases we need something else, another theory or other virtues.

With regard to another theory, he is very sympathetic to Amartya Sen:

> When we attempt to deal with the problem of special medical and health needs, a different or more comprehensive notion than that of primary goods ... will, I believe, be necessary; for example Sen's notion of an index which focuses on persons' basic capabilities may prove fruitful for this problem and serve as an essential *complement* to the use of primary goods.[41]

With regard to the movement towards other virtues, he writes that justice as fairness may fail because

[36] *Id.*, 'Justice as Fairness: Political not Metaphysical', 1999/1985, p. 398 (my italics, YD).

[37] For David Gauthier, people of unusual disability are 'not party to the moral relationships grounded by a contractarian theory'. In D. Gauthier, *Morals by Agreement*, New York: Oxford University Press, 1986, p. 18, speaking of all 'persons who decrease that average (level of well-being in a society)'.

[38] B. Barry, *Justice as Impartiality*, 1996, pp. 42, 60, 272, endn. 28; M.C. Nussbaum, *op. cit.*, 2001, p. 36; *op. cit.*, 2002, pp. 43, 50–51; *Id.*, 'Capabilities as Fundamental Entitlements', 2003, pp. 51, 53; *op. cit.*, 2004, pp. 424, 434, 439–440, 444–446.

[39] J. Rawls, 'Kantian Constructivism', 1999/1980, pp. 332–333; *Id.*, 'Social Unity and Primary Goods', 1999/1982, pp. 368–369; *Id.*, *Political Liberalism*, 1996, p. 272, n. 10.

[40] *Id.*, *Justice as Fairness: a Restatement*, 2001, p. 176.

[41] *Id.*, 'Social Unity and Primary Goods', 1999/1982, p. 369, fn. 8 (my italics, YD). See also: *id.*, *Justice as Fairness: a Restatement*, 2001, p. 176.

> [T]he idea of political justice does not cover everything, nor should we expect it to. Or the problem may indeed be one of political justice but justice as fairness is not correct in this case, however well it may do on other cases. How deep a fault this is must wait to be determined until the case itself can be examined .... In any case, we should not expect justice as fairness, or any account of justice, to cover all cases of right and wrong. Political justice needs always to be complemented by *other virtues*.[42]

A similar line of reasoning is presented in Daniels's writings. According to Daniels, health care is of special moral importance, because it helps us to preserve our status as fully functioning citizens. With regard to the issue of complementary theory, Daniels stresses that both the views of Rawls and Sen converge much more than may appear to at first.[43] With the extension to health care, Rawls is much more clearly concerned with the capabilities of citizens that emerge from meeting their needs, thus entering the same 'space' as Sen's account. Similarly, Sen's discussion of capabilities is most plausibly focused on those capabilities citizens must have to achieve democratic equality.

With regard to the necessity of other virtues, he writes: '[I]t seems to follow that where health care is generally inefficacious ... it loses its status as a special concern of justice and the "caring" it offers may be viewed more properly as a concern of beneficence or charity.'[44] The characterisation of the four layers of health-care institutions reveals the same thought:[45]

> The picture conveyed by [the] four levels ... of health-care institutions should not be taken to imply that each layer corresponds to a different principle of justice, or that the layers are ranked in moral priority. Each corrects in a particular fashion for a type of *departure* from the Rawlsian idealization that all people are functionally normal. It is preferable to prevent than to have to cure, and to cure than to have to compensate for lost functioning. But all these institutions and services are needed if fair equality of opportunity is to be guaranteed. It is only where there is no chance of protecting opportunity, as in the fourth level, where we are concerned with the seriously, *permanently* disabled, that we may be *beyond* measures that justice requires. Here, principles of *beneficence* may be a more important guide to our obligations.[46]

However, both appeals to complementary theory and to other virtues than justice provide no evident solution to our problem. The second extension is not

---

[42] *Id.*, *Political Liberalism*, 1996, p. 21 (my italics, YD).

[43] N. Daniels, 'Justice, Health, and Healthcare', 2001, pp. 5–6; *Id.*, 'Rawls's Complex Egalitarianism', in S. Freeman (ed.), *The Cambridge Companion to Rawls*, Cambridge: Cambridge University Press, 2003, pp. 241–276, esp. pp. 244, 259. Daniels refers to Sen's most recent work, like A. Sen, *Development as Freedom*, New York: Knopf, 1999. See also: E. Anderson, 'What Is the Point of Equality?' in *Ethics* 109(1999), pp. 287–337, p. 319.

[44] N. Daniels, *Just Health Care*, 1985, p. 45, n. 4. Let us return to Daniels's critique on the utilitarian interpretation of the moral importance of health care (meeting health-care needs has – through its reduction of pain and suffering – a definite tendency to promote happiness), which comes down to the fact that the utilitarian must always weigh this reduction of pain and suffering against the satisfaction of all other kinds of preferences that promote happiness. It is the empirical estimate of the strength of the tendency that makes health care special; the special character disappears when the estimate changes. Here, it seems to turn out that Daniels meets similar problems: health care loses its special character when it is no longer efficacious.

[45] Cf. Section 3.3.2.1 *The Special Moral Importance of Health Care?*

[46] N. Daniels, *op. cit.*, 1985, pp. 47–48.

unproblematic. For instance, can the capability approach that easily be added to Rawlsian theory and thus function as a complement? And what are the implications of putting long-term care under the virtues of care and beneficence? It seems to me that this route is not only too easy a way out but moreover hides – as Gauthier frankly acknowledges – an unpleasant feature of bargaining theories that people do not like to mention, namely that the congenitally, lifelong disabled are simply not included in the realm of justice.[47] To clarify this, let me specify a few forceful arguments included in Nussbaum's critique.

4.1.2 Important Qualities of Nussbaum's Critique

4.1.2.1 Continuity

Martha Nussbaum focuses attention on the fact that variations and asymmetries in need are not isolated or easily isolable cases; but a pervasive fact of human life.[48] Dependency is part of the human condition and extreme dependency comes in many forms. As such it has an ontological status. All of us are dependent on others in significant ways:

> Real people begin their lives as helpless infants, and remain in a state of extreme, asymmetrical dependency, both physical and mental, for anywhere from ten to twenty years. At the other end of life, those who are lucky enough to live on into old age are likely to encounter another period of extreme dependency, either physical or mental or both, which may itself continue in some form for as much as twenty years. During the middle years of life, many of us encounter periods of extreme dependency, some of which involve our mental powers and some our bodily powers only, but all of which may put us in need of daily, even hourly, care by others. Finally ... there are many citizens who never have the physical and/or mental powers requisite for independence.[49]

The issue is, Nussbaum urges, that there really is a *continuum* between the cases of lifelong disability that Rawls has postponed on contractarian grounds and the periods of disability imposed by illness, accident and old age, i.e. disabilities that fall within the so-called normal range. Especially as more of us live longer into old age, with its myriad disabilities, the continuity between one group and the other becomes clearly visible. Thus, it seems arbitrary to include the temporary disabled and to exclude the whole class of people with irreversible disabilities on the grounds that they affect only a small number of people. For care is a complicated issue for *every* society. Any real society is a care-giving and care-receiving society,

---

47 'The problem here is not care of the aged, who have paid for their benefits by earlier productive activity. Life-extending therapies do, however, have an ominous redistributive potential. The primary problem is care for the handicapped. Speaking euphemistically of enabling them to live productive lives, when the services required exceed any possible products, conceals an issue which, understandably, no one wants to face'. In: D. Gauthier, *op. cit.*, 1986, p. 18 No. 30. The inhuman implications of Gauthier's theory are also discussed in A. Buchanan, 'Justice as Reciprocity versus Subject-Centered Justice', in *Philosophy and Public Affairs*, 19(1990), pp. 227–252 at esp. 230–232.

48 M.C. Nussbaum, 'Future of Feminist Liberalism', 2000, p. 52; *Id., op. cit.*, 2001, p. 34; *Id., op. cit.*, 2002, pp. 31–34, 48–49; *Id.*, 'Capabilities as Fundamental Entitlements', 2003, p. 51; *Id.*, 'Rawls and Feminism', 2003, pp. 511–514. *Id., op. cit.*, 2004, pp. 421–422, 439–440.

49 *Id.*, 'Capabilities as Fundamental Entitlements', 2003, p. 51. Also in *id.*, 'Future of Feminist Liberalism', 2000, p. 49; *Ibid.*, 2002, pp. 43–44 and in *id.*, 'Rawls and Feminism', 2003, pp. 511–512.

and must therefore discover ways of coping with the facts of human neediness and dependency that are compatible with the self-respect of the recipients and do not exploit the caregivers. In this situation, Nussbaum argues, the simple language of economic efficiency must give way to a more complex, more reflective language of full human development. All too often, economic thought addressing this problem proceeds as if it is only a matter of efficiency and not also a matter of justice and equity. I will consider her proposal in detail in Section 4.2.

4.1.2.2 Knowledge Inconsistency in the Original Position

A second important quality of Nussbaum's critique is that she points to the inconsistent character of knowledge of the parties in the original position. The Rawlsian concept of social cooperation is based on the idea of reciprocity between rough equals and has no place for relations of lifelong extreme dependency.[50] The continuity between dependency and independency is not reflected in the structure of the Veil of Ignorance. This is highly remarkable:

> Anyone might be ... a person [with lifelong mental and/or physical disabilities], so it seems arbitrary for the parties in the Original Position to deny themselves knowledge of their race, class, and sex, but to permit themselves knowledge that their ... abilities fall within the so-called normal range.[51]

That means, we are being asked to imagine ourselves as if we have no needs for care in times of extreme dependency. This is even more remarkable because Rawlsian theory is, from the beginning, strongly committed to removing the unfair influence of various morally irrelevant accidents of life and to promoting equal respect among citizens. Just as race, class, wealth and even sex do not give one person greater worth than another from the perspective of the principles of justice that should underlie society's basic institutions, so too one would have thought the facts that one person's body or mind is more dependent than another's or that one has a dependent aged parent should not be the source of pervasive social disadvantage.[52] If health care and other forms of care are considered as central goods making well-being possible only under the principle of fair equality of opportunity, i.e. as normal health care, then care for the elderly and the mentally and physically handicapped (covering a major part of the work that needs to be done in any society) remains a source of great injustice, both for the cared-for as for the caregiver.

Therefore, it seems that the whole point of the veil of ignorance is reason enough to put dependency behind the veil: [L]et the parties in the original position not know what ... disability they may or may not have: then, and only then, will the resulting principles be truly fair to people with disabilities.'[53] For three reasons, however, Rawls is unable to accept this apparently reasonable suggestion.

---

[50] M.C. Nussbaum, *op. cit.*, 2002, p. 47. See also: E. Feder Kittay, *op. cit.*, 1999, pp. 75–113, esp. p. 81.

[51] M.C. Nussbaum, *op. cit.*, 2004, p. 431.

[52] M.C. Nussbaum, *op. cit.*, 2001, p. 35.

[53] *Id.*, *op. cit.*, 2004, p. 431.

SIMPLICITY. Firstly, like in the case of normal health care, the reason of simplicity returns.[54] By admitting people with disabilities into the calculation, he loses a simple and straightforward way of measuring who is the least well-off in society, the determination that he needs to make for purposes of thinking about material distribution and redistribution, which Rawls makes with reference to income and wealth alone.

STATISTICAL FREQUENCY. Secondly, the parties in the original position know general facts about the world, and thus, they know that certain disabilities – like back trouble or diminished eyesight – are very common and others – like being paralysed or blind – are much less common. In all societies, these facts of statistical frequency determine the shape of public and private space and the general nature of daily life. The real issue for the contractarian is the relative rarity of the non-normal disabilities, and therefore the expensive and difficult character of the arrangements that have to be made to make work and public space fully accessible to people with such rare disabilities, enabling them to be productive. Such expenditures, in general, greatly outweigh the return in economic productivity made possible by the full inclusion of people with disabilities, because they involve redesigning facilities for all, for the sake of the needs of a very small number. Thus, as Gauthier makes explicit, these arrangements are not mutually advantageous in the economic sense.

PRODUCTIVE RECIPROCITY. Although some disabled people may be highly productive in appropriate circumstances, it is implausible to think that this is generally true of all persons with disabilities. Some disabilities greatly interfere with major life functions. Therefore, even if the case for full inclusion of some disabled workers could be made, it would surely not cover all cases of disability. Here, Nussbaum argues, we see the naked face of contractarianism. The whole exercise is one of reaping benefits from cooperation, and all such theorists define the benefits in a quite familiar economic way. David Gauthier turned out to be very clear about this. Rawlsian theory shows tension. On the one hand, one of its central purposes is to give questions of justice priority over questions of efficiency. Once the bargain is under way, it is arranged in such a way that society may not pursue overall well-being in a way that is unfair to any individual. On the other hand, the account of how the contract initially gets going is still a classic contractarian account, with Hume's circumstances of justice as starting point and mutual advantage as goal of the cooperation. Therefore, despite the presence of valuable moral elements in his initial situation, Rawls cannot eliminate the constraint imposed by the fact that it is a bargaining situation.

4.1.2.3 Long-Term Care and the Difference Principle

Gradually, it appears that although the fundamental problem with Rawlsian theory remains twofold, the real issue appears to be less the Kantian interpretation of the person than contractarian reciprocity thinking. Consider the next

---

[54] Cf. Section 3.3.1 *Health Care: an Additional Primary Social Good?*

proposal: wouldn't it be a solution to split the problem of care? Normal health care would then fit under the fair equality of opportunity principle, whereas special and long-term care could be subsumed under the difference principle. In that case the least advantaged would be – 'Sen-like' – the disabled who will never be able to lead an independent 'normal and fully cooperating' life and who will always be in need of care. In *Restatement* Rawls even seems to leave room for this possibility:

> Within the guidelines of the difference principle, provisions can be made for covering these needs up to the point where further provision would lower the expectations of the least advantaged .... This reasoning parallels that in fixing a social minimum. The only difference is that now the expectation of an assured provision of health care at a certain level ... is included as part of that minimum.[55]

However, Rawls is aware of the fact that the difference principle can become unreasonable by transforming itself into a bottomless pit draining away all resources.[56] This becomes clear in the case of special health and medical needs combined with total lack of productivity.[57] In *Restatement* he focuses attention to what I have previously called the external dynamic of scarcity.[58] Consequently, he stresses that the provision of medical care must not be mistakenly understood as to supplement the income of the least advantaged when they cannot cover the costs of the medical care they may *prefer*. On the contrary, medical care must be understood to fall under the fair equality of opportunity principle. It serves to meet the needs and requirements of citizens as free and equal, normal and fully cooperating members of society. Note, however, that this remark refers to the bottomless pit that originates in social hijacking by *preferences*, which comes down to medicine that serves people's wishes and whims, like cosmetic medicine, instead of medicine that serves people's needs to restore normal functioning.[59]

Nevertheless, what about the bottomless pit that originates by placing long-term care for the permanently disabled under the difference principle? Here, we

---

[55] J. Rawls, *A Theory of Justice*, 1971, pp. 284–293; 1999, pp. 251–258; *Id., Justice as Fairness: a Restatement*, 2001, p. 173.

[56] Cf. supra for the critique of G.A. Cohen, 'On the Currency of Egalitarian Justice', in *Ethics*, 99(1989), pp. 906–944. Cf. Section 3.3.2.2 *Which Health Inequalities Are Unjust?*

[57] J. Rawls, 'Social Unity and Primary Goods', 1999/1982, p. 368, answering the objections raised by Kenneth Arrow in K.J. Arrow, 'Some Ordinalist-Utilitarian Notes on Rawls's Theory of Justice', in *Journal of Philosophy* 70(1973)9, pp. 253–254.

[58] 'Observe that what sets an upper bound to the fraction of the social product spent on medical and health needs are the other essential expenditures society must make, whether these are paid for by private or public funds. For example, an active and productive workforce must be sustained, children must be raised and properly educated, part of the annual product must be invested in real capital and another part counted as depreciation, and provision must be made for those who are retired, not to mention the requirements of national defense and a (just) foreign policy in a world of nation-states. The representatives of citizens who view these claims from the point of view of the legislative stage must strike a balance between them in allocating society's resources.' In J. Rawls, *Justice as Fairness: a Restatement*, 2001, pp. 173–174.

[59] Cf. Section 2.3.2 *Needs and Preferences in Health Care.*

understand medical care to supplement the income of the least advantaged when they cannot cover the costs of the special medical care they *need*. This is a problem built deeply into the logic of contractarianism, for it is the idea that people get together with others and contract for principles of justice only in circumstances in which they can expect mutual benefit and in which all stand to gain from cooperation. To include in the initial situation people who are unusually expensive without contributing anything much to the well-being of the group would be contrary to the logic of the whole exercise. For people who in terms of efficiency, are simply a drain on the whole system, compromise its ability to deliver mutual advantage. As Gauthier frankly acknowledges, this is an unpleasant feature of bargaining theories, a feature people do not like to mention.[60] In the writings of Normal Daniels we read:

> What does asking for the restoration of normal opportunity range mean for the terminally ill, on whom we lavish exotic life-prolonging technology, or for the severely mentally retarded? We are not required to pour all our resources into the worst cases, for that would undermine our ability to protect the opportunity of many others. But I am not sure what the approach requires here, if it delivers an answer at all.[61]

In her critique Martha Nussbaum makes clear that the idealising fiction of the 'fully cooperating … over a complete life' is no mere mistake that might be corrected by a longer list of primary goods or by extensions of extensions. It is woven deeply into the very idea of a contract for mutual advantage. We have seen that it is actually quite important for Rawls and fundamental to his entire argument for the difference principle to be able to speak of both social productivity and the well-being of individuals in simple economic terms. The list of primary social goods is a list of needs of citizens possessing the two moral powers within the circumstances of justice. This omits the expensive, non-productive, and severely disabled.

4.1.2.4 Postponement is Not Innocent

At first sight, there seem to be very good reasons for the Rawlsian postponement of both normal health care and long-term care to the legislative stage. For we need detailed knowledge of issues like the complexity of individual cases, probability calculations, alternative uses of the same resources, level of support required, level of productivity and other empirical information. None of this knowledge is available yet in the original position.[62] Furthermore, there seems to be no problem with regard to *normal* health care, for this is actually already integrated in the original position under the fair equality of opportunity principle.

[60] D. Gauthier, *Morals by Agreement*, 1986, p. 18, n. 30: 'The problem here is not care of the aged, who have paid for their benefits by earlier productive activity. Life-extending theories, however, do have an ominous redistributive potential. The primary problem is care for the handicapped. Speaking euphemistically of enabling them to live productive lives, when the services required exceed any possible products, conceals an issue which, understandably, no one wants to face.'
[61] N. Daniels, 'Health-Care Needs and Distributive Justice', 1996/1981, p. 196.
[62] Cf. Section 3.2.4.2 *Actual Choice* and Section 3.3.1 *Health Care: an Additional Primary Social Good?*

However, Nussbaum's critique has shown that the postponement of *inefficacious* health care is a problem that cuts deeply into Rawlsian theory because the parties are, *from the beginning*, characterised by normal reciprocity and social productivity. Nussbaum puts it in harsher terms: 'Rawls is just right in thinking that [the] interests [of the severely disabled] cannot be accommodated in the first stage of the theory, when we are choosing the principles of justice, *for the whole setup does not foster their inclusion*.'[63]

Consequently, it is in effect out of *charity* that these interests will be considered, not out of basic *justice*. With this, Nussbaum shows a serious tension in Rawlsian theory. For justice was supposed to take precedence over efficiency as was specified in the second priority rule, which determines the priority of justice over efficiency and welfare.[64] It does so, Nussbaum argues, but only once the bargain is constituted. In getting the bargain off the ground, something that looks very much like an issue of justice – i.e. taking the needs of the severely disabled into account – is left waiting in the wings.

Can the problem be solved by the idea of the trustee, presupposing that the contracting parties are also trustees for the persons who are incapable of partaking in the process of selecting principles of justice that will form the basic structure of society? In that case, we take the parties in the original position to be trustees for the interests of all independent members of society, as they currently are also trustees for future generations. In confronting the problem posed for a Kantian contract doctrine by people who have little or no reference to normal, rational reciprocity, Thomas Scanlon concludes that the idea of the trustee may be a way in which the facts of extreme dependency may be recognised in such a doctrine.[65]

For two reasons Nussbaum doubts that this will offer a satisfactory solution.[66] Firstly, the idea of the trustee suggests that the dependent are worthy of respect in the design of basic political institutions only in a *derived* way, i.e. only on account of some relationship to the so-called fully cooperating members. With the addition of the trustee idea, the dependent enter the bargain, not because they are equipped to participate, but only because a contracting party *cares* about their interests. Secondly, the contractual bargain remains characterised by productive reciprocity. It is, after all, a bargain for mutual advantage, which assumes rough equality among the parties. Only productivity, or some reference to productive reciprocity, justifies a claim to support. The problem is put most sharply by Gauthier when he says that the elderly dependent have paid for their care by earlier periods of productive activity, but the handicapped have not.[67] The consequence of this is that human neediness in itself has no claim to support.

---

[63] M.C. Nussbaum, *op. cit.*, 2004, pp. 437–438.

[64] Cf. Section 3.1.2 *The Principles of Justice*.

[65] T.M. Scanlon, *What We Owe to Each Other*, Cambridge: The Belknap Press of Harvard University Press, 1998, pp. 177–187.

[66] M.C. Nussbaum, 'Future of Feminist Liberalism', 2000, pp. 52–53; *op. cit.*, 2004, pp. 443–446.

[67] D. Gauthier, *Morals By Agreement*, 1986, p. 18, fn. 30.

Although Rawlsian theory is much more subtle than Gauthier's, it still suffers from something like this problem. Need and animality receive dignity, respect and support only in virtue of their relation to normal, cooperative reciprocity. Rather than recognising that reciprocity has many forms, the trustee solution retains the Kantian split between the rational and the natural.

Another way in which the facts of extreme dependency may be recognised in a Kantian contract doctrine, would be to say that the contract doctrine offers an account of only *one part* of morality and that we will need a different account to cope with the facts of extreme dependency.[68] This means that we would hold, as Rawls and Daniels clearly do, that the original position is *not* a complete device for designing political justice and that we also need other approaches. This reply is fine for Scanlon, because his theory is a moral contract theory, and not a political theory. He employs no hypothetical initial situation, and makes no claims to completeness. However, it creates large problems for the contract doctrine in the area of *political* theory that concentrates on the project of selecting principles that will form the basic structure of society.

I believe Nussbaum rightly criticises the idealisation of the parties in the original position. What is the point of the ideal political theory if it only covers a part of the story and leaves the rest to the tender mercies of charity? Nussbaum agrees with Rawls when he explicitly states that any approach to the design of basic political institutions must aim at a certain degree of completeness and finality.[69] After all, we are designing the basic structure of society, which Rawls defines as those institutions that influence *all* citizens's life chances pervasively and *from the start*. And yet, Rawlsian theory seems unable to defer our problem in hand in the context of basic political theory. Precisely because it is a political and not an ethical theory, setting up the basic institutions, it is not open to us to say: we have done one part of the task, but of course other parts that are equally basic, will come along later. This would leave, thus Nussbaum, 'huge areas of political justice up for grabs and would entail the recognition of much indeterminacy in the account of basic justice as so far worked out'.[70]

4.1.3 Reciprocity and the Problem of Justice: Possible Strategies?

Let us review a few elements. With regard to our problem in hand there are three possible strategies. Firstly, we could take the route of Rawls and Daniels and

[68] M.C. Nussbaum, 'Future of Feminist Liberalism', 2000, pp. 53–54; *op. cit.*, 2004, pp. 445–446.

[69] See: J. Rawls, *A Theory of Justice.*, 1971, pp. 130–136, 175–178; 1999, pp. 112–118, 153–155, where finality is determined as a formal condition of the political principles (1971, p. 135; 1999, pp. 116–117) and where it is made clear that 'the original agreement is final and made in perpetuity' and that 'there is no second chance'. Therefore, 'they cannot enter into agreements that may have consequences they cannot accept'. (1971, p. 176; 1999, p. 153). The formal condition of finality is part of Rawls's opposition to intuitionism. See also: *The Formal Constraints of the Right* in Section 3.1.3.2 *The Original Position.*

[70] M.C. Nussbaum, 'Future of Feminist Liberalism', 2000, p. 53; *Id.*, *op. cit.*, 2002, p. 51; *op. cit.*, 2004, p. 446.

leave it as it is. Long-term care for the permanently disabled is then a matter that has to be solved either by another, more elaborated theory that functions as a complement, or by the virtues of charity.

A second possibility would be to subsume normal health care under the fair equality of opportunity principle and classify care for the disabled that can never be brought to the level of normal functioning under the difference principle. In this case, the Rawlsian presuppositions would still hold.

A third strategy would respond to Nussbaum's critique that it is not a problem of incompleteness but of misdirection from the start. The list of primary goods selected by Rawls's parties omits items that appear central for humans of 'normal' capacity, as well as for the disabled persons, namely dependency as an ontological element in the human condition. The difficulties in Rawlsian contract thinking are deep and cannot be remedied by a mere modification with the help of extensions.[71] Nussbaum argues that we can avoid these difficulties by pointing out that people actually do conceive of their good as including the good of other vulnerable people.[72] A reasonable political conception of the person can take advantage of this fact.

Earlier, I have expressed my concern with the first strategy, as I consider it as too easy a way out, leaving it up to 'the others' – whether these are other virtues or other theories – to solve the problem. Furthermore, I have pointed out the internal incoherency of the second strategy. Rawls and Daniels cannot put long-term care under the difference principle because this would create a bottomless pit, draining away all resources to the special health-care needs for the so-called non-productive.

Therefore, let us consider the third strategy – no modification, but a whole new theory. What we already know is that this will include adopting a richer, more moralised conception of the benefits of social cooperation including 'the good of vulnerable other people' and 'respect for human dignity' because it is 'good in itself'.[73] Benefit, then, should not be understood in 'purely economic terms', 'for there is the great good of justice itself to be considered'. This implies a radical rejection of productive reciprocity and efficiency and a choice for 'complex forms of reciprocity'.[74] Let us consider Nussbaum's proposal in more detail.

[71] *Ibid.*, 2000, pp. 54–59; *Ibid.*, 2002, p. 41. For this reason, Nussbaum rejects the proposal of Feder Kittay. Adding care to the list is useless because the list is a list of needs of citizens who possess *the two moral powers*. Therefore, Nussbaum argues, it is better to opt for a more radical Sen-like proposal. Cf. *ibid.*, 2004, pp. 446 ff.

[72] See also: R.E. Goodin, *Protecting the Vulnerable. A Reanalysis of Our Social Responsibilities*, Chicago: University of Chicago Press, 1985.

[73] M.C. Nussbaum, *op. cit.*, 2004, pp. 436–438.

[74] *Ibid.*, 2004, pp. 441–444.

4.2 NUSSBAUM'S CAPABILITIES APPROACH

4.2.1 Other Virtues?

Would it be a solution to call upon other virtues than justice – like beneficence, humanity or charity – under which long-term care could be classified? I agree with Nussbaum that this would not be the case. Care for the long-term ill or disabled raises distinct issues of social justice.[75] Firstly, there is the issue of *fair treatment and support* of people who *need* a lot of *care* throughout their lives. The just society would support their health, education and participation in social and, when possible, in political life. Furthermore, the just society would also appropriately support the labour of the *caregivers* by providing assistance, both human and financial, and opportunities for rewarding work and for participation in social and political life. Arranging care in a way that protects the dignity of the recipient and does not exploit the caregiver would also seem to be a central job of a just society. To clarify the argument, let us briefly reconsider the basic hallmarks of the concept of justice.

The principles of justice underlie the structure of society's political, social and economic institutions that distribute fundamental rights and duties and determine the appropriate distribution of the benefits and burdens of social cooperation. Thus, this structure profoundly affects people's life prospects and chances, 'what they can expect to be and how well they can hope to do'.[76] If we are committed to removing the unfair influence of various morally irrelevant accidents of life and to promoting equal respect among citizens, i.e. if we do not allow race, class, wealth and sex to give one person greater worth than another from the perspective of the principles of justice that underlie society's basic institutions, so too, the fact that one person's body is more dependent than another's, or that one has a dependent aged parent should not be a source of pervasive social disadvantage in what one can expect to be and how well one can hope to do. As such, justice guarantees security; it provides a fallback framework, a safety net or minimum floor below which nobody is allowed to fall. It offers a structure of rights that people can count on for organising their lives, a structure which stands somewhat *apart* from familial and other *affective* attachments and which can be relied on to survive, no matter what happens to those attachments and no matter how unhappy the situation may turn out: 'Justice makes sense in circumstances in which resources are scarce and human sympathies limited.'[77] Ultimately, it is for reasons of *justice* that we

---

[75] M.C. Nussbaum, *op. cit.*, 2001, p. 35; *op. cit.*, 2002, p. 51; *Id.*, 'Capabilities as Fundamental Entitlements', 2003 p. 40; *op. cit.*, 2004, pp. 420–422, 437–438, 445–446, 451.

[76] J. Rawls, *A Theory of Justice*, 1971, p. 7; 1999, pp. 6–7.

[77] This Humean idea remains upright. My argument here is also supported by the writings of Jeremy Waldron, 'When Justice Replaces Affection: the Need for Rights', in *id.*, *Liberal Rights*, Cambridge: Cambridge University Press, 1993, pp. 370–391. For the quotation, see: p. 383 of this article.

do not allow the problem of the bottomless pit and the language of economic efficiency to bring about stigmatisation and social exclusion of those in need of long-term care.

Following this line of reasoning it is unnecessary and even undesirable to opt for the argument that the failure of Rawlsian liberal theory to deal with the problem of long-term care implies that we should reject the so-called atomistic liberalism, and the cold and impersonal language of justice, and instead should defend a more personal, care-based or attachment-based theory. On the contrary, the structure of justice remains centre stage. It also remains a cold virtue that is, and should be, independent of the affectionate framework of family and friends, because the latter is vulnerable. It can change, or can even be completely absent. Justice, on the other hand, is concerned with offering each a framework of opportunity and fair treatment, no matter what one's personal characteristics or relations are. In this sense, Nussbaum writes:

> is being 'some mother's child' [i.e., someone's relative, being loved, being cared for] a sufficient image for the citizen in a just society? I think we need a lot more: liberty and opportunity, the chance to form a plan of life, the chance to learn and imagine on one's own.[78]

Or, as Waldron puts it, applied to the issue of care for the elderly dependent:

> It is not that the system of rights is the only imaginable way in which needs could be dealt with in a caring society. We could set things up in a way that encouraged old people to rely on the warm and loving support of their families. But even if we did that, I think we would still want to set up a system of rights as a fallback – as a basis on which some assurance of support could be given, without risking the insecurity, resentment, and indignity of leaving the elderly completely at the uncertain mercy of their sons and daughters.[79]

There is certainly no need to leave the language of justice, nor is there any need to believe in the disablement of liberalism. On the contrary, the liberal ideas of personhood, autonomy, rights, liberty, independence, dignity and opportunity remain central in any plausible theory of justice. However, it does mean that we need to recast liberal theory in a way that does justice to temporality

---

[78] M.C. Nussbaum, 'Future of Feminist Liberalism', 2000, pp. 57–58. Here, Nussbaum differs from Feder Kittay in their respective viewpoints on liberalism. Whereas Kittay suggests to distance from the typical liberal notions of individuality and freedom, and to move over to a conception of the state as a parental supporter of its children's needs, Nussbaum proposes to hold on to liberal theory, but to give it a new form, i.e. a form that is more attentive to need and to its material and social conditions. See also: M.C. Nussbaum, *op. cit.*, 2002, with a reply by Feder Kittay in E. Feder Kittay, 'Can Contractualism Justify State-Supported Long-Term Care Policies? Or 'I'd Rather Be Some Mother's Child', in World Health Organization, *Ethical Choices in Long-Term Care. What Does Justice Require?* Geneva: World Health Organization, 2002, pp. 77–83.

[79] J. Waldron, *op. cit.*, 1993, p. 382.

and need and is more attentive to the material and institutional conditions of liberalism.[80]

To further enforce the argument, Nussbaum formulates three issues that seem uncontroversially important in thinking about the quality of life, and particularly in areas that are pertinent to care.[81] These are issues that are situated squarely within liberal theory. Firstly, a just society would minimally be one that offers to all its citizens, regardless of birth, race, sex or disability, *decent life chances* in areas including, though not limited to, health, education, employment and political participation. Secondly, everyone should be given these opportunities on a basis of *equality* with other citizens, i.e. no group should be turned into second-class citizens by accidents of birth, race, sex or disability. Thirdly, everyone should have their *self-respect* protected, so far as it is within the power of the social institutions to do so. Here, Nussbaum argues in the line of Rawls who called this the most important of the primary social goods that all people can be assumed to want, as crucial to the formation and execution of any life plan. These are no matters of beneficence or charity. If a society fails to offer a supporting framework for decent life chances on the basis of equality, supporting the self-respect of both the cared-for and of the caregiver, then this is not a shortcoming of beneficence or charity or a lack of supererogatory action, but a failure of justice.

4.2.2 Complementary Theory?

Secondly, Nussbaum explicitly repudiates the idea of a complementary theory. She argues that the dominant way of thinking about society, its citizens and its

[80] I agree with Waldron's and Nussbaum's critique on some of the feminist and communitarian criticisms on liberalism and the language of justice that are based on a misunderstanding of some of the most appealing conceptions of liberal theory and of the necessity of justice being a cold virtue. See: J. Waldron, 'When Justice Replaces Affection: the Need for Rights', in J. Waldron, *Liberal Rights*, Cambridge: Cambridge University Press, 1993, pp. 370–391; and M.C. Nussbaum, 'The Feminist Critique of Liberalism', in *id.*, *Sex and Social Justice*, Oxford: Oxford University Press, 1999, pp. 55–80; *Id.*, 'The Future of Feminist Liberalism', 2000, pp. 47–79; *Id.*, *Women and Human Development*, Cambridge: Cambridge University Press, 2000, pp. 55–56, en fn. 45; *Id.*, 'Long-Term Care, and Social Justice', 2002, pp. 41–42, and p. 54, fn. 58; and *id.*, 'Rawls and Feminism', 2003. Some influential feminist writings containing the atomistic critique are: A. Baier, 'The Need for More than Justice', in *Canadian Journal of Philosophy* 13(1987) Suppl., pp. 41–56; A. Jaggar, *Feminist Politics and Human Nature*, Towota: Rowman and Allanheld, 1983, repr. 1988, esp. pp. 27–50, 173–206; C. MacKinnon, *Toward a Feminist Theory of the State*, Cambridge: Harvard University Press, 1989, esp. Chapters 3 and 8; *Id.*, *Feminism Unmodified*, Cambridge: Harvard University Press, 1987, esp. Chapters 2 and 8; C. Pateman, *The Problem of Political Obligation: a Critique of Liberal Theory*, Berkeley: University of California Press, 1979; *Id.*, *The Sexual Contract*, Stanford: Stanford University Press, 1988; C. Gilligan, *In a Different Voice*, Cambridge, MA: Harvard University Press, 1982. For a related criticism from communitarian perspective, see: M. Sandel, *Liberalism and the Limits of Justice*, Cambridge: Cambridge University Press, 1982. For an excellent and extensive overview of this discussion, see: L. MacClain, '"Atomistic Man" revisited: Liberalism, Connection and Feminist Jurisprudence', in *Southern California Law Review* 65(1992), pp. 1171–1264. For further discussion, cf. Section 4.4.1 *Rich but Liberal*.

[81] M.C. Nussbaum, *op. cit.*, 2002, p. 35.

goals as it is fostered by the social *contract* tradition and as it is now widely disseminated around the world, is itself a large part of the problem. It has created a biased way of thinking about need, dependency and dignity that makes it hard to place care in a sufficiently prominent place on the agenda of society, and to give it the support it deserves. That way Nussbaum argues against the proposal of addition, complement or second extension and emphasises on the contrary the necessity of a modification *from the start*.

Following this line of reasoning, Nussbaum argues that an approach, based on ideas of human capability and functioning can more adequately deal with issues of social justice raised by the need to provide care for the elderly, the lifelong disabled, and others in a state of extreme and asymmetrical dependency. The capability approach was pioneered in development economics by Amartya Sen.[82] Nussbaum's current version of the approach derives from a period of close collaboration with Sen at the World Institute for Development Economics Research beginning in 1986.[83] Even though they are very similar, there remain substantial differences between both versions. I am unable here to give a full comparison of both approaches, containing a discussion of all the subtle differences. However, in order to sketch Nussbaum's proposal adequately, it is important to first give a concise specification of what is and what is not common between them. This will be a useful addition in support of my main goal, which is to grasp the essence of Nussbaum's perspective on just health care. Sen's approach will appear regularly, but will remain in the background.[84]

---

[82] The initial statement is in A. Sen, 'Equality of What?' in S.M. McMurrin (ed.), *Tanner Lectures on Human Values*, Salt Lake City: University of Utah Press, 1980, pp. 195–220. See also: *Id.*, *Resources, Values and Development*, Oxford: Basil Blackwell, 1984; *Id.*, *Commodities and Capabilities*, Amsterdam: North-Holland, 1985; *Id.*, 'Well-Being, Agency, and Freedom: the Dewey Lectures 1984', in *The Journal of Philosophy* 82(1985), pp. 169–221; *Id.*, *Inequality Reexamined*, New York: Russell Sage, 1992; *Id.*, 'Capability and Well-Being', in M.C. Nussbaum; A. Sen (eds.), *The Quality of Life*, Oxford: Clarendon Press, 1993, pp. 30–61; *Id.*, *Development as Freedom*, New York: Knopf, 1999.

[83] Although there is considerable independency of her work through her studies about Aristotle's ideas of human functioning and Marx's use of them. See: M.C. Nussbaum, 'Nature, Function, and Capability: Aristotle on Political Distribution', in *Oxford Studies in Ancient Philosophy*, Suppl. Vol. I (1988), pp. 145–184; *Id.*, 'Aristotelian Social Democracy', in R.B. Douglas; G.R. Mara; H.S. Richardson (eds.), *Liberalism and the Good*, New York: Routledge, 1990, pp. 203–252; *Id.*, 'Non-Relative Virtues: an Aristotelian Approach', in M.C. Nussbaum; A. Sen (eds.), *The Quality of Life*, Oxford: Clarendon Press, 1993, pp. 242–269; *Id.*, 'Human Functioning and Social Justice: In Defense of Aristotelian Essentialism', in *Political Theory* 20(1992), 202–246; *Id.*, 'Aristotle on Human Nature and the Foundation of Ethics', in J.E.J. Altham; R. Harrison (eds.), *World, Mind, and Ethics: Essays on the Ethical Philosophy of Bernard Williams*, Cambridge: Cambridge University Press, 1995, pp. 86–131.

[84] For further discussion of differences between both approaches, see also: D. Crocker, 'Functioning and Capability: the Foundation of Sen's and Nussbaum's Development Ethic, Part I', in *Political Theory* 20(1992), pp. 584–612. Part II appeared in M.C. Nussbaum; J. Glover (eds.), *Women, Culture, and Development*, Oxford: Clarendon Press, 1995, pp. 153–198.

4.2.3 *Sen and Nussbaum: Main Agreements*

4.2.3.1 *The Capabilities Space*

The first point of agreement between both authors is the use of the capability notion as the proper space within which comparisons of quality of life or standard of living are most rightfully made. Instead of asking about economic growth, people's satisfactions, or how many resources they are able to command, we should ask about what they are really able to do or to be.[85]

Firstly, economic growth is a poor indicator of the quality of life because of the problem of the distributive-aggregative tension. Gross national product (GNP) per capita occludes distributional inequalities; it fails to tell us how deprived people are doing despite of a nation's general prosperity.[86]

Secondly, equality of utility or welfare falls short for reasons of adaptive preferences. We can only have an adequate theory of social justice if we are willing to make claims about fundamental entitlements that are to some extent independent of the preferences that people happen to have, because preferences are often shaped by unjust background conditions; they can be preferences that have adjusted to a second-class status, like those of the contented slave.[87]

Thirdly, equality of resources falls short because it fails to take into account the fact that individuals need differing levels of resources if they are to come up to the same level of capability to function. They also have differing abilities to convert resources into actual functioning. Some of these differences are downright *physical*: a pregnant woman needs more nutrients than a non-pregnant woman to achieve a similar level of healthy functioning. The differences that most interest the authors are also *social*, and connected with discrimination of various types. To cite Sen's famous example, a person in a wheelchair will require more resources connected with mobility than will the person with normal mobility, if the two are to attain a similar level of ability to get around.[88] In this line of reasoning, both

---

[85] Cf. Section 3.2.4.1 *Capabilities*. See also: M.C. Nussbaum, *Women and Human Development*, 2000, pp. 11–15, 34–110; *op. cit.*, 2001, p. 36; *op. cit.*, 2002, pp. 52, 56–65; *Id.*, 'Capabilities as Fundamental Entitlements', 2003 pp. 33–59; *op. cit.*, 2004, pp. 448–457.

[86] *Ibid.*, 2000, pp. 60–61; *Ibid.*, 2003, pp. 33–34.

[87] *Id., op. cit.*, 2002, pp. 61–63, 111–166; *Ibid.*, 2003, pp. 34–35. Keynote works on the subject of adaptive preferences are: J. Elster, 'Sour Grapes – Utilitarianism and the Genesis of Wants', in A. Sen; B. Williams (eds.), *Utilitarianism and Beyond*, Cambridge: Cambridge University Press, 1982, pp. 219–238; see also: J. Elster, *Ulysses and the Sirens: Studies in Rationality and Irrationality*, Cambridge: Cambridge University Press, 1979. On the contented slave: J. Elster, *Solomonic Judgements: Studies in the Limitations of Rationality*, Cambridge: Cambridge University Press, 1989, p. 6.

[88] In: A. Sen, 'Equality of What?' 1980, esp. pp. 217–219. Although Sen tends to treat this example as one of straightforward physical difference, I agree with Nussbaum that it should not be so treated, since the reasons why wheelchair persons cannot get around are also and to a considerably large extent *social*. All societies are typically arranged to cater for the disabilities of the average person, so they do not become handicaps. Thus, Nussbaum argues, we do not have staircases with steps so high that only giants can climb them, nor do musicians play at frequencies inaudible to the human ear. When a person is deaf, or blind, or has to go around in a wheelchair, societies are not so well adjusted to make such persons fully mobile, fully able to occupy public space or to hold a job on a basis of equality. Whereas these

Sen and Nussbaum argue that the Rawlsian theory would be more apt to give an account of the relevant social equalities and inequalities if the list of primary goods were formulated as a list of capabilities rather than a list of things.[89]

4.2.3.2 Priority of Liberty

Another area of strong agreement is the important role that both theories give to political liberties. Sen explicitly endorses the Rawlsian priority of liberty. Although Nussbaum holds that all capabilities are equally fundamental and she does not announce a lexical ordering among them, she strenuously argues that economic needs should not be met by denying liberty.[90]

4.2.3.3 Distinctness of Persons

In contrast with aggregative thinking, human capabilities should be understood to be valuable for each and every person, and it is the capability of each that should be considered in asking how nations are doing and how quality of life is measured. Nussbaum explicitly defends a *principle of each person's capability*.[91] Sen never took such an explicit route but his theory does emphasise the treatment of each person as an end. The ultimate political goal of the capability approach is the promotion of each person's capabilities.

4.2.4 Nussbaum Differs: Towards Her Own Proposal

However, Nussbaum's version of the capability approach is in several ways different from Sen's, mainly in its emphasis on the philosophical foundations (Aristotle and Marx) and scope (guiding social policy) of the approach and in its readiness to take a stand on what the central capabilities are (through providing a list). In the following, I will concentrate solely on those aspects of Nussbaum's own proposal that are in one respect or another relevant for our problem in hand and leave non-related issues undiscussed.

4.2.4.1 Commitment About Substance

Although Nussbaum fully endorses Sen's claim about the capability space and the arguments he has used to support them, his approach, she argues, has the drawback of being *underspecified*. The mere idea of capabilities as a space within which comparisons are made and inequalities assessed is insufficient. It gives

---

persons can be active and participating members of society if only society adjusts its background conditions to include them – by providing well-maintained streets, buildings with ramps, busses with wheelchair access, audio guidance at crossroads and so on. M.C. Nussbaum, *Women and Human Development*, 2000, pp. 65–70; *op. cit.*, 2001, p. 36; *op. cit.*, 2002, pp. 48–49, fn. 41; *Id.*, 'Capabilities as Fundamental Entitlements', 2003, pp. 56–57, endn. 4; *op. cit.*, 2004, pp. 431–434.

[89] *Ibid.*, 2000, pp. 65–70; *Ibid.*, 2002, pp. 48–49; *Ibid.*, 2003, p. 51; *op. cit.*, 2004, pp. 432, 440–441, 446–448.

[90] *Ibid.*, 2000, pp. 12, 54–56.

[91] *Ibid.*, 2000, pp. 12, 74.

us a general sense of what societies ought to be striving to achieve, but without commitment about substance, this guidance remains but a vague outline. To get a view of social justice that will have the requisite critical force and definiteness to *guide* and *direct* social policy we need to go beyond the merely comparative use of the capability space and articulate a *substantive account* of central capabilities that can provide a basis for constitutional principles that citizens have a right to demand from their governments.[92] This commitment entails two things.

Firstly, we need to take a stand, for political purposes, on *which* capabilities are important in our conceptions of justice. For the case for equality of capability depends both on *what* capability we are considering and on how we *describe* it. Thus, equality of capability is important when we consider the right to vote, the freedom of speech and so on. However, when we consider the capability to play tennis, it seems ludicrous to suppose that society should be very much concerned about it.[93] Furthermore, much depends on how we define the relevant capability. With something like health, there is considerable difference between defining the capability as 'having access to the social bases of health' or as 'the ability to be healthy'. The former seems something that a just society should distribute on a basis of equality; the latter contains an element of chance that no just society could, or should, altogether fully control.

Secondly, the substantive account entails the idea of a *threshold* or minimum level of capability that a just society should deliver as a fundamental entitlement of all its citizens.[94] According to Sen's view, nations are compared in areas such as health and educational attainment, but concerning *what level* of health service, or what level of educational provision society should guarantee, his view is suggestive, but basically silent. Sen never says to what extent equality of capability ought to be a social goal, or how it ought to be combined with other political values in the pursuit of social justice.

In dealing with these matters, it is Nussbaum's aim to develop a normative philosophical theory which is abstract and systematising, as well as responsive to empirical facts, i.e. more useful for guiding and bringing about social justice. In this line of reasoning, Nussbaum endorses a specific list of the central human capabilities as a focus both for comparative quality-of-life measurement and for the formulation of basic political principles of the sort that can play a role in fundamental constitutional guarantees.

4.2.4.2 A List of the Central Human Capabilities

EMBODIED HUMAN DIGNITY. The basic idea behind Nussbaum's approach is a conception of human dignity and of a life that is worthy of that dignity: a life that has

---

92 *Ibid.*, 2000, pp. 5–12, 63; *Id.*, 'Capabilities as Fundamental Entitlements', 2003, pp. 34–36, 40–42, 56.

93 Cf. Section 2.3.2 *Needs and Preferences in Health Care*, esp. the arguments considering the objective moral importance of course-of-life needs.

94 *Id.*, *Women and Human Development*, 2000, pp. 6, 70–77.

'truly human functioning' available in it. This idea combines a twofold intuition. Firstly, there is the idea that certain *functions* are particularly central in human life, and secondly, there is the idea that there is something about performing these functions in a *truly human* way, not in a merely animal way. The basic intuition then, from which the capability approach begins, then, is that certain central human abilities exert the moral claim that they should be developed.[95]

This conception, Nussbaum argues, is more Aristotelian and Marxian than Kantian. The idea of human dignity is twofold. It sees a person as both capable and needy; as both having activity, goals and projects – going beyond the mechanical workings of nature – and yet being in need of support for the fulfilment of many central projects.[96] From the *needs viewpoint*, this means that this conception takes seriously the materiality or basic dependency of human beings – their need for food, shelter, friendship, care and so on. The major powers of a human being need material support and cannot be what they are without it. From the viewpoint of the person as being *capable*, the conception does not refer to lower boundary questions – for instance, the question whether we may judge that when the absence of capability for a central function is so acute (as in cases of very severe forms of mental disability, or senile dementia) that the person cannot or no longer be called a truly human being. All in all, Nussbaum is mostly interested in the *higher threshold*, the level at which a person's capability becomes *truly human*. What this approach is after is a society in which persons are treated as each worthy of regard, and in which each has been put in a position to live really humanly.[97]

---

[95] Analyses of the essential characteristics of human nature and human functioning, which serve as the basis of the capabilities can be found in: M.C. Nussbaum, 'Nature, Function, and Capability: Aristotle on Political Distribution', in *Oxford Studies in Ancient Philosophy*, Suppl. Vol. I (1988), pp. 145–184; *Id.*, 'Aristotelian Social Democracy', in R.B. Douglas; G.R. Mara; H.S. Richardson (eds.), *Liberalism and the Good*, New York: Routledge, 1990, pp. 203–252; *Id.*, 'Human Functioning and Social Justice. In Defense of Aristotelian Essentialism', in *Political Theory* 20(1992)2, pp. 202–246; *Id.*, 'Non-Relative Virtues: an Aristotelian Approach', in M.C. Nussbaum; A. Sen (ed.), *The Quality of Life*, Oxford: Oxford University Press, 1993, pp. 242–269; *Id.*, 'Aristotle on Human Nature and the Foundations of Ethics', in J.E.J. Altham; R. Harrison (eds.), *World, Mind, and Ethics. Essays on the Ethical Philosophy of Bernard Williams*, Cambridge: Cambridge University Press, 1995, pp. 86–131. See also: M.C. Nussbaum, *Women and Human Development*, 2000, pp. 71–73; *op. cit.*, 2002, p. 58; *op. cit.*, 2004, pp. 448–452.

[96] For elaboration of this idea, see: M.C. Nussbaum, 'Victims and Agents', in *The Boston Review* 23(1998), pp. 21–24; and *id.*, 'Political Animals: Luck, Love, and Dignity', in *Metaphilosophy* 29(1998), pp. 273–287.

[97] However, she does make a few implicit remarks on the lower boundaries, i.e. in cases where we are inclined to say that beneath a certain level of capability, in each area, a person is not able to live in a *truly* human way: '... we see a human being as having worth as an end, a kind of awe-inspiring something that makes it horrible to see this person beaten down by the currents of chance – and wonderful at the same time, to witness the way in which chance has not completely eclipsed the humanity of the person. As Aristotle puts it, "the noble shines through". Such responses provide us with strong incentives for protecting that in persons that fills us with awe'. In: M.C. Nussbaum, *Women and Human Development*, 2000, p. 73.

UNIVERSAL RELEVANCE. With this idea as starting point, Nussbaum holds that we can arrive at an enumeration of central elements of truly human functioning that can command an overlapping consensus.[98] Consequently, she attempts to identify a list of ten capabilities as central requirements of a life with dignity. For these ten capabilities, she claims universal relevance, both as descriptive and as normative concepts. As descriptive concepts they represent certain basic aspirations to human flourishing that are universally recognisable. As normative concepts they serve as universal norms of human capability that should be central for political purposes in thinking about basic political principles that can provide the basis for a set of constitutional guarantees in all nations.

MINIMUM. Although the abstract list can be further specified by each society in question and thus be modified by context – which sometimes requires that in practical terms *priorities* may have to be set temporarily – the ten capabilities are part of a *minimum* account of social justice. They are to be understood as fundamental entitlements that it is *bad* for citizens to *lack*.[99] This because the capabilities are first held to be important for *each and every person*: each person is treated as an end and no one as a mere means to the ends of others; and second because they are understood as being both *mutually supportive* and *all* of central relevance to social justice. Thus, a society that *neglects* one of them to promote the others has morally harmed its citizens, and there is a failure of justice in this harm. In such cases, the citizens involved are not asked to carry just a big cost; on the contrary, they are forced to bear a burden that no citizen should have to bear. They are forced to carry a cost that is not just disadvantageous, but morally wrong. In short, the ten capabilities form a set of basic entitlements without which no society can lay claim to justice. Here is the most recent version of the list:[100]

The Central Human Capabilities

1. LIFE. Being able to live to the end of a human life of normal length; not dying prematurely or before one's life is so reduced as to be not worth living.
2. BODILY HEALTH. Being able to have good health, including reproductive health, to be adequately nourished and to have adequate shelter.
3. BODILY INTEGRITY. Being able to move freely from place to place, to be secure against violent assault, including sexual assault and domestic violence, and having opportunities for sexual satisfaction and for choice in matters of reproduction.
4. SENSES, IMAGINATION AND THOUGHT. Being able to use the senses, to imagine, think, and reason – and to do these things in a 'truly human' way, a way informed and cultivated by an adequate education, including, but by no means limited to, literacy and basic

---

[98] *Ibid.*, 2000, pp. 34–110; *Id.*, *op. cit.*, 2002, p. 58.

[99] M.C. Nussbaum, 'The Costs of Tragedy: Some Moral Limits of Cost-Benefit Analysis', in M.D. Adler; E.A. Posner (eds.), *Cost–Benefit Analysis: Legal, Economic, and Philosophical Perspectives*, Chicago: University of Chicago Press, 2001, pp. 169–200; *Id.*, *Women and Human Development*, 2000, esp. Chapter 3.

[100] The list has undergone modification over time and no doubt it will undergo further modification in the light of criticism. This is the version as it appeared in M.C. Nussbaum, 'Capabilities as Fundamental Entitlements: Sen and Social Justice', in *Feminist Economics* 9(2003)2–3, pp. 33–59, pp. 41–42.

mathematical and scientific training. Being able to use imagination and thought in connection with experiencing and producing works and events of one's own choice, religious, literary, musical and so forth. Being able to use one's mind in ways protected by guarantees of freedom of expression with respect to both political and artistic speech, and freedom of religious exercise. Being able to have pleasurable experiences and to avoid non-beneficial pain.

5. EMOTIONS. Being able to have attachments to things and people outside ourselves; to love those who love and care for us, to grieve at their absence, in general, to love, to grieve, to experience longing, gratitude and justified anger. Not having one's emotional development blighted by fear and anxiety. (Supporting this capability means supporting forms of human association that can be shown to be crucial in their development.)
6. PRACTICAL REASON. Being able to form a conception of the good and to engage in critical reflection about the planning of one's life. (This entails protection for the liberty of conscience and religious observance.)
7. AFFILIATION.
   A. Being able to live with and toward others, to recognise and show concern for other human beings, to engage in various forms of social interaction; to be able to imagine the situation of another. (Protecting this capability means protecting institutions that constitute and nourish such forms of affiliation, and also protecting the freedom of assembly and political speech.)
   B. Having the social bases of self-respect and non-humiliation; being able to be treated as a dignified being whose worth is equal to that of others. This entails provisions of non-discrimination on the basis of race, sex, sexual orientation, ethnicity, caste, religion, national origin [and disability].[101]
8. OTHER SPECIES. Being able to live with concern for and in relation to animals, plants and the world of nature.
9. PLAY. Being able to laugh, to play and to enjoy recreational activities.
10. CONTROL OVER ONE'S ENVIRONMENT.
    A. POLITICAL. Being able to participate effectively in political choices that govern one's life; having the right of political participation, protections of free speech and association.
    B. Material. Being able to hold property (both land and movable goods), and having property rights on an equal basis with others; having the right to seek employment on an equal basis with others; and having the freedom from unwarranted search and seizure. In work, being able to work as a human being, exercising practical reason and entering into meaningful relationships of mutual recognition with other workers.

4.2.4.3 Distinguishing Characteristics

SEPARATE COMPONENTS. Nussbaum emphasises that the list is a list of separate components. We cannot satisfy the need for one of them by giving a larger amount of another one. All are of central importance and all are distinct in quality. This irreducible plurality of the list limits trade-offs that could be reasonable to make, and thus limits the applicability of quantitative cost–benefit analysis.[102] If one does use cost–benefit analysis in connection with Nussbaum's approach, it will be crucial to represent in the weighing the fact that each of this plurality of distinct

---

[101] My addition between square brackets, YD. I will come back to this in Section 4.3 *Resuming Long-Term Care*.

[102] M.C. Nussbaum, 'The Costs of Tragedy: Some Moral Limits of Cost-Benefit Analysis', 2001, pp. 169–200; *Id.*, *Women and Human Development*, 2000, p. 81.

goods is of central importance. Thus, there is a tragic aspect to any choice in which citizens are left below the threshold of one of the central areas. This tragic aspect could be represented as a huge cost. However, representing it this way Nussbaum argues, cannot hide the fact that a distinctive good has been slighted.

Interrelated Goods. Secondly, the items on the list are related to one another in many complex ways. For instance, one of the most effective ways of promoting people's control over their environment, and their effective right of political participation, is to promote literacy. Reproductive health also is related in many complex ways to practical reason and bodily integrity. Previously, I have analysed the complex framework of social determinants – socio-economic status, educational status and so on – of health.[103] This complex interrelation gives us good reason to avoid promoting one of them at the expense of others.

Social Basis. Some of the items on the list include what John Rawls has called 'natural goods', goods in whose acquisition luck plays a substantial role. Thus, governments cannot hope to make all citizens healthy or emotionally balanced, since some of the determinants of those positive states are natural or luck-governed. Like Norman Daniels, Nussbaum also emphasises that just social policy can do a lot to influence the status of the natural goods (like promotion of health or intelligence). What public policy can aim to deliver is the social basis of these capabilities (by providing access to basic health care and good education for all). Nussbaum's capabilities approach equally insists that this requires doing a great deal to make up for differences in starting point that are caused by natural endowment or by power; however, it is still *the social basis of the good*, not the good itself, that society can reliably provide, and factors we cannot control may still interfere to keep some people from full capability. Therefore, when we use capabilities as a comparative measure of quality of life we must also always examine the *reasons* for the differences we observe. Some differences in health are due to factors public policy can *control*, and others are not. 'Basic political principles have done their job', Nussbaum argues, 'if they have provided people with the full [social] basis of these capabilities.'[104]

Practical Reason and Affiliation. Among the capabilities, practical reason and affiliation stand out as of special importance, since they both organise and cover all the others, making their pursuit truly human.[105] This is not to say that these are the two ends to which all the others can be reduced. Health, for example, is not to be interpreted, then, as a mere means to freedom of choice. What Nussbaum does hold, on the contrary, is that a government that makes available

[103] Cf. Section 3.3.2.2 *Which Health Inequalities Are Unjust?*

[104] M.C. Nussbaum, *Women and Human Development*, 2000, p. 81 (my addition between square brackets, YD). For further discussion, see: *Ibid.*, 2000, pp. 82, 86–96. I will come back to this in Section 4.4.6 *The Tragic and the Unjust*.

[105] *Ibid.*, 2000, pp. 82–83, 92–93. See also: *Id.*, 'Aristotle on Human Nature and the Foundations of Ethics', 1995, pp. 86–131; *Id.*, 'Aristotelian Social Democracy', 1990, pp. 203–252.

only a reduced and animal-like mode of an important item such as health has not done enough. All the items on the list should be made available in a form that involves reason and affiliation. That is why we generally find it so important that all persons, even the very ill or least capable of deciding autonomously, are allowed at least *some* say in the manner of their own treatment and support. In organising health care we try as much as possible to make room for the interests of the person, in the way he himself experiences or wants them, even when his rational decision powers are minor, or absent, or fading away. Only that way does social policy truly recognise his humanity.[106]

At first sight, the list of central human capabilities makes an essentialist and perfectionist impression. As a result, its political justification may well suffer. Being very aware of this Nussbaum urges emphatically that the primary role for the capabilities account remains that of providing political principles that can underlie all just national constitutions. Given the fact of pluralism, the list has to be understood in six additional ways.[107]

OPEN-ENDED. Firstly, the list has undergone modification over time and no doubt it will undergo further modification in the light of criticism. This is all the more to be expected because the list is explicitly to be considered as open-ended and subject to ongoing revision and rethinking. As such, the account of political justification behind the list lies close to the Rawlsian account of argument proceeding towards reflective equilibrium. It demands continued reflection about, testing against and adjustment to the most secure of our intuitions.[108]

ABSTRACT. Secondly, the items on the list ought to be defined in a somewhat abstract and general way, in order to leave room for the activities of specifying and deliberating by citizens and their legislatures. Within certain parameters it is appropriate that different nations should do this somewhat differently, taking their historical, social and other characteristic circumstances into account. As such, each of the capabilities may be concretely realised in a variety of different ways, in accordance with individual tastes, local circumstances and traditions.

FREESTANDING. Thirdly, to use the words of John Rawls, the list is considered to be a freestanding, partially moral conception. It is intended as the moral core of a conception that is explicitly introduced for political purposes only, without grounding in any particular metaphysical view of the world, in any particular comprehensive ethical or religious view. It is the object of a political overlapping consensus. People who otherwise have very different conceptions of the ultimate

[106] See also: Y. Denier, 'Autonomie en afhankelijkheid. Het subject in de medische ethiek', in R. Devos, A. Braeckman; B. Verdonck (eds.), *Terugkeer van het subject? Recente ontwikkelingen binnen de filosofie*, Leuven: Universitaire Pers, 2002, pp. 105–116.

[107] M.C. Nussbaum, *Women and Human Development*, 2000, pp. 105–106; *Id.*, 'Capabilities as Fundamental Entitlements', 2003, pp. 42–43.

[108] *Ibid.*, 2000, pp. 77, 101–103.

meaning and purpose of the good life can endorse it. They can connect it to their comprehensive views of the good in their way.[109]

In this line of reasoning, the importance of the idea of the threshold becomes clear.[110] The list gives us the basis for determining a decent social minimum in a variety of areas on which we can all agree. Nussbaum argues that the structure of political and social institutions should be chosen with a view to promoting at least a threshold level of the human capabilities. However, this proposal is intended to be compatible with several different accounts of distribution above the threshold. Again, it is consequently a partial, rather than a comprehensive theory of just distribution.

Capability as Goal. Fourthly, the appropriate political target is capability and not actual functioning.[111] This again ought to protect pluralism, for it leaves citizens the choice whether to pursue the relevant function or not. Many people, who are willing to support a given capability as a fundamental entitlement, would feel violated if the associated functioning were made basic. When we think of health for example, we should distinguish between the capability or opportunity to be healthy and actual healthy functioning: a society might make the first available and also give individuals the freedom not to choose the relevant functioning.[112]

Legitimate exceptions are children and adults who do not or no longer have full mental and moral powers.[113] Since exercising a function in childhood is frequently necessary to produce a mature adult capability, it seems perfectly legitimate in these cases to require proper education, given the role this plays in all the later choices of an adult life. Similarly, it seems legitimate to insist on the health,

---

[109] *Ibid.*, 2000, pp. 5, 74–77. Nussbaum explicitly urges that her capability approach is to be understood as 'a substantive account of central political goods', (*Ibid.*, p. 8) which, as such, is not to be equalised with a metaphysical theory or a Platonist account of the human good. She defines her own neo-Aristotelian proposal as follows: 'As I interpret Aristotle, he understood the core of his account of human functioning to be a freestanding moral conception, not one that is deduced from natural teleology or any non-moral source. Whether or not I am correct about Aristotle, however, my own neo-Aristotelian proposal is intended in that spirit – and also (clearly unlike Aristotle's) as a partial, not a comprehensive conception of the good life, a moral conception selected for political purposes only'. In: *Ibid.*, 2000, pp. 76–77. See also: M.C. Nussbaum, 'Nature, Function, and Capability: Aristotle on Political Distribution', 1988, pp. 145–184; *Id.*, 'Aristotelian Social Democracy', 1990, pp. 203–252; *Id.*, 'Human Functioning and Social Justice. In Defense of Aristotelian Essentialism', 1992, pp. 202–246; *Id.*, 'Non-Relative Virtues: an Aristotelian Approach', 1993, pp. 242–269; *Id.*, 'Aristotle on Human Nature and the Foundations of Ethics', 1995, pp. 86–131. Note also that there is discussion about her interpretation of Aristotle. See for instance: R. Mulgan, 'Was Aristotle an "Aristotelian Social Democrat"?' in *Ethics* 111(2000), pp. 79–101. I cannot go into this discussion here. For now, it is important to bear in mind that we are dealing here with Nussbaum's neo-Aristotelian proposal.

[110] *Ibid.*, 2000, pp. 70–77.

[111] *Ibid.*, 2000, pp. 86–96.

[112] I will come back to this issue in detail in Section 5.5.3 *On Personal Responsibility and the Right to Health Care*.

[113] *Ibid.*, 2000, pp. 89–90.

emotional well-being, bodily integrity and dignity of children in a way that does not take their fully individual choices into account. Furthermore, we will often be justified in restricting the scope of choice for mentally disabled adults, promoting actual functioning (e.g. in the areas of health, shelter and bodily integrity) rather than simple capability. The argument for this is the simple expression of respect for their worth and dignity.

Another important exception is provided by cases in which we may feel – even where normal adults are concerned – that some of the capabilities are so important, so crucial to the development or maintenance of all the others, that we are sometimes justified in promoting functioning rather than capability, however, this should happen within *limits* set by an appropriate concern for *liberty*.[114] This can be done either by direct effective policy – as is mostly the case in matters of *public health* – or by providing incentives to avoid hazardous health behaviour – which is the case in matters where *personal responsibility* becomes relevant.[115]

Most nations treat *public health and safety measures* as not to be left altogether to people's individual choices. Building codes, control of infectious diseases, vaccinations, sanitation regulations, regulation of food, medicine and environmental issues, all these restrict individual liberty in a certain sense. However, they are understood to be justified because of the difficulty of making informed choices in all these areas, and the burden of inquiry such choices would impose on citizens, as well as by the thought that health and safety are simply too basic to be left to people's choices. We may also feel that health is a human good that has value in itself, independent of choice, and that it is not unreasonable for government to take a stand on its importance in a way that to some extent (though not totally) bypasses individual choice.[116]

What should we say when adults knowingly and consistently risk signing away a major capability in a significant way, for instance through hazardous health behaviour? Frequently, though certainly not always, we will judge that interference is justified to protect the capability. Laws against drug use typically reflect a judgement that drugs impair capabilities in a long term and frequently irreversible way. Seat belt and helmet laws, although more controversial, reflect a widespread view that it is appropriate to protect people's long-term capabilities against the consequences of momentary carelessness. In other areas – like alcohol and tobacco use – governments offer disincentives to health damage, rather than preventing choice outright. All these issues are controversial because they do raise concerns about paternalism. This concern, however, can be weakened by the argument that health and bodily integrity are so important to all the other

---

114 *Ibid.*, 2000, pp. 89–96.

115 Again, cf. Section 5.5.3 *On Personal Responsibility and the Right to Health Care* for more details on this matter.

116 See also: Y. Denier, 'Public Health, Well-Being and Reciprocity', in *Ethical Perspectives* 12(2005)1, pp. 41–66.

capabilities that they are legitimate areas of interference with choice to some extent and up to some point.[117]

Liberties and Practical Reason. Fifthly, the content of the capabilities list gives a central role to citizen's powers of choice and to the major political and civil liberties that protect pluralism. They have a central and non-negotiable place on the list. However, the importance attached to the basic liberties is permanently interwoven with the understanding that the major powers need material support and that they cannot be what they are without it.[118]

Justification and Implementation. Finally, Nussbaum insists on a strong separation between issues of justification and issues of implementation. The approach is designed to offer the philosophical grounding for constitutional principles. As such, the list can be justified as a good basis for political principles for all just societies. However, the implementation of such principles must be left to the internal politics of the nations, although international agencies are justified in using persuasion to promote such developments.[119]

4.2.5 Some Counterreactions

Nussbaum's universal framework to assess person's quality of life faces a few respectable criticisms that deserve a serious answer. Among the ones that should be borne firmly in mind are the arguments of democratic deliberation, freedom as all-purpose-good, pluralism and paternalism. Let us consider concisely the way in which the above characteristics serve to answer these counterreactions.

4.2.5.1 Democratic Deliberation

Firstly, what does Amartya Sen think of the questions raised by Nussbaum? There appears to be a tension in his writings at this point. On the one hand, he speaks as if certain specific capabilities are absolutely central and non-negotiable. His discussions of health, education, political and civil liberties seem to be fully in agreement with Nussbaum's view that these human capabilities should enjoy strong priority and should be made central by all just societies, as fundamental entitlements of each and every citizen.[120] On the other hand, Sen explicitly refuses

[117] Although there will rightly be much disagreement about where that point is in each area. Much of this debate does not fall within the scope of basic political principles, and should be left to the democratic processes of each nation. See also: Y. Denier, 'On Personal Responsibility and the Moral Right to Health Care', in *Cambridge Quarterly of Health Care Ethics* 14(2005)2, pp. 224–234.

[118] M.C. Nussbaum, *Women and Human Development*, 2000, pp. 82–83, 92–93; *Id.*, 'Capabilities as Fundamental Entitlements', 2003, p. 43.

[119] *Ibid.*, 2000, pp. 101–105; *Ibid.*, 2003, p. 43.

[120] Some writings that concentrate explicitly on health are: A. Sen, 'Objectivity and Position: Assessment of Health and Well-Being', in L. Chen et al. (eds.), *Health and Social Change in International Perspective*, Boston: Harvard University Press, pp. 115–128; and A. Sen, 'Health in Development', in *Bulletin of the World Health Organization* 77(1999)8, pp. 619–623; *Id.*, 'Why Health Equity?' in *Health Economics* 11(2001), pp. 659–666. See also: S. Anand; F. Peter; A. Sen (eds.), *Public Health, Ethics, and Equity*, Oxford: Oxford University Press, 2004.

to endorse any full account of the central capabilities. The examples mentioned above clearly are examples of what he thinks is very important, but it is not clear to what extent he is prepared to *recommend them* as universally important goals connected with the idea of social justice itself. It is equally unclear, Nussbaum urges, whether there are *other capabilities* not mentioned so frequently that might be equally important, and if so, *what* those capabilities might be. The reason for this appears to be his respect for democratic deliberation. People should be allowed to settle these matters for themselves. Although Nussbaum makes a clear distinction between justification and implementation, Sen suggests that democracy is inhibited by the endorsement of a set of central entitlements in international political debate.[121] It is most likely that his refusal to make commitments about substance is based on the central importance he attaches to the idea of freedom as a general good.

4.2.5.2 Freedom

Throughout Sen's work 'the perspective of freedom', suggesting that freedom is a general all-purpose social good, and that capabilities are to be seen as instances of this more general good of human freedom, is very much present. However, Nussbaum urges, this approach cannot offer valuable normative guidance because the perspective of freedom is much too vague. Freedom, she argues, can have both good and bad dimensions, and not all freedoms are of equal worth.

Firstly, it is unclear whether the idea of promoting freedom is a coherent political project, for some freedoms limit others. Any particular idea of freedom involves the idea of constraint, for a person is only free to do a certain action if other people are constrained from interfering with it. We cannot coherently frame the notion of increase or decrease in freedom without specification of *whose* freedom, and freedom to do *what*.[122] No society that pursues equality or even an ample social minimum can avoid curtailing freedom in many ways – for instance, through taxation of some form.

Secondly, not all freedoms are of equal worth. The above examples show that any political project that is going to protect the equal worth of certain basic liberties for everyone, and to improve the living conditions of the least advantaged, needs to say that some freedoms are central for political purposes, and some are distinctly not. Some freedoms involve basic entitlements and others do not. Among the ones that are not part of a core group of entitlements required by

---

[121] M.C. Nussbaum, *op. cit.*, 2003, pp. 44–47.

[122] Nussbaum refers to the triadic definition of freedom as used by Rawls: 'Liberty can always be explained by a reference to three items: the agents who are free, the restrictions or limitations, which they are free from, and what it is that they are free to do or not to do'. In: J. Rawls, *A Theory of Justice*, 1971, p. 202; 1999, p. 177. This reflects the argument of the triadic structure of political freedom, an argument that was first developed by Gerald MacCallum, in his article 'Negative and Positive Freedom', in *Philosophical Review* 76(1967), pp. 312–334, as a reaction against the famous dual conception of Isaiah Berlin. Cf. Section 3.1.5.4 *Real Equality of Opportunity*.

the notion of social justice, some are simply less important (e.g. the freedom of motorcyclists to drive without helmets), but others may be positively bad (e.g. the freedom to non-consensual intercourse within marriage (marital rape)) and may in many ways subvert the core entitlements.

In other words, Nussbaum holds that all societies that pursue a reasonably just political conception have to *evaluate* human freedoms, saying that some are central and some trivial, some good and some actively bad. This is a qualitative difference that affects the way we will assess abridgements of freedom. Certain freedoms are taken to be *entitlements* of citizens based upon justice. When any one of these is abridged, it is an especially grave failure of the political system. The abridgement is then not just a huge cost to be borne, but also a cost of a distinctive kind, involving a violation of basic justice. When, on the other hand, some freedoms *outside the core* are abridged, it may be a small cost – or a large cost to some persons – but it is a cost of a different kind. It is not a cost that from the viewpoint of justice no citizen should be asked to bear.[123]

Therefore, the reason why it would not be better to leave the list-making up to the national democratic process is simple: some human matters are too important to be left to whim and caprice, or to the dictates of a cultural or national tradition.[124] The list represents a set of universal norms, benchmarks that we believe to be justified by good philosophical argument. What is on the list has intrinsic importance and should not be left to sole national discussion. For instance, just because a country does not recognise an equal right to education or to adequate health care for all does not make that country's policy morally justified.[125] A very important part of public discussion, Nussbaum rightly urges, is radical moral statements and the arguments supporting them. Such statements may be justified long before they are widely accepted. If we want to take the issue of social justice seriously, and to use a norm of justice to critically assess the various nations of the world and their practices, then the endorsement of some *content* of a conception of justice is necessary.

Nevertheless, it is important to bear in mind that Nussbaum's approach is designed to offer the *philosophical grounding* for constitutional principles. As

---

[123] M.C. Nussbaum, 'The Costs of Tragedy: Some Moral Limits of Cost-Benefit Analysis', 2001. See also: *Id.*, 'Capabilities as Fundamental Entitlements', 2003, pp. 44–47.

[124] *Ibid.*, 2003, pp. 47–48.

[125] Cf. 'The fact that many third world countries do not have a similar system [i.e. a well-organised health care system] that meets the health-care needs of their citizens does not mean that these persons do not *have* these needs. Having the need does not depend on the existence of the system. Being in good health is of fundamental value for every person and eliminating or reducing barriers that undermine this value – such as disease, illness or injury – is a positive obligation for every just society.' In: Y. Denier; T. Meulenbergs, 'Health Care Needs and Distributive Justice. Philosophical Remarks on the Organisation of Health Care Systems', in R.K. Lie; P.T. Schotsmans, *Healthy Thoughts. European Perspectives on Health Care Ethics*, Leuven: Peeters, 2002, pp. 265–297, p. 272. Cf. Section 2.4.4.2 *Reassessing Scarcity in Health Care*.

such, the list can be justified as a good basis for political principles for all just societies. Further implementation and precise specification of such principles must be left to the internal politics of the nations, although international agencies can be justified in using persuasion to promote such developments.

4.2.5.3 Concerns of Pluralism

Furthermore, one may be reluctant to endorse a list because of concerns about pluralism. We need to ask whether a universal framework is sufficiently flexible to enable us to do justice to the human variety. In this respect, Nussbaum makes three points that pertain to the norm of respect for pluralism.[126] Firstly, pluralism and respect for difference are themselves universal values that are not everywhere observed. They require a normative articulation and defence.

Secondly, the most meaningful kind of pluralism entails that all persons are worthy of concern and respect, deserving all necessary support for their equal opportunity to lead a life of one's own and to make their own choices regarding it. Real respect for that goal requires a non-negotiable commitment to some universal principles as fundamental entitlements – like guaranteeing an equal right to the basic rights and liberties, including freedom of thought and association, freedom of speech, and so on. The same goes for primary goods such as basic health care and education for all. These entitlements set the stage on which valuably different forms of human activity can flourish.

Thirdly, against accusations of perfectionism Nussbaum repeatedly urges that her list has to be understood within the spirit of political liberalism.[127] Respect for pluralism is fostered both by making capability and not functioning the appropriate political goal and also by endorsing a relatively small list of core capabilities for political purposes. To begin with, this means that to ask people to endorse the capabilities list does not require them to endorse the associated functioning as a good in their own lives. Moreover, as with Rawls's list of primary goods, so with the list of central capabilities: it is not meant to be an exhaustive account of what is worthwhile in life; on the contrary, it leaves plenty of room to value other things in mapping out one's life plan. The just society is required to endorse the list of capabilities for political purposes and as applicable to all citizens. Then everyone may get on with the lives they prefer. Genuine respect for persons involves respect for choice. It is in this sense that the list is a partial, and not a comprehensive conception of the good.

---

126 M.C. Nussbaum, *Women and Human Development*, 2000, pp. 31–33, 50–51, 59, 40, 88, 95–96; *Id.*, 'Capabilities as Fundamental Entitlements', 2003, pp. 48–49.

127 *Ibid.*, 2000, pp. 5, 8, 59–60, 74–77, 101–105. For further discussion, see: R. Arneson, 'Perfectionism and Politics', in *Ethics* 111(2000), pp. 37–63; and M.C. Nussbaum, 'Aristotle, Politics, and Human Capabilities: a Response to Antony, Arneson, Charlesworth, and Mulgan', in *Ethics* 111(2000), pp. 102–140.

4.2.5.4 Paternalism

Finally, the argument of paternalism says that when we use a set of universal norms as benchmarks telling people what is good for them, we show a lack of respect for people's freedom as agents. People are themselves the best judges of what is good for them and if we prevent people from acting on their own choices we treat them as children. To this critique Nussbaum gives a threefold answer.[128]

Firstly, a commitment to respecting people's choices hardly seems incompatible with the endorsement of universal values. Indeed, it appears to endorse explicitly at least one universal value: the value of having the opportunity to think and choose for oneself. Consider the following: many traditional value systems are highly paternalistic, treating certain groups of people as unequal. When we encounter a system like this, it is in one sense paternalistic to say that it is unacceptable under the universal norms of equality and liberty. For to say, that is to tell people how to conduct their lives with one another, in a way that may run counter to their actual preferences. If paternalism means telling people that they cannot behave in some way that they want to behave, then, any system of law and any bill of rights is paternalistic with respect to practices that treat people with insufficient or unequal respect. It is clear that this is hardly a good argument against the rule of law, or, more generally, against opposing the attempts of some people to tyrannise over others. We dislike paternalism because we like each person's liberty of choice in fundamental matters. Therefore, it is fully consistent to reject some forms of paternalism while supporting those legitimate forms that underwrite these universal values, on an equal basis.

Next, we should note that the various liberties of choice have material preconditions. Liberty is not just a matter of having rights on paper, it requires being in a position to *exercise* those rights. And this requires material and institutional resources, including legal and social acceptance of claims. A state that is going to *guarantee* people rights *effectively*, is going to have to take a stand about more than the importance of these basic rights themselves. It will have to take a stand on the distribution of resources to guarantee citizens what John Rawls has called the 'fair value' of the various liberties – for example, by raising revenue through taxation in sufficient quantity to make education and basic health care available to all. Such redistributive measures are paternalistic, meaning interference with activities that some people choose. The question is, 'is it acceptable or not?' The argument of paternalism indicates that we should prefer a universal normative account that allows people the liberty to pursue their own conceptions of value, within limits set by the protection of the equal worth of the liberties of others. It gives us no good reason to reject all universal accounts, and some strong reasons to construct one, including in our account not only the liberties themselves, but also forms of economic empowerment that are crucial to making liberties *truly* available.

And finally there is the *principle of each person as end.* If we agree that citizens are worthy of concern and respect, and grant that they live separate lives, we

---

[128] *Ibid.*, 2000, pp. 51–60.

ought to conclude that politics should treat each of them as ends, as sources of agency and worth in their own right, with their own plans to make and their own lives to live, therefore as deserving of all necessary support for their equal opportunity to be such agents. To do this implies that we have to take a stand on some values that will be made central for political purposes, and *against* some ways of treating persons disrespectfully. However, taking a stand in this way should not raise the charge of illegitimate paternalism, since we do so in order to treat *each* person as an end and permit *all* citizens to search for the good in their own ways. In that sense, the list proposes universals that are facilitative, rather than tyrannical; that create space for choice rather than obliging people to a desired total way of functioning.

Understood at its best, the paternalism argument is not an argument against universals. For it is all about respect for the dignity of persons as choosers. This respect requires us to universally defend a wide range of liberties, plus their material conditions; and it requires us to respect persons as separate ends. If we are really to show respect for all citizens in a pluralistic society, universal values are not just acceptable but badly needed.

4.3 RESUMING LONG-TERM CARE

Having set forth the major strains of Nussbaum's theory, we now have more tools to integrate the full and complex framework of health care (including issues of prevention, cure *and* care) into an adequate theory of justice together with a clearer view on how to do it. Before going into the issue of full integration of health care and justice, let us first return to the problem of long-term care, which is, lest it be forgotten, an important aspect of the integration question.

It is Nussbaum's firm belief that the issue of care for persons who are dependent on others (e.g. children, the disabled and the elderly) helps us to see both why the capabilities approach is superior to other approaches to social justice within the liberal tradition (like Rawls's), and why a list of definite entitlements is required if the approach is to deliver an adequate conception of justice (vs Sen). Firstly, care for children, the elderly and the mentally and physically disabled is a major part of the work that needs to be done in any society and which can be a source of great injustice. Any adequate theory of justice, Nussbaum urges, needs to think about the problem *from the beginning*, in the design of the most basic levels of institutions, and particularly in its theory of primary goods.[129] But how exactly does she solve the problem? What can be done to give the problem of care and dependency sufficient prominence

---

[129] In this, she follows Eva Kittay who writes: 'Dependency must be faced from the beginning of any project in egalitarian theory that hopes to include all persons within its scope', in E. Feder Kittay, *Love's Labor: Essays on Women, Equality, and Dependency*, New York: Routledge, p. 77. See also: M.C. Nussbaum, 'Future of Feminist Liberalism', 2000, pp. 51–52; *Id.*, *op. cit.*, 2002, p. 48; *Id.*, 'Capabilities as Fundamental Entitlements', 2003, p. 53; *op. cit.*, 2004, p. 440.

in a theory of justice? In a second move, she proposes four changes in the Rawlsian list of primary goods.[130] Having the major strains of Nussbaum's theory in mind, we can discuss these changes concisely.

4.3.1 Add Care to the List

Firstly, Nussbaum proposes to follow Eva Kittay's suggestion of adding the need for care during periods of extreme and asymmetrical dependency to the Rawlsian list of primary goods, thinking of care as among the basic needs of citizens whose fulfilment profoundly affects people's life prospects and chances.[131]

4.3.2 Redesign to a List of Capabilities

However – and Nussbaum is quick in asserting this – such a suggestion would require making another modification, for care is not a commodity like income and wealth, to be measured by the sheer amount of it citizens possess.[132] Even though the Rawlsian list is already quite heterogeneous – including thing-like items such as income and wealth along with capability-like items as liberties, opportunities, the moral powers and the social basis of self-respect – we would do well, Nussbaum proposes, in the line of Sen to understand the *entire* list of primary goods as a list not of things, but of central capabilities. This change would not only enable us to better deal with people's needs for various types of *care* as elements of the list, but would also answer the point that Sen has repeatedly made about the unreliability of income and wealth as indices of well-being. The well-being of citizens will now be measured not by the amount of income and wealth they have, but by the degree to which they have the various capabilities on the list. Looking at capabilities rather than at resources gives us a much more accurate sense of what people are actually able to do and be.

4.3.3 To a Richer Account of Primary Goods

If we accept the two changes suggested above, we should also add a third, relevant to the earlier thoughts analysed about dependency, both physically and mentally, of the baby at the breast, of the growing child, of the mentally disabled children and adults, of the senile demented elderly and of the physically dependent persons who are perfectly sane and painfully aware of their unequal and in some cases increasing limitations. We then would broaden the list of primary goods and add other pertinent capability-like things to it, for example, the social basis of health, affiliation, adequate working conditions, bodily integrity and the social basis that stimulates imagination and emotional well-being. Such

---

[130] M.C. Nussbaum, *ibid.*, 2000, pp. 54–55; *Id.*, *op. cit.*, 2001, pp. 36–37; *Id.*, *op. cit.*, 2002, pp. 51–65; *Ibid.*, 2003, pp. 50–56; *Ibid.*, 2004, pp. 446–452. For the general framework of these changes, see: M.C. Nussbaum, *ibid.*, 2000, pp. 34–110.

[131] E. Feder Kittay, *op. cit.*, 1999, pp. 102–103.

[132] This is a very important move, for otherwise, Nussbaum would meet the same problems as Ronald Green. Cf. Section 3.3.1 *Health Care: an Additional Primary Social Good?*

broadening of the list enables us to hold and foster a *rich concept of reciprocity* between people. With this concept, the narrow language of economic efficiency must give way to a more complex, more reflective language of full human development *for all.*

4.3.4 Redesign the Political Concept of the Person

Finally, we would have to shift to another concept of the person. The Rawlsian account of primary goods is introduced in connection with the Kantian political conception of the person, as an account of what citizens characterised by the two moral powers need. Thus, we attribute basic importance to care only from the point of view of our own current independence. It is good to be cared for only because care serves our moral personality, understood in a Kantian way as conceptually quite distinct from need and animality. The idea is that because we are dignified beings capable of political reciprocity, we had better provide for times when we are not, so that we can regain that status as quickly as possible. This leads us in the direction of a contemptuous attitude towards dependency, elderly disability and congenital disability. Somehow or other, care is supposed to be valuable only with reference to the 'fully cooperating' condition. As such, Nussbaum argues, the full humanity of people so affected is being denied.

Therefore, Nussbaum urges that we need to delve deeper, and redesign the political conception of the person, thus bringing the rational and the animal into a more intimate relation with one another, and acknowledging that there are many types of dignity in the world, including the dignity of the long-term disabled, and of the senile demented elderly. We want a political conception of the person that does justice to temporality and need. Nussbaum wants to bring in a kind of reciprocity in which we humanly engage, being confronted with variations in need, with periods of symmetry and independence as well as with extreme necessity and irreversible dependency. When we think of designing just institutions, this kind of *rich and complex reciprocity* has to be brought in *from the start.*

Nussbaum proposes an Aristotelian and Marxian inspired political conception of the person; one that sees 'the person from the start as both capable and needy', to be shaped both by practical reason and affiliation, in 'need of a rich plurality of life-activities ... whose availability will be the measure of well-being'. This conception, which builds in 'both growth and decline' into the characteristics of human life, will put us better on the road to thinking well about what society should design.[133] The basic idea is that we do not have to contract for what we need by producing; we have a claim to support in the dignity of our human need itself. Since this is an idea that corresponds well with human experience, there is good reason to think that it can command a political consensus in

---

[133] M.C. Nussbaum, *op. cit.*, 2002, p. 58; also in: *op. cit.*, 2004, pp. 448–452.

a pluralistic society. If we begin with this conception of the person and with a suitable list of the central capabilities as primary goods, we can begin designing institutions by asking what it would take to get citizens up to an acceptable level on all these capabilities.

All things considered – the idea of central capabilities as the richer and more complex analogue of Rawlsian primary goods, the conception of the human being as a being in need of a rich plurality of life activities, to be shaped both by practical reason and affiliation, all within the sphere of liberal political theory, which aims at securing capability only, thus giving citizens a great deal of space to pursue their own comprehensive conceptions of the valuable and the good, but which is at the same time more attentive to the variations in need that characterises all citizens, and to the various material and institutional conditions of being able to lead a truly human life – we can conclude that Nussbaum's proposal is better adjusted to integrate *care* within an adequate theory of justice. Care is to be subsumed under the idea that the public conception must design the material and institutional environment so that it provides the requisite affirmative support for *all* the relevant capabilities.

Thus, care – both in situations with reference to (re)gaining normal functioning as in situations of physical and mental dependency – will enter into the conception *at many points*, as part of what is required to secure to citizens *as much as possible* of the capabilities on the list. On closer inspection, one could reasonably link the issue of care for the dependent – whether temporary or irreversible – to the capabilities of life, bodily health, bodily integrity, imagination, emotional attachments, practical reason, affiliation and play – which are eight out of ten. Within the same line of reasoning, it was necessary to add the morally irrelevant character of *disability* to the capability of having the social bases of self-respect and of non-discrimination.[134]

4.4 ON THE CONTRIBUTION OF NUSSBAUM'S CAPABILITIES APPROACH

What are the merits and limitations of Nussbaum's capabilities approach in dealing with the problem of just health care? Does her analysis do justice to Rawlsian theory? And does her approach provide a useful and valuable answer to what we should understand as 'just health care', and to pressing problems in contemporary health care – like the issue of cost control, the role of personal responsibility and of high-tech medicine? Does she, all in all, indeed provide a better answer to such issues than Rawls and Daniels?

The starting point of my analysis of Nussbaum has been the problem of integrating long-term care into a general theory of justice; a problem that neither Rawls, nor Daniels could tackle in a satisfactory manner. After having considered in detail both Nussbaum's critique on the matter and her proposal to transform our dominant way of thinking about social justice, introducing a whole new

[134] Cf. Section 4.2.4.2 *A List of the Central Human Capabilities*.

approach, we have good reasons, I believe, to think that Nussbaum's capabilities approach is superior in providing an adequate framework of justice that is able to *integrate* the full complexity and heterogeneity of contemporary health care, which I have defined as essentially being threefold in character, including the triple structure: prevention, cure and care. There will, however, also remain some important questions, especially regarding the contemporary problems, which turn out to be only partially answered for the time being. Let us go through these qualities and questions.

4.4.1 Rich but Liberal

The primary differences between Nussbaum's capabilities approach and Rawlsian theory are the length and definiteness of the list of primary goods; the refusal to make thing-like items, such as income and wealth, goals in their own right, and in particular, Nussbaum's determination to place the social basis of several goods that Rawls has called 'natural goods', like 'health and vigor, intelligence and imagination' on the list.[135] Nussbaum's theory is much broader, more complex and is based on the rich continuum of varieties in need between persons. One is therefore, in philosophical literature and discussion, easily inclined to categorise Nussbaum's theory as 'thick', by analogy with Rawls's 'thin' theory of the good. Two reasons show why this is a somewhat misguiding designation, although it is important to note that Nussbaum uses this designation herself.[136] The long and the short of the issue is that Nussbaum's approach is positioned exactly between the classic feminist and communitarian critique on liberalism and the accusations of perfectionism.

4.4.1.1 Broad but Not Comprehensive

Firstly, although the list is considerably different both in structure and in substance from Rawls's list, it is offered in a similar political-liberal spirit: as a list that can be endorsed for *political* purposes, as the moral basis of central constitutional guarantees, by people who have otherwise very different views of what a complete good life for a human being would be.[137] The core of the political conception is endorsed for political purposes only, giving citizens *space* to pursue their own comprehensive conceptions of value. More room for reasonable pluralism in conceptions of the good is secured by insisting that the appropriate political goal is capability only. The list provides, like Rawls's list, an objective basis with which we can all agree. Nussbaum's list is just a richer account of the primary goods that a just society should distribute; it is broader and much more complex. However, if 'thick' is meant

---

[135] J. Rawls, *A Theory of Justice*, 1971, p. 62; 1999, pp. 54–55.

[136] See: M.C. Nussbaum, 'Aristotelian Social Democracy', 1990, p. 216; *Id.*, 'Human Functioning and Social Justice. In Defense of Aristotelian Essentialism', 1992, pp. 217.

[137] M.C. Nussbaum, *Women and Human Development*, 2000, pp. 5, 74–75.

to equal 'comprehensive' in the Rawlsian sense we can confidently hold that Nussbaum's capabilities approach is not a 'thick theory'. It is a partial, not a comprehensive, perfectionist, or metaphysical conception.[138]

4.4.1.2 Embodied Liberalism

In a second move, she opposes the classic feminist and communitarian critique: the liberal tradition is not inherently linked to a pernicious form of egoism or denial of communal and relational values.[139] Nussbaum stresses that, though broader and more complex, her approach is liberal in very important respects. For that reason, she does not assent to Eva Kittay's suggestion that we should depart from all forms of the liberal tradition and reconfigure political theory to put the fact of dependency at its heart. That we are all 'some mother's child', Kittay says, existing in intertwined relations of dependency, should be the guiding image for political thought.[140] Such a care-based theory will be likely to be very different from any liberal theory, since the liberal tradition is deeply committed to goals of independence and liberty. However, what is so wrong then with liberty and independence? Surely, nobody is ever self-sufficient; the independence we enjoy is always both temporary and partial and it is good to be reminded of that fact by a theory that also stresses the importance of care in times of dependency. However, I agree with Nussbaum that 'being some mother's child' is not a sufficient image for the citizen in the just society. For this, we need a lot more: liberty and opportunity, the chance to form a life plan, the chance to learn and imagine.

Nussbaum's theory lies squarely *within* the liberal tradition, although it involves a more social and interrelational conception of the person than do many types of liberalism, and although it insists more than many on the material prerequisites. Ideas of respect for human dignity are as central to the capabilities approach as they are for Kant and his modern descendents, although dignity is now conceived of in a subtly different and more 'embodied' way. Nussbaum defends a form of *embodied* political liberalism: a form of liberalism that is more attentive to need and its material and institutional conditions, and which therefore can integrate the element of *care* as one of the major things that need to be supported and secured in a just society.[141]

4.4.2 Complex Reciprocity

A second advantage is that this embodied liberalism contains a rich and complex conception of reciprocity. This is the second reason why Nussbaum's approach is

---

[138] *Ibid.*, 2000, pp. 66–70, 76–77; See also: *Id., op. cit.*, 2002; p. 59, fn. 67; *op. cit.*, 2004, p. 448. For further discussion, see: R. Arneson, 'Perfectionism and Politics', 2000, pp. 37–63; and M.C. Nussbaum, 'Aristotle, Politics, and Human Capabilities: a Response', 2000, pp. 102–140.

[139] M.C. Nussbaum, 'Feminist Critique of Liberalism', 1999, pp. 55–80; *Id.*, 'Future of Feminist Liberalism', 2000, pp. 47–79; *Id.*, 'Long-Term Care and Social Justice', 2002, pp. 60–63. Cf. Section 4.2.1 *Other Virtues?*

[140] E. Kittay, *Love's Labor*, Chapter1, Part III, on political strategies is titled 'Some Mother's child'.

[141] M.C. Nussbaum, 'Future of Feminist Liberalism', 2000, pp. 58–59; *op. cit.*, 2002, p. 62.

better adjusted to integrate the complex framework of health care. Three things deserve emphasis.[142]

4.4.2.1 Human Functioning and Independence

Firstly, it is important to stress that independence remains a central value. Although dependence is inherent to the human condition, we do not *aim* for it. On the contrary, our social institutions like health care are thus organised that we strive to back away from it; we try as much and as long as possible to reach and enlarge the period of strength and independency in a human life. When a human being is given a life that in some way lacks the powers of independent human functioning and expression, it gives us a strong sense of tragedy. Independence and activity, i.e. being able to lead a life of one's own, being one's own advocate, one's own author, one's own best representative, has very strong normative value in a way that dependency has not. Nevertheless, this may not lead us into myopia. It should not lead us into a practice of denigrating dependency by neglecting care as a social problem, devaluing the work of caregiving, and shortchanging the needs of the dependent.

4.4.2.2 Rich and Complex Reciprocity

To lead us away from myopia, Nussbaum's approach distances from the idea of sheer productive reciprocity and draws attention to the complexity of interpersonal relationships and the rich character of reciprocity. She emphatically opposes a narrow focus – concentrating on the state's role in producing adults who have all the capabilities on the list, focusing on independency, activity, literacy and other basic skills that are important for productive, technical and economic development, and on political skills understood in a narrow sense. Nussbaum argues that, in order to be doing what they should be doing, states must be concerned with *all* the capabilities on the list, even when these do not seem so useful for economic growth, or even for political functioning.

At this point, we need to question whether the idea of society as cooperation for mutual advantage is a complete account of social cooperation. The point of a cooperation that includes educating children with severe disabilities and supports their development with appropriate care cannot be seen in terms of mutual economic advantage. The *benefit* for society of including them, interacting with them, and fully supporting them is much more complex, versatile and diffuse. It includes, for instance, the advantage of respecting the dignity of the disabled and developing their human potential, whether or not this potential is socially 'useful' in the narrow, productive sense. It includes, as well, the advantage of understanding humanity and its diversity that comes from associating with mentally disabled people on terms of mutual respect and reciprocity. It includes new insight about the dignity of the aging and of ourselves as we age. It includes the value of the

---

142 *Id.*, *Women and Human Development*, 2000, pp. 83, 90; *op. cit.*, 2001, pp. 36–37; *op. cit.*, 2002, p. 55; *op. cit.*, 2004, pp. 441–448, 450–451.

interactions for people with disabilities themselves, who without special support would live, as they once did, isolated and stigmatised lives. Of course, this includes the value of justice in including the good of vulnerable to other people.

The case of the long-term disabled shows us to what extent the idea of society as cooperation for mutual advantage biases the whole idea of the benefits of social cooperation. If we are to deliberate well about these problems of justice, a rich and supple sense of what it is to be human is crucial, including complex forms of reciprocity and social interaction.[143] Justice, Nussbaum says, does not begin with the idea that we have something to gain from bargaining together. We have a claim to support based on justice in the idea of our human need itself.[144] Society is held together by a very wide range of attachments and concerns, only some of which relate to productivity and profitability. Productivity is necessary, and even good, but it is far from the main end of life. In some cases, 'the simple language of economic efficiency must give way to a more complex, more reflective language of full human development'.[145]

4.4.2.3 Social Inclusion

Thirdly, all this allows us to treat the asymmetrical needs for care of children, elderly and the mentally and physically disabled as part of their human dignity, rather than as heavy social costs to be borne. And this respect for the long-term asymmetrical needs contains reference to activity, to participation in social life, to as much independence as one's condition allows. People with physical

[143] To illustrate what she means by complex forms of reciprocity and social interaction, Nussbaum refers to Jamie Bérubé, who is born with Down syndrome and will therefore need extensive care throughout his life, but who is also a boy who loves B.B. King, Bob Marley and The Beatles, who can imitate a waiter bringing all his favourite foods, and who interacts in a loving, playful and generous way both with his family and with other children. Nussbaum also refers to Sesha. Because of congenital cerebral palsy and severe mental retardation, Sesha will never be able to walk, talk or read and will always be profoundly dependent on others. But she is also a young woman in her early thirties, who loves music, hugs those who care for her, dances with joy and shows appreciation for the care she is given. These forms of reciprocity have nothing to do with the two moral powers, with the capacity for forming a life plan and an overall conception of the good, or with being a 'self-authenticating source of valid claims' (*Political Liberalism*, 1996, p. 32). Nevertheless, they involve complex forms of reciprocity that form a challenge to our dominant theories of social justice. The story of Jamie is written down by his father in: Michael Bérubé, *Life as we Know it: a Father, a Family, and an Exceptional Child*, New York: Pantheon, 1996. The story of Sesha is written by her mother, Eva Feder Kittay, *Love's Labor: Essays on Women, Equality, and Dependency*, New York: Routledge, 1999. See also: M.C. Nussbaum, *op. cit*., 2001, pp. 34–37; *op. cit*., 2002, pp. 61–62; *op. cit*., 2004, pp. 418–420, 441–442.

[144] This idea is also supported by the writings of A. Buchanan, 'Justice as Reciprocity versus Subject-Centered Justice', in *Philosophy and Public Affairs* 19(1990), pp. 227–252; T.M. Scanlon, 'Preference and Urgency', 1975; *Id*., *What We Owe to Each Other*, 1998, pp. 177–187; H. Brighouse, 'Can Justice as Fairness Accommodate the Disabled?' in *Social Theory and Practice*, 27(2001)4, pp. 537–560. R.E. Goodin, *Protecting the Vulnerable: a Reanalysis of Our Social Responsibilities*, Chicago: University of Chicago Press, 1985.

[145] M.C. Nussbaum, *op. cit*., 2002, p. 34.

impairments can usually be highly active and productive members of society in the usual economic sense, performing a variety of jobs at a sufficiently high level. People in wheelchairs can get around quite well and do their work, so long as buildings have ramps, busses have wheelchair access and so on. The blind can work more or less anywhere in these days of varied audio-technology and tactile signage; the deaf, too, can take advantage of e-mail in place of the telephone, and of many other visual technologies, so long as workplaces and society structure themselves so as to *include* such persons. And for people with mental disabilities, we do not allow them to be isolated and locked away. On the contrary, we feel duty-bound to guarantee them their space in the world, as members of society alongside everyone else, for 'doesn't a full human life involve a kind of freedom and individuality, namely a space in which one can exchange love and enjoy music, light and sound, free from confinement and mockery?'[146]

4.4.3 All Human Beings

The above-analysed elements of embodied liberalism, rich and complex reciprocity, full social inclusion, departure from the Kantian split between the rational and the animal together illustrate a third important quality, namely that the concept of human dignity resides in the sole and simple fact of being human, of being 'child of two human parents'.[147] It is rooted in human nature and in the general conditions of human life that appear to be common to all – that we are mortal, of human born, in need of support, that we are capable of human functioning.

The tie of birth is an important argument in the question of integrating the full complexity of health care, i.e. in speaking about what justice requires with regard to long-term care. It is an argument that is also used by Thomas Scanlon.[148] Although it has sometimes been characterised as a prejudice called 'speciesism', it is not prejudice, he argues, to hold that our relation to the severely disabled

---

[146] *Ibid.*, 2002, p. 62.

[147] Note that this is a significant departure from Aristotle whose ideas on dignity are based on criteria (i.e. sex, status and merit) which lead to focus solely on the free male population of the Greek polis and their rights of citizenship, and to exclusion of the women and slaves from their political rights, because of their supposed natural inferiority. For slaves, the main discussion is in Aristotle, *Politics*, 1.4–7, 13, and in *Nichomachean Ethics*, 8.11.1161a32–b8. For his argument on women, see Aristotle, *Politics*, 1.12–13, 2.5.

Not surprisingly, Nussbaum has stressed the stupidity and unacceptability of Aristotle's arguments on women and slaves from the beginning of her work on his political thought. Cf. M.C. Nussbaum, 'Aristotle, Politics, and Human Capabilities: a Response', 2000, p. 108. See also: *Id.*, 'Nature, Function, and Capability', 1988, esp. pp. 164–172; *Id.*, 'Aristotelian Social Democracy', 1990, pp. 203–252; and R. Mulgan, 'Was Aristotle an "Aristotelian Social Democrat"?' in *Ethics* 111(2000), pp. 79–101. Her departure has lead Richard Arneson to characterise her approach as a liberal perfectionist egalitarian approach to politics in R.J. Arneson, 'Perfectionism and Politics', in *Ethics* 111(2000), pp. 37–63.

[148] T. Scanlon, *What We Owe to Each Other*, 1998, p. 185.

humans who will never develop even limited capacities of rationality gives us reason to accept the requirement that our actions should be justifiable. The tie of birth gives us very good reason to treat them as human, despite their limited capacities.

The list of central human capabilities, Nussbaum stresses, is the same for the mentally disabled as for the so-called normal citizen. It is important that this be so, although the same level of functioning may not always be possible, in order to stress that people with mental disabilities are full human beings and citizens. This integrating move reminds us of the element of tragedy that persists in many such lives. Sesha and Jamie are not non-human just because their capabilities are tragically out of step with those of most members of their species community who possess normal species-typical functioning. Severely disabled humans are crucially unlike happy animals, even though their abilities are less developed than some animal that flourishes in its own way.[149] Moreover, in many mentally disabled lives there is a disharmony that does not exist in the life of a flourishing animal: some abilities are developed, others are not. In these cases, life does not fully fit together, certainly not without special support. Using the capabilities list also for people with disabilities reminds us of the strong reasons we have to deal with obstacles that prevent their full functioning.[150]

4.4.4 Self-Respect

Fourthly, an important quality has to do with the social bases of self-respect. Both Rawls and Nussbaum stress that everyone should have their *self-respect* protected, so far as it is within the power of the social institutions to do so. Perhaps this is, beside the fact that they both represent a form of political liberalism, another point where both authors are most related, and where the tension between them tones down. Rawls called the social basis of self-respect 'the most important of the primary goods'.[151] For Nussbaum, it is one of the three issues that are uncontroversially important in thinking about a society's quality of life, and particularly in areas that are pertinent to care.[152]

The issue of self-respect is of crucial importance, Nussbaum argues, both on the side of the cared-for and on the side of the caregiver, *and*, I would add, on the side of the just society as a whole as well. On the side of the cared-for, it would be disrespectful to limit their use of public space on account of their disability only; to limit their capacity to participate in society on that account; to simply lock them away as was common practice in previous times and centuries. On the side of the caregiver, it is very important that the general atmosphere in society

[149] Cf. the issue of embodied human dignity in Section 4.2.4.2 *A List of the Central Human Capabilities*.
[150] M.C. Nussbaum, *op. cit.*, 2004, p. 452.
[151] J. Rawls, *A Theory of Justice*, 1971, pp. 396, 440; 1999, pp. 348, 386.
[152] Cf. Section 4.2.1 *Other Virtues?* See also: M.C. Nussbaum, *Women and Human Development*, 2000, pp. 66–67.

is one of respect for their work by providing the necessary aid and support for those persons with burdens of care at home as well as sufficiently appropriate working conditions for the professional caregivers. All in all, I believe it is also a matter of self-respect on the side of society itself. For the degree of moral refinement and justice of a particular society is reflected in the way it *truly* includes the least advantaged.[153]

4.4.5 Strong Connection with Human Experience and Sensibility

A fifth important quality is that Nussbaum's embodied liberalism is closely linked to general human experience and sensibility, in a way that Rawlsian theory is not. Firstly, the *list* of central human capabilities is more familiar than the Rawlsian list of primary goods, which can therefore make it easier to reach an overlapping consensus on what is of fundamental importance in every human life. Furthermore, we generally see that the social goal of promoting *human flourishing* and capabilities development is very much present not only in the impressive frameworks of support of the welfare states, but also as a prominent goal of development programmes for the Third World countries. Thirdly, the Aristotelian and Marxian inspired political conception of the person – that sees the person as both capable and needy, to be shaped both by practical reason and affiliation, in need of a rich plurality of life activities, knowing growth and decline – also corresponds largely to common human experience. Finally, the capabilities approach allows us to integrate care into the framework of justice as one of the primary goods that has a prominent influence on people's prospects and life chances. Previously, I have shown that Rawls did feel uneasy with the fact that he could not incorporate care from the beginning into his theory. It was an uneasiness that he, nor Daniels, were able to remove. It is the merit of Nussbaum's theory that she was able to do it.

4.4.6 The Tragic and the Unjust

Furthermore, I would like to underline an important distinction that, I believe, is generally and implicitly present in Nussbaum's theory, but remains insufficiently carried out. This is the distinction between the tragic and the unjust.

Previously, I have commented on the fact that Norman Daniels draws attention to the relationship between *efficacy* and justice. In the same line of reasoning, David Miller argues that justice has to do with things we can *control*; with things we can distribute.[154] As such, the state of the weather is not a matter of justice or injustice. In the same line of reasoning, increasing possibilities and efficacy of diagnostic and therapeutic techniques stimulate questions of just distribution and allocation of medical resources such as intensive care units, organ transplantation

---

[153] Again, the relevance of adding disability to the list of capabilities, especially where the social basis of self-respect is concerned, becomes relevant. Cf. Section 4.2.4.2 *A List of the Central Human Capabilities*.

[154] D. Miller, *Social Justice*, Oxford: Clarendon Press, 1976, pp. 17–18.

and artificial reproduction techniques. As long as these techniques were still in an experimental stage, issues of just allocation were out of the question.

With regard to health, we have also dealt with the fact that health is to a certain extent a 'natural good', since luck plays a substantial role in every person's health condition, and to a certain extent a 'social good', since just social policy can do a lot to influence the status of a natural good like health. Related to this, the distinction between the tragic and the unjust becomes relevant. When a person, through bad luck, is given a life that lacks the powers of human action and expression, this gives us a sense of *tragedy*. However, when we find that a person is kept, or left or even pushed beneath the threshold on any one of the ten capabilities because of irresponsible or discriminating social policy, it gives us a strong sense of *injustice*.[155]

Nussbaum has stressed throughout that it is *the social basis of the good*, not the good itself, that society can reliably provide and factors we cannot control may still interfere to keep some people from full capability. Therefore, when we use her framework of capabilities as a comparative measure of quality of life, we must also always examine the *reasons* for the differences we observe. Some differences in health are due to factors public policy can *control*, and others are not.

In that sense, Daniels was right in relating justice to effectiveness, to the fact that social policy can make a difference; that the result is in some way 'in our hands'. The difficulty, however, resides in the fact that he determines efficacy in health care by the extent to which there is a return to 'normal species-typical functioning'. Nussbaum's merit is that she enlarges the domain of justice: alongside with the element of efficacy and normal functioning she takes in the idea of the moral importance of the matter without reference to returning to normal functioning. Care for the dependent, for the least advantaged, just as it is, is a matter of such moral importance that every respectable and just society should guarantee it. That means that 'efficacy' and 'things we can control' receive a new meaning that is no longer exclusively linked to a narrow return to normal productive functioning. On the contrary, it opens the possibility to incorporate a variety of things a society can do that greatly influence *all* people's prospects, life chances and sense of self-respect, i.e. not only the lives of the fully cooperative but of *all* human beings. Providing equal access to decent health care for all is one of those things a society can do.

Taking the above-mentioned qualities together, we have good reasons to say that she succeeded in her goal to provide a theory that is both responsive to experience and reality – able to come to grips with the daily reality of people's lives – and still general and systematic – able to provide a strong normative philosophical theory.[156] Translated to our problem in hand, her theory is able to

---

[155] Cf. Section 3.3.2.2 *Which Health Inequalities Are Unjust?* Those that are avoidable by just and responsible social policy.

[156] M.C. Nussbaum, *Women and Human Development*, 2000, pp. xv–xvi, 10–11.

provide normative guidance for integration of the full complexity of health care within a theory of justice.

4.4.7 *Scarcity and the Language of Limits*

A pressing problem, however, remains. After due consideration, it seems that Nussbaum ignores what is considered by many as a basic fact of human life and society, i.e. the fact of scarcity and the ensuing language of limits. Does this mean that Nussbaum's philosophical theory claims that the sky is the limit; that she is not at all concerned with the fact of limited resources? Certainly not. However, it is difficult to get a clear understanding of what the contribution of her approach would be in the matter of just allocation of *scarce* resources. Let us try to list the relevant points on this matter.

On the one hand, there seems to be some room left for the issue of limits, and the circumstances of scarcity in Nussbaum's capabilities approach. First of all, her idea of truly human functioning is based on certain general conditions of human life that appear to be common to all human societies: 'that we are mortal; that we have bodily desires that are difficult both to control and to satisfy; that there is *scarcity* of material resources, with the *distributional problems* attendant on that, and so on'.[157] Furthermore, Nussbaum emphatically declares that the capabilities approach includes only the basic political framework and thus is only responsible for the *social basis* of the natural goods, and not for the individual happiness of a perfect life. She is a liberal, not a Platonist, not a perfectionist. 'Basic political principles have done their job', she says, 'if they have provided people with the social basis of the central capabilities.'[158] Moreover, she does say at certain points that there are obviously *limits* to any programme of social benefits; and that there are indeed questions to be asked about how much the state should invest in which domains.[159] Finally, the capabilities list and the threshold idea function as a social *minimum*. And it is this social minimum that has universal relevance. Further discussion about what should be guaranteed above the threshold, and which priorities should then be set, should be left, she argues, to national democratic discussion.[160] These elements, however, do not provide a truly satisfying answer.

For, on the other hand, it seems that her initially just critique on the Humean character of Rawlsian theory – i.e. on the idea of the 'circumstances of justice' that, among others, builds in the concept of scarcity from the beginning – ultimately risks throwing out the child with the bathwater. There seems to be no place left for the language of limits, of economic efficiency, priority setting and cost-containment.

[157] *Id.*, 'Nature, Function, and Capability', 1988, pp. 177–178.

[158] *Id.*, *Women and Human Development*, 2000, p. 81.

[159] *Id.*, *op. cit.*, 2004, pp. 441.

[160] M.C. Nussbaum, *Women and Human Development*, 2000, pp. 77, 95; *Id.*, 'Capabilities as Fundamental Entitlements', 2003, pp. 42–43.

Even though the capabilities list and the threshold idea function as a social minimum, it is a very demanding minimum. The richness and complexity of the list, the emphasis on human *flourishing*, the primary interest in the *higher* threshold which does not refer to the lower boundary but to the level at which a person's capability becomes *truly* human functioning and so on, all suggest that the general framework is one of abundance: 'A society that does not guarantee these to all its citizens, at some appropriate threshold level, falls short of being a fully just society, *whatever its level of opulence*.'[161] This gives the impression that social policy has never done enough. There are no substantive or guiding answers to the fact that scarcity of resources is a general condition of human life and that there are limits to what society can spend besides the proposal to leave the problem of setting limits on the national democratic discussion.

However, it is important to be cautious and not to jump to conclusions. In a recent article on the moral legitimacy of cost–benefit analysis, Nussbaum concentrates on the issue of public choice and argues – rightly I believe – that there are moral limits to trade-offs.[162] We have to be cautious, she says, not to forget that in all situations of choice there are two sorts of questions. The first is *the obvious question*: what should we do? What should we decide? The second is *the tragic question*: is any of the alternatives open to us free from moral wrongdoing? It is important to see that the moral understanding of the capabilities list entails that there is always a potential for a 'no' answer to the second question, i.e. for tragedy to arise. This is the case whenever necessary public choice unavoidably results in leaving some citizens beneath the threshold of any one of the ten capabilities. What are the possibilities in such situations? These are threefold, Nussbaum argues.[163]

Either we can fail to see the force of the tragic question altogether or recommend a policy of deliberately not facing it, in order to just get on with one's duties. This would be to deny that there is an issue of importance that needs to be addressed.

Or we could, secondly, try to find a solution through a method such as cost–benefit analysis, understood as a strategy for choice in which weightings are allocated to the available alternatives, arriving at some kind of aggregate figure for each major option.

However, and this is the third and most essential aspect of her argument, it is important to be aware of the fact that while methods like cost–benefit analysis offer an attractive way of approaching the obvious question, they offer *no good*

---

[161] *Ibid.*, 2003, p. 40 (my italics, YD).

[162] M.C. Nussbaum, 'The Cost of Tragedy: Some Moral Limits of Cost–Benefit Analysis', in M. Adler; E. Posner, *Cost–Benefit Analysis: Legal, Economic, and Philosophical Perspectives*, Chicago: Chicago University Press, 2001.

[163] Note that this partially corresponds with the three strategies for coping with the gap between supply and demand. Cf. Section 2.1.2 *Setting Limits: Coping with the Gap*.

way at all of registering the force of the *tragic* question or of representing a situation in which the answer to that question is 'no'. Too much reliance on cost–benefit analysis as a general method to deal with trade-offs can distract us, it can *obscure* the presence of a tragic situation by suggesting that the obvious question is the only pertinent one. This is most unwise, for the tragic question registers not the *difficulty of solving* the obvious question but a *distinct difficulty*, namely the possibility that all the answers to the obvious question, including the best one, can be morally bad, involving serious wrongdoing. There is the possibility that there is no 'right answer'. The point of the tragic question is that we have to bear in mind that some choices are simply wrong. It also reminds us of the deep importance of some spheres of life that are in conflict, and of the dreadful results when they are opposed and we have to choose between them. The sense of tragedy should inform decent moral choice.[164] In this regard, she argues, there are *moral limits* to trade-offs. If one does use cost–benefit analysis in connection with the capabilities approach, it will be crucial to represent in weighing the fact that each of this plurality of distinct goods is of central importance. Thus, there is a tragic aspect to any choice in which citizens are left below the threshold of one of the central areas. This tragic aspect could be represented as a huge cost. But representing it this way, Nussbaum argues, cannot hide the fact that a distinctive good has been slighted.

What does this interesting analysis imply for the question: how should we deal with just distribution of scarce health-care resources; how should we deal with the health-care trade-off? It cannot be denied that issues of cost-containment and economic efficiency remain very important elements within a sensible theory of just health care. To deny this would show a lack of sense of reality in the opposite way. Efficiency is as important to a just and responsible health-care system as is long-term care. However, it remains reasonable to say that Nussbaum's theory cannot provide a satisfying answer to contemporary health-care problems like issues of cost control, questions regarding the role of personal responsibility, regarding limits to high-tech medicine and so on. Further specification of the threshold, of the social minimum that every just society is obliged to provide, and of possible priorities, should be left to the national democratic discussion. As I said earlier, this is a somewhat disappointing answer. Are there no relevant principles conceivable that can serve as a normative guide to the democratic discussion on rendering the goals of good quality care, equal access for all, freedom of choice and economic efficiency more coherent?

---

[164] Bear in mind that 'tragedy' has a somewhat different meaning here than in the previous mentioning of the term. Whereas 'tragic' and 'tragedy' previously referred to the realm of luck, in which things just happen, as opposed to the realm of 'justice' or 'injustice', referring to things that are in some way in our hands, things we can control, do something about. Now, the tragic question comes *within* the realm of justice. This makes matters more complex.

In Chapter 5, I will concentrate on these questions by putting the fact of scarcity and the issue of limits distinctly back on the menu. Let us see whether we can legitimately provide more substantive guidelines for trying to answer questions like 'Where should we set the limits?' and 'What deserves priority?' than a mere 'We should leave this to the democratic discussion process', an answer that has, until now, been the dominant strategy. In the search for a more substantive strategy in the realm of trade-offs, scarcity and limits, Ronald Dworkin's proposal proves to be very promising. Let us consider his proposal, while bearing Nussbaum's analysis of the tragic question firmly in mind.

CHAPTER 5

SETTING LIMITS: DWORKIN'S PROPOSAL

5.1 TWO INTUITIONS ON LIMITS

Let us, for the sake of the argument, recapitulate: how have the previously-analysed theories dealt with the fact of scarcity and the issue of limits? Two basic intuitions are helpful to answer this question in a coherent way.

5.1.1 The General Liberal Intuition

Firstly, there is the liberal intuition of political, non-comprehensive theory. Whatever the differences between individuals' life plans, we are all committed to something like pursuing a conception of the good life, and certain things (resources, institutional frameworks) are needed in order to pursue these commitments, whatever their more particular content. In providing these means, however, the task and duty of the just society is *limited* in the sense that it is not responsible for securing and guaranteeing personal happiness but only for providing the material and institutional framework in which a person can pursue his own ideal of the good life.

In Rawls's theory, these things and frameworks are called the 'primary social goods'. We have seen that Rawls expressly chooses to keep the framework narrow. The primary goods serve as a small-scale objective measure of well-being, not only for the main reason that the concept has to serve a *political* theory, but also because Rawls is aware of the *external dynamic of scarcity*.[1] In the same line of reasoning, Norman Daniels has developed an objective small-scale basis for justice in health care in which the moral importance of meeting health-care needs is located in the objective impact on normal functioning and opportunity rather than in the subjective impact on happiness. As such, both theories take the issue of limited resources into account, which leaves room for economic efficiency. Unfortunately, this happens, as we have seen, at the expense of justice for long-term care and support.

The capabilities approach of Sen and Nussbaum also provides a liberal perspective. The theory is not about the complete fulfilment of a happy life. Sen's inflexibility critique, however, has shown that the theories of Rawls and Daniels are too narrow and that it is better to concentrate on capabilities, that is, on what persons are able to do and be. In Nussbaum's approach this results in a theory

[1] J. Rawls, *Justice as Fairness: a Restatement*, Cambridge, MA: The Belknap Press of Harvard University Press, 2001, pp. 173–174. Cf. Section 4.1.2.3 *Long-Term Care and the Difference Principle*.

of human flourishing that presupposes a realm of abundance. Even though she proposes a liberal theory, according to which the duty of society is limited, the issue of scarcity is kept aside and the language of efficiency is fiercely criticised.[2] Although her criticism on sacrificing long-term care is most just and right, the aversion to any constructive talk of efficiency is a flaw in her theory.

5.1.2 Personal Responsibility

Is there a possibility within *just* health care to (1) take economic *efficiency* into account, by setting limits on the high cost of health and (2) not having to sacrifice long-term *care* in doing so? Ronald Dworkin seems to offer a way out by elaborating a second basic intuition on limits, which holds that society should not be accountable for people's personal choices. Although founded on the liberal intuition in the first place, the second intuition carries the discussion further because it brings into much sharper focus the issue of personal responsibility, which appears only embryonically in the first intuition and in the works of Rawls, Daniels, Sen and Nussbaum.[3]

Dworkin tries to take efficiency into account while saving long-term care by (1) distinguishing more carefully than other theorists do between a person's ambitions (which have to do with our personality-manifesting choices) and his endowments (which are a matter of unchosen circumstances) and (2) by elaborating his idea of the hypothetical insurance market. By doing this, his strategy occupies a middle position between scarcity and abundance, between 'too narrow' and 'too broad' a theory.[4]

To give a sketch of Dworkin's theory that is both thoroughly elaborated in relating the arguments to the previously analysed theories of Rawls, Daniels, Sen and Nussbaum and avoids elements that are irrelevant for our problem in hand, I shall start the discussion again with a dialogue with Rawlsian theory. Next, I shall present some of the central ideas of Dworkin's very abstract theory, involving different forms of luck, the use of auctions, hypothetical insurance schemes and

---

[2] This critique is most forcefully present in M.C. Nussbaum, *op. cit.*, 2001; *op. cit.*, 2002.

[3] Let us recapitulate shortly: the idea of citizens taking responsibility for their ends, being capable of adjusting their aims and aspirations in the light of what they can pursue by the means one can reasonably expect to acquire given one's prospects and situation in society, has been one of the reasons why Rawls's account of justice measures people's share of primary goods, not their level of welfare. In the same line of reasoning, Daniels argues against the problem of social hijacking by expensive tastes, and proposes a small-scale objective measure of well-being. In the capability approach of Sen and Nussbaum, the idea of personal responsibility turns up in the distinction between capabilities and functioning.

[4] The initial statement is in: R. Dworkin, 'What is Equality? Part 1: Equality of Welfare', in *Philosophy and Public Affairs* 10(1981)3, pp. 185–246; repr. in: R. Dworkin, *Sovereign Virtue: the Theory and Practice of Equality*, Cambridge, MA: Harvard University Press, 2000, pp. 11–64; R. Dworkin, 'What is Equality? Part 2: Equality of Resources', in *Philosophy and Public Affairs* 10(1981)4, pp. 283–345; repr. in R. Dworkin, *Sovereign Virtue*, *op.cit.*, pp. 65–119. References below include both the 1981 and the 2000 editions of the articles.

envy tests. Consequently, I shall discuss the implications of his theory for the issue of justice in health care and move the discussion to the prudent insurance principle as an extended example of the hypothetical insurance idea. Finally, I shall examine his position in relation to three questions. Firstly, what is the value of Dworkin's contribution to the problem of limits to health care? Secondly, is his conception of health – being no more special than other resources, containing no special status – tenable within the light of his own theory? And thirdly, how should we understand Dworkin's proposal in relation to the luck-egalitarian proposal defended by Cohen and Arneson? I will consider this discussion in the light of the following question: to what extent does justice require society to try to restore the health of those who voluntarily put their own health at risk?

5.2 THE DIALOGUE WITH RAWLS: TWO PROBLEMS WITH THE SECOND PRINCIPLE

The second Rawlsian justice principle holds that social and economic inequalities between individuals are justified provided that they result from choices, ambitions and efforts in life, open to all under conditions of fair equality of opportunity (the fair equality of opportunity principle); and only if it benefits the least advantaged members of society to do so (the difference principle).[5] This embryonically reflects the second intuition concerning the responsibility distinction between choices and circumstances; ambitions and endowments, preferences and (natural and social) resources: we should not leave too much room for our *undeserved* circumstances and endowments.[6]

According to this principle, it is fair for individuals to have unequal shares of social goods if those inequalities are earned and deserved by the individual, i.e. if they are the product of the individual's ambitions, actions and choices. However, it is unfair for individuals to be disadvantaged or privileged by arbitrary and undeserved differences in their *social circumstances*. If I am pursuing some personal ambition, my success or failure should be determined by my performance, not by my race, class or sex. Hence, whatever success I achieve is 'earned'

---

[5] J. Rawls, *A Theory of Justice*, 1971, p. 62; 1999, p. 53; *Id., Political Liberalism*, 1996, p. 291.

[6] This idea is part of his critique on the view of formal equality of opportunity that, according to Rawls, leaves too much room for the influence of our natural endowments (he calls it the 'system of natural liberty', in which the initial distribution is regulated according to the conception of 'careers open to talents'). Cf. Section 3.1.5.4 *Real Equality of Opportunity*. According to the equality of opportunity view inequality of income and prestige are assumed to be justified if and only if there was fair competition in the awarding of the offices and positions that yield those benefits. Everyone has a fair go. While Rawls also requires equality of opportunity in allotting positions, he denies that the people who fill the positions are thereby entitled to a greater share. In the Rawlsian society, such people may have more than the average, only if it benefits those who have the least. See J. Rawls, *A Theory of Justice*, 1971, pp. 65–67, 71–72, 106–107; 1999, pp. 57–58, 62–63, 91–92. See also W. Kymlicka, *Contemporary Political Philosophy*, Oxford: Clarendon Press, 1990, pp. 55–57; in the 2nd ext. edn., 2002, pp. 53–101.

by me rather than merely 'endowed' on me. In a society that provides equality of opportunity, unequal income is fair because success is 'merited'; it goes to those who 'deserve' it.

In the same line of reasoning, there is another source of undeserved inequality that has to be taken into account. It is true that social inequalities are undeserved, and hence it is unfair to be made worse off by that undeserved inequality. However, the same thing can be said about inequalities in *natural talents*. No one deserves to be born handicapped or very intelligent, any more than one deserves to be born into a certain class, sex or race. If it is unfair to be made worse off in life because of the latter causes, it is equally unfair when people's life is unequally determined by the former causes.

The central intuition in each case is the same: distributive shares should not be influenced by factors that are arbitrary from the moral point of view. Natural talents and social circumstances are both matters of brute luck, and people's moral claims should not depend on brute luck. If we are genuinely interested in removing undeserved inequalities, then we must take both social contingencies and natural talents into account. How then, should we deal with these undeserved differences? As Rawls says:

> [While] no one deserves his greater natural capacity nor merits a more favourable starting place in society ... it does not follow that one should eliminate these distinctions. There is another way to deal with them. The basic structure can be arranged so that these contingencies work for the good of the least fortunate. Thus we are led to the difference principle if we wish to set up the social system so that no one gains or loses from his arbitrary place in the distribution of natural assets or his initial position in society without giving or receiving compensating advantages in return.[7]

What the second principle says is that while no one should *suffer* from the influence of undeserved natural or social inequalities (fair equality of opportunity for all), there may be cases where everyone *benefits* from allowing such an influence. And although no one deserves to benefit from their natural talents and social circumstances, it is not unfair to allow such benefits when they work to the advantage of those who were least fortunate in the natural and social lottery (the difference principle).[8] The higher expectations and income of the naturally talented and socially fortunate are just if and only if they are part of a scheme that improves the expectations of the least advantaged members of society.

As such, the difference principle gives some weight to the consideration singled out by the *prima facie principle of redress*.[9] This is the principle that undeserved inequalities (of birth and natural endowments) are to be somehow compensated for. The idea is to redress the bias of contingencies in the direction of equality. In pursuit of this principle society must give more to those with fewer native assets and

---

[7] J. Rawls, *A Theory of Justice*, 1971, p. 102; 1999, p. 87.

[8] *Ibid.*, 1971, p. 75; 1999, p. 65.

[9] *Ibid.*, 1971, pp. 100–102; 1999, pp. 86–88. See also: N. Daniels, 'Rawls's Complex Egalitarianism', 2003, pp. 251, 255.

to those born into less favourable social conditions. However, it counts as a prima facie principle that has to be weighed in the balance with other principles – like the principle of efficiency and of the advance of the common good. The difference principle differs from the principle of redress in the fact that the latter requires society to even out handicaps as if all were expected to compete on a fair basis in the same race. This is not required by the former. According to the difference principle, one should not *eliminate* the interpersonal differences. Justice is about mitigating, not about levelling. As long as the differences improve the situation of the least advantage, the setting is fair.

However, two problems crop up, the first of which will sound familiar to the reader. Although the second justice principle expresses the intuition that we should not leave too much room for the influence of our undeserved social circumstances and natural endowments, it turns out that the principle is, on the one hand, not endowment-insensitive enough and, on the other hand, too ambition-insensitive. After due consideration, it turns out that Rawls leaves too much room for the influence of natural inequalities (only social primary goods are distributed by the second principle) and too little for the influence of choice (the difference principle holds that *all* inequalities must work to the benefit of the least advantaged). Let us consider these issues in more detail.

5.2.1 Natural Primary Goods

According to Rawls, the second justice principle is the best way to ensure that endowments do not have an unfair influence. The talented and well-endowed do not deserve any greater income, and they should only receive more income if it benefits the least well-off. This, however, is misleading for remember that Rawls distinguished two kinds of primary goods that we all need in order to pursue our conception of the good life: social primary goods and natural primary goods. The first are the goods that are directly distributed by social institutions, like rights and liberties, powers and opportunities, income and wealth. The second are goods like health, intelligence, vigour, imagination and natural talents, which are affected by social distributions, but are not directly distributed by them.[10]

The second principle still allows too much room for the influence of our undeserved *natural* endowments because the benchmark for assessing the justice of social institutions lies in the prospects of the least advantaged in terms of *social* goods. We are only led to the difference principle if gains or losses are interpreted as social goods. The second principle ensures that the *well-endowed* do not get *more* social goods *solely* because of their arbitrary place in the distribution of natural assets, and that the *handicapped* are not *deprived* of social goods just because of their place. Nevertheless, this does not mitigate the effects of both natural accident and social circumstance. On the contrary, for the handicapped person still faces an *undeserved natural disadvantage* that creates causing both special, additional and extra needs for social resources

[10] J. Rawls, *A Theory of Justice*, 1971, p. 62; 1999, p. 54.

and a burden in this person's ability to lead a satisfactory life, a burden caused by his endowments and circumstances. The difference principle allows rather than removes that burden for it does not make room for the fact that the *handicapped* need *more* social resources to compensate for their natural disadvantage. In order to be in a position of equal opportunity the naturally handicapped should not only have a claim to *non-discrimination*, but also to *compensation* for their disadvantage (subsidised medicine, transportation possibilities, adjusted working areas, etc.). Only then the principle can claim to be endowment-insensitive.

In this preamble to the analysis of Dworkin's theory, we need to stress an important difference with the seemingly similar capability critique on the use of primary goods. This critique blames the problem on Rawls's commitment to using *primary goods* to define the least well-off position.[11] This, however, Dworkin urges, is a mistake. As we shall see, he rather blames the problem on the *incomplete* use of primary goods, i.e. on the arbitrary exclusion of natural primary goods from the index. We have to concentrate on compensation as a way of eliminating undeserved inequality in *overall* primary goods.[12] Consequently, his theory includes not just external, transferable goods like income, but also natural endowments like talents, mobility, vision; goods that cannot be redistributed per se. Disabilities and talents are treated as opposites: disabilities count as negative internal resources, while talents count as positive internal resources. This means that to be born disabled or susceptible to major health problems is to have an internal resource deficit that does make a call on justice. External health-care resources may be able to reduce this internal resource deficit, or to cure or alleviate the poor health condition. However, even if no reduction in an internal health-related resource deficit is possible, external resources can be redistributed to compensate for it. In addition, it is important to mention that internal resources do not include people's tastes, ambitions or preferences. Nor should they strictly include health conditions or disabilities that represent option luck, i.e. that are the result of freely chosen risks.[13] This brings us to the second problem, that of choice and ambition.

---

[11] A. Sen, *op. cit.*, 1980, pp. 215–216; B. Barry, *The Liberal Theory of Justice*, Oxford: Oxford University Press, 1973, pp. 55–57.

[12] W. Kymlicka, *op. cit.*, 1990, pp. 91–92; J. Roemer, *Theories of Distributive Justice*, Cambridge, MA: Harvard University Press, 1996, pp. 246–247. An important implication of this critique is that it shows Dworkin's viewpoint of holding on to the resource-egalitarian approach (like Rawls's), instead of moving over to the capabilities approach (of Sen and Nussbaum). However, an important aspect of his critique on Rawlsian resource egalitarianism comes to the same point as the capabilities critique: the Rawlsian primary goods approach is too narrow. It excludes things that should not be excluded.

[13] I will consider the notion of brute luck and option luck in detail in Section 5.3.2 *Brute Luck, Option Luck and Insurance*.

Before going into this, however, I must mention a second interesting point. Compensating people for the unchosen costs of their natural disadvantages should be done, not so that they can *compete* with others on an equal footing, but so that they can equally have the same ability to *lead a satisfying life according to one's own idea of the good.*[14] As for this, it turns out that Dworkin's point is not so wholly different from Nussbaum's. The story is not about productive reciprocity and competition for mutual advantage, but about 'the right to be in the world' and to lead a rewarding life of one's own, according to what best serves one's beliefs about what gives value to life.[15]

5.2.2 Choice and Ambition

The second problem concerns the flip side of the distinction between choices and circumstances. If people do not deserve to bear the burden of *unchosen* costs, how should we then respond to people who *choose* to do costly things? A theory that wants to avoid insensitivity to the unchosen nature of natural and social inequalities consequently has to be able to answer the question of how to be sensitive to people's choices.

Let us consider the following example.[16] Imagine two persons, Adrian and Bruce, with equal natural talents, sharing the same social background and starting with an equal distribution of resources: a similar amount of land. Bruce wants to play tennis all day and only works long enough to sustain his desired lifestyle. Adrian wants to be a gardener, producing and selling vegetables. While they both began with equal shares of resources, soon Adrian the gardener will generate a larger income through larger amounts of work and will thus have more resources than Bruce the tennis player.

According to the difference principle, this inequality is only allowed if it benefits the least well-off, i.e. if it benefits the tennis player who has much less income. Intuitively, there is something peculiar about this way of mitigating inequality. Given that the differences in lifestyle are freely chosen, how can one reasonably say that the tennis player is treated unequally just because the gardener is allowed to have the income earned through a working lifestyle that the tennis player did not want? In such situations, the difference principle leads to counterintuitive results. Rawls defends the difference principle by saying that it counteracts the inequalities of natural and social contingencies. However, these are not relevant here. Rather than removing a disadvantage,

---

[14] R. Dworkin, 'Equality and the Good Life', in *id., Sovereign Virtue*, 2000, pp. 237–284.

[15] M.C. Nussbaum, *op. cit.*, 2002, p. 32.

[16] R. Dworkin, 'Equality of Resources', 1981, pp. 304–309; 2000, pp. 83–87. The example is challenged, for instance, by Philippe Van Parijs in his *Real Freedom for All: What (If Anything) Can Justify Capitalism?* Oxford: Clarendon Press, 1995. I cannot go into this discussion here.

the difference principle makes Adrian subsidise for Bruce's expensive desire for leisure. As Will Kymlicka points out:

> [Adrian] has to pay for the costs of [his] choice – i.e. [he] forgoes leisure in order to get more income. But [Bruce] does not have to pay for the costs of his choice – i.e. he does not forgo income in order to get more leisure.[17]

This means that the gardener has to pay twice: for the costs of his own choices, and for the tennis player's choice. They are treated unequally for no legitimate reason. This does not promote equality but rather undermines it. When inequalities in income are the result of *choices*, not circumstances, the difference principle creates, rather than removes unfairness. Treating people with equal concern requires that people pay for the costs of their own choices. Paying for choices is the flip side of our intuition about not paying for unequal circumstances.

Taken together, it turns out that although Rawls's second justice principle *appeals* to the choices–circumstances distinction – by emphasising that his conception of justice regulates inequalities that affect people's life chances, not inequalities that arise from people's life choices, which are the individual's own responsibility – the principle at the same time *violates* this distinction in two important ways.

Firstly, it mitigates too little. Rawls wants the principle to mitigate the unjust effects of one's place in the distribution of natural and social assets. However, because he excludes the *natural* primary goods from the index which determines who is least well-off, there is in fact no compensation for those who suffer from undeserved natural disadvantages.

Secondly, it mitigates too much. It also mitigates the legitimate effects of personal *choice*, effort and ambition. From the fact that natural and social inequalities are arbitrary; it should follow that only those kinds of inequalities should influence distribution when it would benefit the least well-off. However, the difference principle says that *all* inequalities must work to the benefit of the least advantaged and not just those, which stem from morally arbitrary factors. As such, the difference principle requires that some people subsidise the costs of other people's choices, thus making the ants pay for the grasshoppers. Can we make a better job of being ambition-sensitive and endowment-insensitive? This challenge is taken up by Ronald Dworkin.

5.3 DWORKIN'S RESOURCE EGALITARIANISM

Dworkin begins by supposing that justice requires equal treatment of individuals, and by asking in what dimension that treatment should be measured. He considers two classes of possibility for the equalisandum – welfare and resources – and argues that equality of resources is the right way of defining

[17] W. Kymlicka, *op. cit.*, 1990, p. 74.

distributive justice. Justice requires equality of resources but not equality of welfare.[18]

Two principles together shape and support his resource egalitarian account. All ideas and devices further developed in his theory can be seen as shaped by these twin demands.[19] The first is the *principle of equal importance*, which holds that it is objectively and universally important that human lives are successful, that they come to something, rather than being wasted. This is equally and impartially important for *each* human life. The equality in question is not attached to any property or merit of people. Consequently, this principle requires governments to adopt laws and policies that ensure that its citizen's life chances are, as far as can be achieved by a government, insensitive to who they otherwise are (which is determined by their economic backgrounds, gender, race or particular set of skills or handicaps).

The second is the *principle of special responsibility*. It holds that although we all must recognise the equal objective importance of the success of a human life, one person has a special and final responsibility for that success, that is, the person whose life it is. Dworkin hastens to say that this principle is neither to be understood as a metaphysical statement about the truly free character of the human will; nor does it intend to deny that psychology, sociology or even biology can provide persuasive causal explanations of why different people choose to live as they do, or that culture, education or social class influences such choices. The principle is rather *relational*: it insists that so far as choices are to be made about the kind of life one wishes to live, within whatever range of choice is permitted by resource and culture, a person himself is responsible for making those choices. Consequently, it demands that governments work – again: as far as they can achieve – to make people's lives *sensitive* to the *choices* they make.[20]

---

[18] The essence of Dworkin's analysis on welfare egalitarianism comes down to fact that equality of welfare is ultimately either an empty theory in need of an independent, external justification or a self-defeating theory leading to conclusions that do not do credit to an egalitarian theory: 'people cannot be treated as equals by making them equal in some dimension they value unequally' in R. Dworkin, 'Equality of Welfare', 1981, p. 245; 2000, p. 63. One of the counterarguments is the famous one on Tiny Tim. Against understanding justice in terms of equality of welfare, Dworkin invokes the case of Tiny Tim from Dickens's *A Christmas Carol*. The aim to equalise welfare would supposedly give more resources to the disabled, in order to raise them to the same level of welfare as others. However, Tiny Tim has such a positive outlook and cheerful disposition that he already is at a higher level of welfare than most people, despite being disabled. If justice requires that we equalise welfare, it would not require us to compensate Tiny Tim for his disability by giving him more resources. Since Tiny Tim is already so well off, the aim to equalise welfare will regard his disability per se as irrelevant to justice. Surely, it is not irrelevant. And Tiny Tim's good cheer does not dilute the requirement of just compensation for disability at all. Therefore, justice cannot be understood in terms of equality of welfare. See: R. Dworkin, 'Equality of Welfare', 1981, pp. 240–244; 2000, pp. 59–62.

[19] R. Dworkin, *Sovereign Virtue*, 2000, pp. 4–7.

[20] Here, the relevance of public incentive measures crops up, cf. Section 4.2.4.3 *Distinguishing Characteristics*. I will consider this issue in detail in Section 5.5.3 *On Personal Responsibility and the Right to Health Care*.

It does not assume that people choose their convictions, preferences or their personality more generally, any more than they choose their race, or physical or mental abilities. However, it does relate to an ethic, which supposes – as almost all of us in our own lives do suppose – that we are in *some* way and to *some* extent responsible for the consequences of the choices that we make out of those convictions, preferences or personality.

As such, Dworkin aims to achieve an account of justice that unifies a political community's *collective responsibility* to show equal concern for all its citizens with each citizen's *personal responsibility*.[21] The principle of special responsibility seizes the idea of allowing space for personal responsibility that, as I have said, occurs both in Rawlsian theory and in the capabilities approach but remains incompletely developed in these theories. This can be illustrated with Dworkin's acceptance of the endowment-insensitive and ambition-sensitive goal that motivated Rawls's second justice principle:

> On the one hand we must, on pain of violating equality, allow the distribution of resources at any particular moment to be (as we might say) *ambition-sensitive*. It must, that is, reflect the cost or benefit to others of the choices people make so that, for example, those who choose to invest rather than consume, or to consume less expensively rather than more, or to work in more rather than less profitable ways, must be permitted to retain the gains that flow from these decisions in an equal auction followed by free trade. But on the other hand, we must not allow the distribution of resources at any moment to be *endowment-sensitive*, that is, to be affected by differences in ability of the sort that produce income differences in a laissez-faire economy among people with the same ambitions.[22]

However, different from Rawls's resource egalitarianism Dworkin thinks that *another* scheme of distribution can make a better job in living up to that ideal. His theory is complicated, involving the use of auctions, insurance schemes, markets and taxation models. In the following sections, I will focus on some of his central ideas that help us toward an answer to our problem of combining efficiency and long-term care within a theory of just health care and leave others undiscussed.

5.3.1 The Ambition-Sensitive Auction

To begin with, Dworkin argues that an equal division of resources presupposes an economic market of some form, mainly as an analytical device, but also, to a certain extent, as an actual political institution. Contrary to the perspective that pictures equality as the victim of the values of efficiency and liberty supposedly served by the market, Dworkin argues that the idea of an economic market, as a device for setting prices for a variety of goods and services, must be at the centre of any feasible theoretical development of equality of resources. This point is most clearly shown by his example of the ambition-sensitive auction.[23]

---

[21] R. Dworkin, *op. cit.*, 2000, p. 7.
[22] R. Dworkin, 'Equality of Resources', 1981, p. 311; 2000, p. 89.
[23] *Ibid.*, 1981, pp. 283–290; 2000, pp. 65–71.

Dworkin proposes an imaginary auction as a model whereby equality of resources may be achieved. He asks us to imagine that all of society's resources are up for sale in an auction, to which every person is a participant. Everyone is given an equal amount of some form of currency – 100 clamshells, in his example – and uses the clamshells to bid for those resources that best suit their life plan.

Against the initially equal stock of purchasing power, each person plays an equal role in determining the bundle of resources that will be chosen. Therefore, if the auction works out, no one will envy another's set of purchases because, by hypothesis, he could have bid for that bundle of goods, rather than the goods he did bid for. Everyone will thus be happy with the result. Dworkin calls this *the envy test*:

> No division of resources is an equal division if, once the division is complete, any [person] would prefer someone else's bundle of resources to his own bundle.[24]

If this is met, then people are treated with equal consideration, for differences between them reflect their different ambitions, their different beliefs about what is valuable. This generalises the case of the tennis player and the gardener who, starting with the same amount of resources, use them in the way they need for their desired activities. A distribution could not be challenged as unequal on the ground of varieties of taste. A successful auction meets the envy test and makes each person pay for the costs of their own choices.

The market character of the auction is not a coincidence in the example. In fact it proposes an important Dworkinian idea; namely that the true measure of the social resources devoted to the life of one person is fixed by asking how important, in fact, that resource is for him, and for others. As such, the auction proposes what the envy test assumes, namely that the market is the best means of enforcing equal division of resources, as measured by the *opportunity cost* of such resources to others.[25] Previously, I have pointed out the notion of opportunity cost as a term used in economics to mean the cost of something in terms of an opportunity foregone (and the benefits that could be received from that opportunity). The role of opportunity cost in Dworkin's theory illustrates his attention for the external aspect of scarcity, the aspect, which also refers to scarcity as a natural condition of life. In economics, the notion of opportunity cost plays a crucial part in the issue of efficient use of scarce resources. We are not in the Garden of Eden, we cannot have everything, and our choice for one thing implies giving up something else.[26]

5.3.2 Brute Luck, Option Luck and Insurance

The envy test is important. It checks whether people are being treated as equals because it disallows any division which leaves a person preferring another bundle

---

[24] *Ibid.*, 1981, p. 285; 2000, p. 67.

[25] *Ibid.*, 1981, pp. 285–290, 338; 2000, pp. 67–71, 112.

[26] Cf. Section 2.1.1.1 *The Twofold Dynamics of Scarcity*.

to his own. In reality, however, equality of resources does not hold because people will have many reasons to prefer the bundles that others have. Once the auction is completed, and after some time, the envy test will shortly fail.[27] People will be advantaged or disadvantaged in terms of natural assets or in the occurrence of events. Some may be handicapped. Some may be more skilful than others, may like to work hard and earn more money than others who do not like to work or prefer a job that will bring them less. Or put inversely, Lazy might become very rich after having bought the winning lottery ticket, while Crazy falls sick and desperately won't be able to continue his rewarding but demanding career.[28] For such, and dozens of other reasons, people will prefer the bundle others have. Consequently, we must ask *which* such developments are consistent with equality of resources, that is, which developments count as legitimate grounds for challenging the distribution and which do not.

Crucial in this regard is the character and impact of luck on a person's fortune, which is more specifically illustrated by Dworkin's distinction between brute luck and option luck.[29] *Option luck* is the sort of luck we might have in gambling, whereby we take a deliberate risk in relation to something. It is a matter of gaining or losing through accepting a risk one should have anticipated and might have declined. If someone develops lung cancer after a life of heavy smoking, that person has chosen an option that turned out to be an unsuccessful gamble. *Brute luck* on the other hand is a matter of how risks turn out without being a deliberate gamble. It refers to a risk one could not anticipate or did not choose to run. It happens to someone without being the result of choice. An instance of brute bad luck is lung cancer developing in the course of a normal non-smoking life.

Dworkin holds that option luck is consistent with equality of resources. In fact, the main point of the auction and the envy test is to support the argument that people should pay the price of the life they have decided to lead, measured in what others give up so that they can do so. If I am made worse off because the gambles I have made have turned out badly, i.e. because I have had poor option luck, then egalitarian concerns are not triggered. People should be permitted to take risks of certain sorts and to reap both the lucky and unlucky consequences. The gambler who wins made a deliberate choice to take a risk and the true cost of that risk is to be measured against what it costs those who have chosen a safer life. The price of that safer life is forgoing any chance of gains. The gambler who loses has been in the position to take the risk, with consequent possible gains. He thus has paid the price, viz. losing, for having had the chance of gambling. From the viewpoint of equality of resources there is no reason why risk-taking should

---

[27] *Ibid.*, 1981, pp. 292–293; 2000, pp. 73–74.

[28] I borrow Crazy and Lazy from Philippe Van Parijs. See: P. Van Parijs, *Real Freedom for All: What (If Anything) Can Justify Capitalism?* Oxford: Clarendon Press, 1995, pp. 92–96.

[29] R. Dworkin, 'Equality of Resources', 1981, pp. 293–298; 2000, pp. 73–77.

be limited, although Dworkin thinks that there could be qualifications such as those based on paternalistic reasons, or reasons of political morality (like forbidding someone to gamble with his freedom or political rights).[30]

However, Dworkin does not think that brute luck is consistent with equality of resources, simply because it is not a matter of deliberate choice.[31] If I fare worse than others because of matters outside my control, then I am a victim of brute bad luck, and egalitarian concerns appropriately come to the fore. However, this problem has not yet been confronted by the auction scheme. If two people lead roughly the same lives, but one suddenly goes blind or loses the use of his limbs, then we cannot explain the resulting differences in their incomes either by saying that one took risks that the other chose not to take, or that we could not redistribute without interfering in the lives both prefer. For the accident has, let us assume, nothing to do with choices. The handicapped person now faces extra burdens in leading a good life, burdens that cut into his income. This will leave him less well-off than the other person. It might make us wonder whether the ambition-sensitive auction can be seen as an endowment-insensitive scheme. What should we do with natural disadvantages that are the result of brute bad luck? Dworkin has a complex answer to that question.

The problem may be dealt with through the possibility of insurance that will provide a link between brute and option luck.[32] Suppose insurance against blindness or loss of limbs is available in the initial auction. Also suppose that two sighted people have an equal chance of suffering an accident that will blind them, and they both know that they have this chance. Now, if one of them chooses to spend part of his initial resources for such insurance and the other does not, or if one of them buys more coverage than the other, then this difference will reflect their different opinions about the relative value of different components (use of limbs, being able to see) of their prospective lives. Apart from paternalist additions, the bare idea of equality of resources would not argue for redistribution from the person who had purchased insurance to the person who had not, if they were both blinded or paralysed in the same accident. For the availability of insurance would mean that, although both have had brute bad luck, the difference between them was a matter

---

[30] *Ibid.*, 1981, pp. 293–295; 2000, pp. 74–77. This qualification is an important addition because it allows society to guide personal preferences and options to some extent. Thus, it creates the possibility of reconciling personal responsibility for one's health condition and the impact of social incentives to healthy behaviour, like health campaigns, for legitimate paternalist reasons. I will consider this issue in detail in Section 5.5.3 *On Personal Responsibility and the Right to Health Care*. See further: R. Dworkin, *op. cit.*, 2000/1994, p. 319, endn. 6, p. 492. See also Y. Denier, 'On Personal Responsibility and the Human Right to Health Care', in *Cambridge Quarterly of Healthcare Ethics* 14(2005)2. Note moreover that the argument shows similarities with Nussbaum's argument for legitimate paternalism. Cf. Section 4.2.5.4 *Paternalism* and Section 4.2.4.3 *Distinguishing Characteristics*, especially in the paragraphs on 'Capability as Goal'.

[31] R. Dworkin, *op. cit.*, 1981, p. 296; 2000, p. 76.

[32] *Ibid.*, 1981, pp. 296–297; 2000, pp. 76–77.

of option luck. By not having insured against this bad luck, one person decided to have a particular bet in the gamble.[33] This means that if the condition were met that everyone ran an equal risk of suffering a handicapping catastrophe, roughly knew the chances of being struck and had ample opportunity to insure themselves, then handicaps would pose no special problem for equality of resources.

Of course, this condition is not met. The problem with much brute luck is that it occurs before anyone is near the position of being able to take insurance to convert it into option luck. Some people are born with handicaps, or develop them before they even have sufficient knowledge or funds to insure on their behalf. In such cases, one is stuck in a cruel catch-22 situation. Those who are born handicapped would not be insured because there is no insurable risk, and even if there were, they would not be able to afford the premium.[34]

5.3.3 Compensation before the Auction?

Suppose then we say that any person's physical and mental powers must count as part of his resources so that someone who is born handicapped starts with fewer resources than others.[35] Why not allow this person to catch up, by way of transfer payments out of a general stock of social resources, *before* the auction and then divide up the remaining resources equally through the auction? The auction results would then meet the envy test. Compensation *before* the auction would ensure that each person is equally able to choose and pursue a valuable life plan; equal division of resources *within* the auction ensures that those choices are fairly treated. This means that the distribution would be both endowment-insensitive and ambition-sensitive.

Nevertheless, the suggestion of an initial compensation to alleviate differences in physical or mental resources is troublesome in a variety of ways.[36] Firstly, it requires some standard of normal powers to serve as the benchmark for compensation. However, as was pointed out previously, a standard of normal functioning is not unproblematic. Whose powers should be taken as *normal* for this purpose? Nussbaum's *complex* reciprocity proposal has shown that the concept of normal functioning has to be treated with caution.

Secondly, it suffers from the defect of resource transfer. Although extra resources can compensate for some natural disadvantages – some physically handicapped people can be as mobile as able-bodied people if we provide the best

[33] Although the distinction between brute luck and option luck seems to fade away once the possibility of insurance is introduced, insurance will not erase the distinction, Dworkin holds: 'Someone who buys medical insurance and is hit by an unexpected meteorite still suffers brute bad luck, because he is worse off than if he had bought insurance and not needed it. But he has better option luck than if he had not bought the insurance, because his situation is better in virtue of his not having run the gamble of refusing to insure.' In *ibid.*, 1981, p. 293; 2000, p. 74.

[34] *Ibid.*, 1981, p. 297; 2000, p. 77.

[35] *Ibid.*, 1981, pp. 240–242, 300; 2000, pp. 59–60, 79–80. See also J. Roemer, *op. cit.*, 1996, pp. 245–249.

[36] *Ibid.*, 1981, p. 300; 2000, pp. 79–80.

framework available – no amount of initial compensation could make someone born blind or mentally incompetent equal in physical and mental resources with someone who does not suffer such disadvantages. Full equality of circumstances is impossible. Although powers are internal resources, they are not resources in the way external material resources like income and wealth are. They cannot be transferred.[37] In this way, the initial compensation solution would misdescribe the problem of handicaps in saying that equality of resources must strive to make people equal in physical and mental condition as far as possible.

Thirdly, the initial compensation proposal collides with the bottomless pit problem, which Dworkin calls the problem of 'no upper bound to initial compensation'.[38] Even if we would try to equalise circumstances as much as possible, it would seem unacceptable. Since each additional bit of money might help the severely disadvantaged person, yet is never enough to fully equalise the circumstances, we might be required to use *all* our resources for the initial compensation of one person, leaving nothing for everyone else. If resources had to be used to equalise circumstances first, *before* the auction starts, there would be none left for us to act on our choices *in* the auction.[39] In this regard, Kymlicka adds an interesting thought to the bottomless pit argument:

> one of our goals in equalizing circumstances was precisely to allow each person to act on their chosen life-plans. Our circumstances affect our ability to pursue our ambitions. That is why they are morally important, why inequalities in them matter. Our concern for people's circumstances is a concern to promote their ability to pursue their ends. If in trying to equalize the means we prevent anyone from achieving their ends, then we have failed completely.[40]

Given these difficulties, Rawls's refusal to compensate for natural disadvantages seems to make sense. Although Nussbaum's analysis has shown that the moral critique on the problem of natural primary goods is just and considerable, actually including natural disadvantages in the index that determines the least well-off seems to create an insoluble problem. We do not want to ignore such disadvantages, nor can we equalise them, and what could be in between, other than ad hoc acts of compassion or mercy? If care for the long-term disabled and congenitally handicapped should be, for moral reasons, a matter of justice and not just mercy or compassion, *how* then might we face the issue and avoid the classical problems that were repeatedly indicated above?

---

[37] Remember that Dworkin's theory holds that a person's comprehensive bundle of resources consists of both transferable, external resources that are distributed by society, and resources that cannot be easily transferred from person to person (such as genes, parents, handicaps and talents). Cf. Section 5.2.1 *Natural Primary Goods*. See also J. Roemer, *op. cit.*, pp. 246–247.

[38] R. Dworkin, *op. cit.*, 1981, p. 300; 2000, p. 80.

[39] R. Dworkin, 'Equality of Welfare', 1981, p. 242; 2000, pp. 60–61; *Id.*, 'Equality of Resources', 1981, p. 300; 2000, pp. 79–80; See also: C. Fried, *Right and Wrong*, Cambridge, MA: Harvard University Press, 1978, pp. 120–128; *Id.*, 'Distributive Justice', in *Social Philosophy and Policy* 1(1983)1, pp. 45–59.

[40] W. Kymlicka, *op. cit.*, p. 79.

Seemingly paradoxical at first sight, Dworkin argues that the idea of a *hypothetical* insurance market by way of supplement to the auction provides a guide, a workable baseline through which equality of resources might face the problem of handicaps in the real world. As such he carries the debate an important step further. Dworkin's innovation is that he proposes a *new scheme* that allows us to actually deal with an issue that is acknowledged to be a problem by many, but to which a constructive and coherent answer remained forthcoming. *How*, according to Dworkin, should we face the issue of natural disadvantages *within* a theory of justice and *without* presupposing abundance of resources?

5.3.4 Reaching Endowment-Insensitivity: the Hypothetical Insurance

Initially, Dworkin's proposal is similar to the Rawlsian idea of the original position. Imagine people behind a modified veil of ignorance: no one knows his place in the future distribution of natural talents, and everyone is assumed to be equally susceptible to the various natural disadvantages that might arise. Each person is given an equal share of resources – 100 clamshells – and asked how much of the share he is willing to spend on insurance coverage for the different natural disadvantages one may suffer. How much coverage would the average member of the community purchase?[41]

Dworkin argues that this counterfactual question would give us a workable baseline from which to work out a premium. This will be so, even though particular persons differ in the risks they are willing to take and the insurance premiums they would be prepared to pay. People would, he says, make roughly the same assessment of the value of insurance against handicaps such as blindness or loss of limbs. On average, people might be willing to spend 30% of their bundle of resources, for instance. Consequently, we could compensate those who develop handicaps accordingly, out of some fund collected by income taxation or some other compulsory insurance at a fixed premium for everyone, designed to match the fund that would have been provided through speculation about what the average person would have purchased by way of insurance had antecedent risk of various handicaps been equal (e.g. 30%). The various welfare, health care and unemployment schemes would be ways of paying out the coverage to those who turned out to suffer from the natural disadvantages covered by the insurance. Those who develop handicaps will then have more external resources at their disposal than others, but the extent of their extra resources will be fixed by market decisions that people would have made if circumstances had been more equal than they are.

5.3.5 A Middle-Course Proposal

According to this scheme, individuals insure themselves against not having the natural assets they believe to be valuable in life. Ultimately, this will lead to a

---

[41] R. Dworkin, 'Equality of Resources', 1981, pp. 297–298, 301–302; 2000, pp. 77–78, 80–82; *Id.*, 'Justice, Insurance, and Luck', in *id.*, *Sovereign Virtue*, 2000, pp. 331–340.

middle-course solution between ignoring unequal natural assets and trying in vain to equalise circumstances.[42] It would not lead to ignoring the problem, for everyone would buy *some* insurance. It is short-sighted not to provide *any* protection against the misfortunes that might befall one. However, no one would spend *all* of their clamshells on insurance, since they then would have nothing left to pursue their goals with. As such, the average person will not be inclined to insure against not having the voice of Maria Callas, the pop star talent of Mick Jagger or the health condition of a top athlete like Michael Jordan. The insurance premium for not having such talents would be very high, because very few persons are able to achieve that level. The amount of society's resources that we dedicate to compensating for natural disadvantages is limited to the coverage people would buy through premiums paid out of their initial bundle.[43] The result will be a middle course.

Does this system make sense? Does it provide for progress in dealing with the problems that turned up in the initial compensation proposal? Firstly, in order to compensate handicaps in people it is not necessary, Dworkin argues, to have some independent idea of what constitutes 'normal' powers.[44] The enormous difficulties inherent in that idea are not necessary in the hypothetical insurance system because the latter allows the hypothetical market to determine which infirmities are compensable. There, the market determines which handicaps – i.e. instances of brute bad luck – people would be prepared to insure against because their preferences will be linked to the cost.

This brings us to an answer to the bottomless pit problem. No one would purchase no insurance at all, while at the same time no one would spend all of its resources on insurance since then nothing would be left to pursue one's life projects. The market will therefore determine an upper limit on the handicaps that are compensable. The *limit* will be determined by factors such as the risks people would be prepared to take and the premiums they would be prepared to pay given the resources available, measured by the *opportunity cost* of such resources to others.[45] It would, perhaps rather unexcitingly, be a matter of sole actuarial calculation.

However, the defect of full equality of circumstances remains. On close reading, the hypothetical insurance scheme does not appear to account for the very feature it intended to, since we have not found the *pure* ambition-sensitive and endowment-insensitive distribution that we were looking for. Some people will still be disadvantaged in undeserved ways under the insurance scheme. There will, therefore, be a failure in the envy test: people will still envy the capacities of the most talented.[46] Dworkin's answer to this is that the envy test cannot be expected to eliminate differences between people in the way they are actually

---

[42] *Id.*, 'Equality of Resources', 1981, pp. 329–331; 2000, pp. 104–106.
[43] *Ibid.*, 1981, pp. 296–9; 2000, pp. 76–79.
[44] *Ibid.*, 1981, fn. 8, p. 300; 2000, endn. 6, p. 478.
[45] *Ibid.*, 1981, pp. 304–305, 338; 2000, pp. 84, 112.
[46] *Ibid.*, 1981, pp. 304–308; 2000, pp. 84–86.

born to be. Neither does equality of resources mean that all wealth differences should be eliminated.

5.3.6 A Second-Best Theory?

It might seem that Dworkin's unwillingness to try as best we can to mitigate the effects of natural disadvantages, shows an inadequate regard for the well-being of the disadvantaged. After all, they did not choose to be disadvantaged. However, the attempt to provide the highest possible coverage to the disadvantaged would *enslave the talented*.[47] If the level of coverage were too high, then those who were fortunate in the natural lottery would be forced to be as productive as possible in order to pay the high premiums they had hypothetically bought against natural disadvantage. As such, their talents would be a burden that restricted their options, rather than a resource that expanded their options. Hence, equal concern for both the advantaged and the disadvantaged requires a middle-course solution – installing a middle course between the slavery of the talented and the exclusion of the handicapped – even though it will leave the well-endowed with less resource than in a non-insurance situation and the disadvantaged envying the well-endowed.[48] At bottom, both parties will continue to have reasons to be dissatisfied in some way.

Consequently, Dworkin's theory could be called 'a second-best theory'.[49] His scheme does not and cannot fully compensate for undeserved inequalities. There will always be a failure in the envy test. Nevertheless it is, he urges, the best we can do to live up to our convictions of justice in dealing with the arbitrariness of one's place in the distribution of natural and social circumstances. He suggests that a society run along his lines, with its wealth differences, reflecting differences in tastes and ambitions, would be more nearly equal in resources than all other alternatives, even though full equality will never be possible. Wealth differences will continue to partially reflect also innate differences like intelligence, talents and strength.[50]

Bearing Dworkin's proposal of the hypothetical insurance in mind, it is now worth asking what his proposal implies in the field of health care. Can he provide an acceptable and workable account of limits to care? Can he develop a theory that is sufficiently endowment-insensitive and ambition-sensitive, also in health

---

[47] *Ibid.*, 1981, p. 322; 2000, p. 98.

[48] As we have seen, a way to avoid this would be to concentrate on equalisation of basic resources, like Sen's basic capabilities, and Rawls's basic primary goods, rather then the full-fledged equalisation of circumstances, which Dworkin rightly rejects as impossible. This has also been the essence of Rawls's rejection of the principle of redress as being both impossible and undesirable. J. Rawls, *A Theory of Justice*, 1971, pp. 100–102; 1999, pp. 86–88. See also: W. Kymlicka, *op. cit.*, endn. 7, p. 93. The same line of reasoning motivated Nussbaum to concentrate on the threshold of capabilities, and Norman Daniels to point at the fact that in such cases 'Justice is always rough around the edges'. Cf. Section 3.3.2.3 *When Are Limits to Health Care Fair?* and Section 4.2.4 *Nussbaum Differs: Toward Her Own Proposal*, respectively.

[49] W. Kymlicka, *op. cit.*, p. 80.

[50] R. Dworkin, *op. cit.*, 1981, p. 331; 2000, p. 106.

care? In the following I shall take a close look at his proposal of the prudent insurance principle, which is an extended example of the hypothetical insurance idea.

5.4 ON HEALTH CARE: RECONCILING QUALITY, EQUALITY, LIBERTY AND EFFICIENCY

Dworkin begins by arguing that the contemporary crisis in health care implies a crisis in our *conception* of what a just health-care system would be; that is to say, a crisis in the answers we should give to two of the biggest questions of justice in health care.

HOW MUCH? The first is the question of a society's *aggregate* health care budget: How much money should a society like ours spend on health care in the aggregate? This question refers to the external dynamic of scarcity: money spent on health care is money that might be spent on education, employment and housing policy, culture, recreation and a host of other goods and opportunities that make up the value of life.

ON WHAT? The second question, which is in fact part of the first, is the question of *distribution*: What kind of health care should a decent society make available *for everyone*, i.e. what should be included in the *basic package* of health care to which *all* individuals should have equal access? This question refers to the internal dynamic of scarcity. This question refers to the internal dynamics of scarcity: in principle, there is no limit to the amount and variety of health-care services that could be included.

Behind these two questions lies an explicitly philosophical one. What is the right standard to use in answering these questions? What should we take as our *ideal* of justice in medical care? In a number of articles, Ronald Dworkin has developed an extended approach of the hypothetical insurance idea that provides a means to begin to answer both questions in a way that brings together the goals of quality, equality, liberty and efficiency in health care.[51]

5.4.1 Rethinking the Ideal of Insulation

Dworkin begins by describing a powerful ideal of justice in health care: *the ideal of insulation*.[52] This ideal has three features. The first involves the idea that life

[51] R. Dworkin, 'Justice and the High Cost of Health', 1994, repr. in *id.*, *Sovereign Virtue. The Theory and Practice of Equality*, Cambridge, MA: Harvard University Press, 2000, pp. 307–319; R. Dworkin, 'Justice in the Distribution of Health Care', 1993, repr. in M. Clayton; A. Williams (eds.), *The Ideal of Equality*, Basingstoke: Palgrave Macmillan, 2002, pp. 203–222. For critical analysis of Dworkin's model, see: J. Butler, *The Ethics of Health Care Rationing*, London: Cassell, 1999, pp. 21–23; M. Powers, 'Hypothetical Choice Approaches to Health Care Allocation', in J. Humber; R. Almeder (eds.) *Allocating Health Care Resources, Biomedical Ethics Reviews*, Totowa: Humana Press, 1995, pp. 55–84; R. Smith, 'Being Creative About Rationing', in *British Medical Journal* 312(1996), pp. 391–392.

[52] R. Dworkin, *op. cit.*, 2002/1993, pp. 205–208; also partially analysed in R. Dworkin, *op. cit.*, 2000/1994, pp. 309–310.

and health are the *summum bonum*, or as René Descartes put it, chief among all goods, which need to be protected and promoted by all means.[53] Everything else is of minor importance besides them.[54] The second component is *equality*. The ideal supposes that even in a society which is otherwise very inegalitarian medical care should be distributed in an egalitarian way so that no one is denied the care he needs simply because of an inability to pay. The third component, which in fact flows from the other two, is the old *rescue principle*, and holds that it is unacceptable when people die, although their lives could have been saved, because the necessary resources were withheld on economic grounds.

This ideal of insulation has exerted great power throughout history. It has served medical practice for millennia and although critical voices crop up from time to time, it is still instinctively accepted by most people and widely supported in political rhetoric.[55] In contemporary political philosophy the ideal is represented for instance by those theories who hold that the provision of medicine

---

[53] Cf. where I have mentioned the idea before: Section 2.2.2.2 *Modern Medicalisation of Life*.

[54] In Discourse on the Method, Descartes writes: '[T]he preservation of health is ... without doubt the first good and the foundation of all the other goods of this life; for even the mind depends so much on the temperament and on the disposition of the organs of the body that, if it is possible to find some means that would render men more wise and more skillful than they have been up until now, I believe that it is in medicine that one ought to be searching for it. It is true that the medicine that is now practiced contains few things of which the utility be so remarkable; but not that I had any intention of disparaging it, I am sure that there is no one, even among those who make a profession of it, who would not admit that all that which one knows in medicine is almost nothing in comparison with that which remains to be known there, and that one might rid oneself of an infinity of maladies, as much of the body as well as of the mind, and maybe even of the enfeeblement of old age, too, if one had sufficient knowledge of their causes and of all the remedies that nature has provided us.' In: R. Descartes, *Discourse on the Method*, ed. and trans. G. Heffernan, Notre Dame: University of Notre Dame Press, 1994, p. 87. 'la conservation de la santé ... est sans doute le premier bien et le fondement de tous les autres biens de cette vie; care même l'esprit dépend si fort du tempérament, et de la disposition des organes du corps que, s'il est possible de trouver quelque moyen qui rende communément les hommes plus sages et plus habiles qu'ils n'ont été jusqu'ici, je crois c'est dans la médecine qu'on doit le chercher. Il est vrai que celle qui est maintenant en usage contient peu de chose dont l'utilité soit si remarquable; mais, sans que j'aie aucun dessein de la mépriser, je m'assure qu'il n'y a personne, même de ceux qui en font profession, qui n'avoue que tout ce qu'on y sait n'est presque rien, à comparaison de ce qui reste à y savoir, et qu'on se pourrait exempter d'une infinité de maladies tant du corps que de l'esprit, et même aussi peut-être de l'affaiblissement de la vieillesse, si on avait assez de connaissance de leur causes, et de tous les remèdes dont la nature nous a pourvus.' In: R. Descartes, 'Discours de la Méthode, Sixième Partie', in A. Bridoux, (ed.), *Descartes: Oeuvres et Lettres*, Paris: Gallimard, 1952, 168–169.

[55] See: R. Porter, *The Greatest Benefit to Mankind: a Medical History of Humanity*, New York: Norton & Company, 1997. For critical voices, see: M. Foucault, *Naissance de la clinique*, Paris, Presses Universitaires de France, 1963; I. Illich, *Medical Nemesis – The Expropriation of Health*, London: Calder & Boyars, 1975; *Id.*, 'Body History', in *The Lancet* (1986)2, pp. 1325–1327; P. Skrabanek, *The Death of Humane Medicine and the Rise of Coercive Healthism*, London: Social Affairs Unit, 1994; H. Häfner (hrsg), *Gesundheit – unser höchstes Gut?* Berlin: Springer, 1999; R. Hanson, 'Why Health is not Special: Errors in Evolved Bioethics Institutions', in *Social Philosophy and Policy* 19(2002)2, pp. 153–179.

constitutes a separate sphere of justice within which decency, community, solidarity and equality must reign; like Michael Walzer's theory, or in those theories that focus on the special character of health care, e.g. the early writings of Norman Daniels.[56] The power of the insulation ideal is so great that people think that it might easily be thought to provide the right standard for answering the two questions of justice in health care: 'How much should we spend?' and 'What should we spend it on?' However, so Dworkin urges, this is a serious mistake. For we face the intellectual crisis in our conception of what a just health-care system would be, that is, what answers we should give to the questions of justice just set out, precisely because it has become clear that the insulation ideal, for all its ancient popularity, has become irrelevant. What is more, it is almost wholly useless for answering these questions and the assumption that it sets the proper standard has done more harm than good.[57]

Consider, for instance, the first question posed: the problem of the aggregate expense a decent society should commit to health care, compared to competing needs and values. The ideal of insulation would give the advice of spending *all* the society can on health care until it has reached the level at which no more gain in health or life expectancy is to be expected. Of course no society ever did organise its household in that way, any more than a sane individual organises his life with the sole goal of making that life as long and healthy as possible, at the expense of many other valuable things – after all, we have seen that most people do trade health for other goods, like when they commute longer distances for a better job, practice certain sports, go on a skiing holiday or on an exotic hiking tour.[58] In the past centuries, however, there was not such a significant gap between the rhetoric of the insulation ideal and the various medical possibilities. However, with the increasing supply of medical technology during the last decennia, meaning that we have so much more to buy, it is unreasonable that society should treat health as lexicographically prior to all other values and treat longer life as a good that must be protected at all costs.[59]

Once the ancient ideal is confronted with this problem, it has nothing more to say. As Dworkin puts it: 'It simply falls silent.'[60] As a result, philosophers, theorists and medical specialists who nominally subscribe to the ideal of insulation all despair of attacking the first question. After some reflection, one concludes

[56] M. Walzer, *Spheres of Justice. A Defense of Pluralism and Equality*, New York: Basic Books, 1983, esp. ch. 3; M.J. Trappenburg, 'Defining the Medical Sphere', in *Cambridge Quarterly of Healthcare Ethics* 6(1997), pp. 416–434; N. Daniels, 'Health-care Needs and Distributive Justice', 1996/1981, pp. 179–207; *Id., Just Health Care*, Cambridge: Cambridge University Press, 1985. Although Daniels in his recent works increasingly concentrates on the social determinants of health. See for instance: N. Daniels; B. Kennedy, I. Kawachi; A. Sen, *Is Inequality Bad for Our Health?* Boston: Beacon Press, 2000; N. Daniels, 'Justice, Health, and Healthcare', in *American Journal of Bioethics* 1(2001)2, pp. 2–16.

[57] R. Dworkin, *op. cit.*, 2000/1994, p. 309; *op. cit.* 2002/1993, pp. 206–207.

[58] Cf. Section 3.3.2.3 *When Are Limits to Health Care Fair?*

[59] Cf. Section 2.2.2 *Various Causes.*

[60] R. Dworkin, *op. cit.*, 2000/1994, p. 309; *op. cit.*, 2002/1993, p. 206.

that the size of the overall medical budget will be 'decided in politics', which is a way of saying that abstract considerations of justice have nothing much to contribute to this part of health-care discussion, except for mere formal instructions for fair decision procedures in the democratic discussion.[61] When it comes to the issue of setting priorities in health care, the dominant strategy has generally turned out to be a retreat to procedural justice. Ronald Dworkin urges both that this strategy is undesirable, and that the dominance of the insulation ideal has been a hindrance, and not just a positive contribution to achieving just health care.

Regarding the second question, the question of distribution, the insulation ideal adds that justice will require that the budget be spent in a fair way. However, how does it help us to define a fair distribution? It tells us something negative and undoubtedly very important: that medical treatment should not be distributed according to ability to pay. Nevertheless, we need a more positive advice. If health care rationing is necessary, what then should the principle of rationing be, if it is not our purse? Once again, the ancient ideal has very little to say. The egalitarian impulse of the ideal seems to recommend that medical care should be distributed according to some principle of need.[62] However, this concept, as we have seen, is multiply ambiguous and cannot solve the matter, Dworkin argues.[63] For how should we balance needs? Does someone need an operation if it might save his life but is highly unlikely to do so? Is someone's need for life-saving treatment affected by the quality his life would have after successful treatment? Does someone need treatment less at 70 than at 40? Therefore, the old ideal of insulation fails to answer our second question, as well as our first. Its proposal is fatally ambiguous. We need a different, more helpful statement of ideal justice in health care.

5.4.2 The Prudent Insurance Principle

Dworkin's alternative approach to justice in health care is based, not on the insulation of health care as a separate sphere of justice or activity, but on the contrary, on the *integration* of health care into competition with other goods. The central idea is:

> We should aim to make collective, social decisions about the quantity and distribution of health care so as to match, as closely as possible, the decisions that people in the community would make for themselves, one by one, in the appropriate circumstances, if they were looking from youth down the course of their lives and trying to decide what risks were worth running in return for not running other kinds of risks.[64]

Dworkin's *prudent insurance ideal* argues that we should allocate resources between health and other social needs, and between different patients who need

[61] *Ibid.*, 2002/1993, p. 207.
[62] Cf. Section 2.3.2 *Needs and Preferences in Health Care.*
[63] R. Dworkin, *op. cit.*, 2000/1994, p. 310; *op. cit.*, 2002/1993, pp. 207–208.
[64] *Ibid.*, 2002/1993, pp. 208–209.

treatment, by trying to imagine what health care would be like if it were left to a free and unsubsidised market that would be corrected in three ways.[65] In this imaginary situation, each individual is free to purchase health-care insurance as much or as little as he wishes to do.

The first correction is an analogy to the ambition-sensitive auction. The economic structure, including the distribution of income and wealth, is as fair as it possibly can be. In Dworkin's view, this means that the economic structure treats all members of the community with equal concern when it divides resources equally, and then leaves each member free in principle to spend those resources designing a life that each believes valuable. This conception will not make people equal in the amount of money or goods each will have at any particular time, nor will it mean that everyone will lead the same kind of life. Some people will have invested and some will have consumed. Some will have spent early and some will have saved for later. The result will nevertheless be egalitarian, because the choices people will have made will answer to their own conceptions of a good life. As such, equality of resources is measured by the opportunity costs of each particular resource.

Secondly, the public at large has full and good information about the value, cost and effectiveness of different medical treatments. In other words, everybody knows what very good doctors know.

Thirdly, no one – including insurance companies – has any knowledge about the genetic, cultural or social influences on our health. No one would be in a position to say that a person has higher probability to contract sickle-cell anaemia, diabetes or some other disease than another person.[66]

The question that Dworkin invites us to ponder is: which treatments would we prudently choose to be insured for, and which would we regard as not worth the cost of the insurance? What kind of health-care arrangements would develop in such a community? How much of its aggregate resources would end up devoted to health care? And how would medical treatment be distributed?[67]

Dworkin claims in advance that whatever that imaginary society will spend on health care through collective institutions governed by individual decisions

---

[65] R. Dworkin, *op. cit.*, 2000/1994, pp. 310–312; *op. cit.*, 2002/1993, pp. 209–210; R. Smith, 'Being Creative about Rationing', in *British Medical Journal* (1996)312, pp. 391–392.

[66] Note that these latter two corrections neutralise Kenneth Arrow's problems of both uncertainty of the quality of the product, and of informational inequality. Cf. Section 2.3.1.2 *Equal Welfare?* Note moreover that an important issue, which is nevertheless raised by the responsibility argument, is ignored in the prudent insurance ideal. What is also excluded is information relating risk of disease to voluntary chosen behaviour (should insurance companies be in a position to charge smoking, or mountain climbing higher premiums?). If so, what then counts as voluntary behaviour? I will consider such issues in detail in Section 5.5.3 *On Personal Responsibility and the Right to Health Care.*

[67] R. Dworkin, *op. cit.*, 2000/1994, pp. 312–313; *op. cit.*, 2002/1993, pp. 210–211.

would be just – both in the aggregate and in its distribution. This follows from the central assumption that Dworkin formulates as follows:

> [A] just distribution is one that well-informed people create for themselves by individual choices, provided that the economic system and the distribution of wealth in the community in which these choices are made are themselves just.[68]

No one would resist this claim if everyone accepts the conception of social justice that assigns individuals responsibility for making the ethical choices for their own lives against a background of competent information and a fair initial distribution of resources. If one accepts that conception of justice, then one will accept this claim. It cannot be faulted on grounds of justice.

5.4.3 Implications

Carrying the model through, he discusses its implications for our own society. Although it is impossible here to provide elaborate policy measures in full detail and with practical precision, he offers some speculations regarding the question what health-care arrangements the hypothetical society would make:

> Of course, what is prudent for someone depends on that person's own individual needs, tastes, personality, and preferences, but we can nevertheless make some judgements with confidence that they would fit the needs and preferences of most [individuals in industrialised societies].[69]

It is important to consider what arrangements the hypothetical society would generally make, Dworkin argues, because these decisions can serve as a *guide* to what we should do to improve justice in our own real, imperfect and often unjust circumstances.

5.4.3.1 A Limited Moral Right

Firstly, the prudent insurance strategy presumably allows one to determine what justice would require in the way of a *decent minimum*. Dworkin speculates that private insurance would develop into large collective insurance arrangements, which might result in something close to a comprehensive public health insurance scheme for a *basic level* of provision, with supplemental private insurance possibilities.[70]

---

[68] *Ibid.*, 2000/1994, pp. 312–313.

[69] *Ibid.*, p. 313.

[70] R. Dworkin, *op. cit.* 2000/1994, pp. 315–319; *op. cit.*, 2002/1993, pp. 211–212, 215–216, 219–221; *Id.*, 'The Place of Liberty,' 1987, repr. in *id.*, *Sovereign Virtue*, 2000, pp. 124, 127, 130–131, 171–172. Dworkin believes that a universal and mandatory insurance scheme for comprehensive basic health care would probably be the best way to approach the hypothetical insurance idea. For reasons of justice, a progressive tax system would be the correct implementation method. Suppose for instance, that the standard medical package nearly everyone would purchase through collective insurance arrangements would include a particular set of benefits. This set of benefits should make up the basic package that must be made available to everyone in our community. However, through this, some relatively low-income people may end up paying a higher share of their actual income for medical care premium (for themselves and, through taxes, for others) than they would have chosen to pay. They might have less left over for other expenses, which may not seem, particularly to them, an improvement in justice. A progressive tax system, Dworkin argues, would be the best way to avoid this.

As such, Dworkin's position combines an argument for a moral right to health care with a limitation on that right, both based on the idea of the prudential insurer. Establishing precise entitlements invariably involves trade-offs, and no right to health care will trump all competing claims of social utility or the common good when larger questions of macro allocation are at stake.

5.4.3.2 Probably Not Including…

Consequently, he identifies a number of disparities between the choices that people would probably make in this hypothetical world and the decisions of health-care providers in the real world.[71] Let us consider some of his proposals.

LIFE-EXTENDING TREATMENT AND PERSISTENT VEGETATIVE STATE. Dworkin suggests, for example, that few people would insure themselves for life-sustaining treatment if they fell into a persistent vegetative state (PVS); yet thousands of people are kept alive in such a condition at any time today.[72] The substantial sum spent year after year in insurance premiums to provide that coverage would be at the expense of education, job training, culture, investment or travel experience, and such sorts of things that would have enhanced someone's actual, conscious life. The opportunity cost of such insurance would be irrationally high.

LIFE-SAVING TREATMENT, ALZHEIMER'S DISEASE AND OLD AGE. Furthermore, it is reasonable to think that almost no one would purchase insurance providing for expensive medical intervention, even of a life-saving character, after he entered the late stages of irreversible dementia. Almost everyone would regard that the money spent on premiums for such insurance would have been better spent making life before dementia more worthwhile. Although most prudent people would want to buy insurance to provide decent-quality custodial care, in conditions of dignity and adequate comfort, if they became demented, no one would insure for expensive, life-saving intervention in this situation. The same goes for relatively old age. The prudent insurer might not be inclined to insure for expensive technology whose main results benefit people in relatively *old age*, like life-saving treatment over the age of 85.[73] Paying all our lives to secure such kind of technology, if we would need it at all, might seem a poor decision when it means that we run a higher risk than we need to an otherwise less satisfactory life in general.

EXPENSIVE TREATMENT AND TERMINAL ILLNESS. A further suggestion is that very few people would choose to insure for very expensive medical treatment in the last months of a terminal illness; treatment which would lengthen their lives

Note, however, that Dworkin also argues that a government's choice for one specific arrangement or another is more likely to depend on considerations other than justice. Nor is justice decisive of that issue in one way or another as long as the various possibilities include the idea of a basic package equally available to all. The main issue of justice consolidates in the question of what should be in that basic package. In R. Dworkin, *op. cit.*, 2002/1993, pp. 215–216.

[71] *Id.*, *op. cit.*, 2000/1994, pp. 313–315; *op. cit.*, 212–214.

[72] J. Butler, *op. cit.*, 1999, p. 22.

[73] R. Dworkin, *op. cit.*, 2000/1994, p. 315; *op. cit.*, 2002/1993, p. 214.

for a few additional months. When we imagine the condition in which most such patients undergoing that intervention live, compared with the value the premiums necessary to purchase that insurance could add to their earlier lives if spent in other ways, it is reasonable to believe that most people would not think it prudent to insure for such treatment. Yet some 40% of medical expenditure during the last year is on people in the last four months of their lives.[74] This is not to say, Dworkin emphasises, that most people would not *want* those additional months. For indeed many people want to remain alive as long as possible, provided they remain conscious and alert, and provided the pain is not too great and the quality of life reasonably good. The point is rather that they would not want those additional months *at too great a cost* of sacrifices in their earlier, vigorous life; a cost that would be necessary if they had to make that choice. On the other hand, they would certainly want insurance to provide the much less expensive care that would keep them as comfortable and as free of pain as possible.[75] In this regard, the relevance of my previous defence of a contemporary Socratic perspective becomes clearer.[76] As the gap between all that is technically possible and what should be equitably accessible increases, our moral responsibility and the answer to the question what we owe to each other at the bar of just health care has less to do with doing all we can in the name of health as the *summum bonum*, and ever more with providing *equal access* to the *abundance* of respectful, good, decent, attentive and qualitative basic *care* for all.

Speculative High-Tech Medicine. How much further can we go down this road? How much more insurance can we be reasonably confident people would not buy in the circumstances we are imagining? Dworkin raises one further issue that is of major importance and will become increasingly more critical in the next decades. That is, how far would people in the imagined community go in making provision for access to the ultra-expensive high-tech medical equipment now in use or being developed, like the various forms of research

---

[74] J. Lubitz; G. Riley, 'Trends in Medicare Payments in the Last Year of Life', in *New England Journal of Medicine* 328(1993), pp. 1092–1096.

[75] Note, however, that the abstract ideas of age-based rationing and limiting expensive treatment in the terminal stages of life are contested. Critics contend that age-based rationing of life-extending technologies would not save substantially on resources, in part because the provision of care, including long-tem care and support services, is expensive and cannot always be sharply differentiated from the care that prolongs life. Experts argue that saving the costs of the last few weeks of life would not produce large reductions of costs overall, and they note great difficulties in predicting the final weeks of life for many patients. See: P. Zweifel; S. Felder; M. Meiers, 'Ageing of Population and Health Expenditure: a Red Herring', in *Health Economics* 8(1999), pp. 485–496; D.W. Jahnigen; R.H. Binstock, 'Economic and Clinical Realities: Health Care for Elderly People', in R.H. Binstock; S.G. Post; L.S. Mills (eds.), *Too Old for Health Care? Controversies in Medicine, Law, Economics, and Ethics*, Baltimore: Johns Hopkins University Press, 1991, pp. 13–43; J.M. O'Connell, 'The Relationship between Health Expenditures and the Age Structure of the Population in OECD Countries', in *Health Economics* 5(1996), pp. 573–578.

[76] Cf. Section 2.4.4 *A Contemporary Socratic Perspective*.

in molecular biology?[77] Undoubtedly these technologies will save *some* lives, but just as undoubtedly at a cost that would seem very high when we consider how a community might use the funds in other ways (like for instance enhancing economy and providing more jobs and a higher standard of living conditions for more people, which are, lest it be forgotten, important social determinants of health). In this line of reasoning, people might not spend to insure for highly expensive speculative technology even though it could save some lives, like for instance separating a Siamese twin when there is only a minute chance of their survival.

5.4.3.3 But Prudently Providing

Inversely, we might use our speculations about what people in the imaginary community would consider prudent to provide for themselves, as a guide to help us define what justice demands everyone should have. What should be in the basic package of health-care coverage that should be available to everyone, at a reasonable cost and be supplied without charge to those who cannot carry that reasonable cost themselves?

THE BASIC PACKAGE. Informed and reflective people in the imagined society, ultimately deciding for themselves how to allocate their resources, might make the following decisions. They might pay to provide for life-saving techniques for diseases that tend to occur relatively *early in life*, particularly when these techniques have a high probability of success. As such, most people would consider it prudent to insure for immediate and expert treatment for handicapping conditions in childhood, including treatments which are traditionally in short supply such as those for children with speech or learning difficulties.[78]

The point of the prudent insurance principle is that if *most prudent* people would buy a certain level of medical coverage in a free market if they had average means – i.e. if nearly everyone would buy insurance covering primary medical care, hospitalisation when necessary, standard prenatal and paediatric care, routine examinations, inoculations and other preventative medicine, and finally, respectful, decent and attentive long-term care – then the *fairness* or unfairness of our real-time society can be measured according to the amount of people that do not have such coverage now. The above quoted elements of medical coverage would constitute the basic package that any responsible health-care system would establish.

On the other hand, if very *few prudent* people would want to buy insurance covering a much higher level of coverage – like some heroic medical technologies – it would be unjust to force everyone to have such insurance through a mandatory scheme. There are of course exceptions to the prudent insurance principle: some people have special preferences and would make decisions different from those of most others. It seems fair however, to construct a mandatory coverage scheme

[77] See also R. Dworkin, 'Playing God: Genes, Clones, and Luck', in *id.*, *Sovereign Virtue*, 2000, pp. 427–452.

[78] J. Butler, *op. cit.*, 1999, p. 22.

on the basis of assumptions about what all but a small number of people would consider appropriate, allowing those few who would be willing to spend more on special care to do so, if they can afford it, through supplementary insurance.[79]

5.5 TOWARD EVALUATION: QUALITIES AND DRAWBACKS

It is time to take stock and consider some qualities and drawbacks of Dworkin's contribution to the debate on justice in health care. In the following sections, I will consider three main issues. Firstly, I will concentrate on Dworkin's dealings with the issue of limits and the fact of scarcity. Secondly, I will concentrate on his statement that health has no special status and examine whether this is a tenable position in the light of his own theory. Whether he can consistently maintain this position will turn out to be doubtful. Finally, it is time to consider the debate between Dworkin's resource egalitarianism and the luck-egalitarian critique of Cohen and Arneson. I will consider this discussion in the light of the following question: does justice require society to try to restore the health of those who voluntarily put their own health at risk?

5.5.1 Reassessing the Fact of Scarcity and the Issue of Limits

An important quality of Dworkin's contribution is that he puts the fact of scarcity and the issue of limits back on the menu. Thus, he provides a possible answer to contemporary health-care problems like issues of cost control, questions regarding the role of personal responsibility, regarding limits to high-tech medicine and so on. His proposal serves as an interesting alternative to the dominant strategy of retreat to procedural justice.

5.5.1.1 The Internal Health-Care Trade-Off

Firstly, his prudent insurance principle provides a step forward in trying to render the goals of quality care, equal access, freedom of choice and economic efficiency more coherent. Most importantly, his critique on the insulation ideal has made clear that this ideal increases the tension between the various health-care goals rather than provide the possibilities to reconcile them. As such, Dworkin is right in claiming that the assumption that it sets the proper standard has done more harm than good.[80] Modern medical technology, which produces more and more ways to spend large sums on health care, puts this ancient ideal to the test and reveals it as no longer tenable.[81]

As we have seen, the prudent insurance principle expresses a *middle course*. It balances the anticipated value of medical treatment against other goods and

---

[79] R. Dworkin, *op. cit.*, 2000/1994, p. 315, endn. 10, p. 492, *op. cit.*, 2002/1993, pp. 219–221.

[80] R. Dworkin, *op. cit.*, 2000/1994, p. 309; *op. cit.* 2002/1993, pp. 206–207.

[81] J. Butler, *op. cit.*, 1999, p. 17.

risks, and it presupposes that people might think they lead better lives overall when they invest less in doubtful or superior quality medicine and more in making life successful or enjoyable, or in protecting themselves against other risks, like unemployment, that might also blight their lives. Health policy makers might well decide that while most prudent people would provide themselves and their families with the basic care mentioned above, they would forego heroic treatment of improbable value in return for more certain benefits like education, housing and economic security. If this is reasonable, then justice demands that a universal health scheme should not provide such enhanced treatment

It is reasonable to think that the Dworkinian middle-course leads to the following conclusions about the trade-off questions of quality, equality, liberty and efficiency. Firstly, the basic package of health-care coverage, to which everyone should have *equal* access, would include decent *quality* primary care – like hospitalisation when necessary, standard prenatal and paediatric care, routine examinations, inoculations and other preventative medicine, decent long-term care and life-saving treatment with high probability of success for diseases that tend to occur relatively early in life. However, it would be limited to this because of otherwise unreasonably high opportunity costs. No speculative high-tech medicine would be included, nor expensive life-saving or life-extending treatment at the clear end of life.

As for the goal of *liberty*, it has turned out that the initial hypothetical insurance idea set up within a free market framework ultimately leads to a collective mandatory scheme for the basic package. This, however, does not count for the additional private insurance package.[82]

And finally, against picturing *equality* as the antagonist or victim of the values of *efficiency* and *liberty* that are served by the market, Dworkin holds that the idea of efficiency and of an economic market, as a device for setting prices for a variety of goods and services, must be at the centre of any attractive theory of equality of resources.[83] Efficiency, he argues, should not be taken as an enemy of justice, but rather as required by justice. He emphasises that his proposal is not 'an argument from efficiency as distinct from fairness', but rather an argument that in 'the circumstances described ... efficiency simply is fairness, at least as fairness is conceived under equality of resources'.[84] The same goes for his view on limits to care. In presenting the prudent insurance principle, Dworkin argues that if we would take this principle as 'our abstract ideal of justice in health care, we would ... accept certain limits on universal coverage, and we would accept these not as compromises with justice but as required by it'.[85]

---

[82] R. Dworkin, *op. cit.*, 2000/1987, pp. 120–183; *op. cit.*, 2000/1994, p. 315; *op. cit.*, 2002/1993, pp. 219–221.

[83] *Ibid.*, 2000/1994, p. 66.

[84] *Ibid.*, p. 84.

[85] *Ibid.*, p. 315.

5.5.1.2 Efficiency and Long-Term Care

By arguing that justice requires limits, Dworkin rightly takes the Humean notion of the circumstances of justice as implying scarcity into account, but without allowing its harsh implication of excluding the non-profitable disadvantaged. For similar to Nussbaum, Dworkin claims that long-term care for the chronically ill and disabled is a matter of justice, and not of charity or compassion. In Dworkin's view, disabilities count as negative internal resources. This means that to be born disabled or susceptible to major health problems is to have an internal resource deficit that according to his resource-egalitarianism makes a call on justice.[86]

Instead of concentrating mainly on the bottomless pit problem that is related to long-term care, he focuses his efficiency thinking on another drain on social resources, namely that of medical technology, creating a technological bottomless pit. Whereas the first is related to the problem of it being impossible to raise certain people above a basic level of independent functioning, the second arises from the absence of an upper bound. Dworkin's prudent insurance principle clarifies that excluding long-term care is much more problematic than excluding some possibilities of high-tech medicine. Setting limits in the field of high-tech medicine produces much less objections of justice than setting them in the field of long-term care.

5.5.1.3 Additional Private Insurance

This, however, does not mean that we have to prohibit people to privately purchase more expensive high-tech medicine or better quality care in addition to the basic package. Although the prudent insurance principle holds that if only very *few* prudent people would want to buy insurance with a much higher level of coverage – including some heroic medical technologies – it would be unjust to force *everyone* to have such insurance through a mandatory scheme, it nevertheless seems fair to allow those few who would be willing to spend more on special care to do so, if they can afford it, through supplementary insurance.[87]

This would mean, however, that some people would have better medical care – some people would live longer and healthier lives – only because they have more money. Although many people, backed up by the insulation ideal, would feel that this comprises solidarity, Dworkin argues that this should not pose any problems of justice. This should not surprise us, since the spirit of the Dworkinian argument has been all along that no one can complain *on grounds of justice* that he has less of something than someone else does, so long as he has all he would have if society were overall just. By hypothesis, the poor are not denied what they would have if economic justice were perfect. Dworkin is very much aware of the fact that many people will hate this argument and think it intolerable. Two reactions serve as an answer.

---

[86] See also J. Roemer, *op. cit.*, 1996, pp. 246–249.

[87] R. Dworkin, *op. cit.*, 2000/1987, pp. 124, 127, 177–178; *op. cit.*, 2000/1994, p. 315, endn. 10, p. 492; *op. cit.*, 2002/1993, pp. 219–221.

Firstly, he invites us to bear in mind that the inequality in medical care that is now being considered is, in an important respect, relatively benign compared to other inequalities in society. If health care were rationed in the way he proposes, then everyone would have at least the medical care he would have in a just society, and this would not be true of most other cases of resource allocation. In education, employment, culture, recreation, and a host of other goods and opportunities that for most of us make up the value of being alive, the poor would continue to have far less than they would have if we had reformed not just health care but our economic and social life more generally.

Secondly, if we somehow manage to succeed in providing the poor with the medical care, justice requires, it would be perverse, given that a rich man can spend whatever he likes on more comfortable housing or better education for his children, not to allow him to spend whatever he likes on more expensive health care. Instead, we would do better to put an excise tax on special health care, and use the proceeds of that excise tax to improve public education, the economic infrastructure or other areas that would make the community distinctly more egalitarian. This in its turn, as we have seen, would have important effects on the distribution of health.[88] It may be thought, after all, that medical advances and technologies driven by demand from those at the highest income levels in a given society will in due course trickle down to improve medical care for those at lower levels. Perhaps these medical advances and technologies would not be developed and realised in the first place if it weren't for the rich and/or private medical insurance who pay for them. Once established, it may be possible that their costs may fall or they may have spin-off benefits for the general quality of medical care. Note, however, that this is very much an empirical issue and that the facts may be counterintuitive. Leaving empirical aspects aside, the point of the argument is that allowing this form of inequality – by setting limits on the basic health-care package and allowing additional private coverage – does not create problems of justice as long as it involves forms of care that would, according to the prudent insurance principle, not be included in the basic package.

5.5.1.4 Reflective Equilibrium

A fourth quality is that setting limits according to the prudent insurance principle is related to the reflective equilibrium. Dworkin's suggestions about how people would behave are speculative and made within an idealised situation. In real life, resources are unjustly distributed and people do not have state-of-the-art information about medicine. In addition, our insurance companies do

[88] Cf. N. Daniels et al., *Is Inequality Bad for Our Health?* Boston: Beacon Press, 2000; R.G. Wilkinson, *Unequal Societies: The Afflictions of Inequality*, London: Routledge, 1996; Section 3.3.2.2 *Which Health Inequalities Are Unjust?* It is interesting to see how this remark brings Dworkin's position closer to that of Norman Daniels. Instead of investing all of our resources directly into health care it is important to remain attentive to the social determinants of health. It is not reasonable to expect from the health care system to immediately react on all developments in the field of acute care and medical technology. In this respect, one can safely say that not all health inequalities are unjust.

know that risks are higher among certain groups. However, Dworkin argues, these speculations can offer some guidance toward public opinion research. If information of this kind were put into the public domain, and challenged and debated there, the resulting discussion would be at least minimally informative about how much people value what kind of care.[89]

As such, the prudent insurance test is not to be understood as a technical cost-benefit test, but as including people's sense of priorities:

> no matter how much information an agency seeking to apply the prudent insurance test is able to gather, its results must be provisional, open to revision on the basis of further evidence of public preference as well as of medical technology and experience.[90]

This means that further refinement and adjustment must be possible. Any judgement that the officials of a society might make would be speculative and open to a variety of objections.[91] If, after a society has established a basic coverage package, it would turn out that a substantial amount of people of average income would buy supplementary insurance, in spite of its expense, the basic package should be expanded.

5.5.2 Health: One Resource Among Others?

A second issue that deserves careful consideration because it might turn out to be problematic is the status of health and health care in Dworkin's theory. Although Dworkin's resource egalitarianism treats health, in reaction to Rawls, as an *internal resource* it refuses to give health any special role or status. Health and health care, Dworkin argues, should not be insulated in a separate sphere of justice or be given priority over other goods. In itself, they do not raise special issues of justice and should thus be integrated into competition with other goods. The idea of putting health and health care on a par with other goods is, as we have seen, an important part of his critique on the insulation ideal. I have three remarks on this point.

5.5.2.1 Conception of Just Health Care

Firstly, Dworkin's critique on the insulation ideal is important and valuable. Spending all on health care, trying to protect life and health by all possible means, whatever the opportunity cost of doing so has become an untenable ideal. On the one hand, his critique has shown that an important part of the contemporary crisis in health care is due to the insulation-based conception of a just health care system. The gap between the rhetoric of that ideal, and all that is medically possible for a community to do, has become too significant. On the other hand, however, it remains to be seen whether this critique necessarily implies that health and health care have no special value whatsoever.

---

[89] R. Dworkin, *op. cit.*, 2002/1993, pp. 217–218.
[90] *Id.*, *op. cit.*, 2000/1994, p. 318.
[91] *Ibid.*, p. 79.

5.5.2.2 *The Argument of Fair Income Shares*

A first argument that questions the above conclusion on the status of health and health care can be found in Daniels's critique on the argument of fair income shares. The argument holds, similar to Dworkin's starting point, that if we could agree on a theory of distributive justice that gives us a notion of a fair income share, individuals could then protect themselves against the risk of needing health care by prudently buying the relevant insurance, provided appropriate conditions obtain. This way, there would be protection against expensive but rare needs for health care, for which relatively inexpensive insurance can be bought. So too, common but inexpensive services can either be risk-shared through insurance or paid out of pocket without great sacrifice. However, expensive and potentially common needs – for example, life-saving treatment or expensive technological intervention in the final stage of life – would not become a drain on social resources since individuals who want protection against them would have to buy expensive insurance out of their own fair shares. Consequently, meeting health-care needs would not create a bottomless pit into which we are forced to drain all available resources.[92] However, in Daniels's view the argument will not do:

> we cannot just finesse the question of whether there are special issues of justice in the distribution of health care by assuming that fair shares of primary goods will be used in part to buy reasonable health insurance.[93]

This is because a share of resources is *fair* only if it is enough to buy a *reasonable* health insurance package. Such a package can meet the health care needs it is reasonable for people to want to be protected against. A share that is too small to do this is inadequate, and there is something unacceptable about it. The key assumption underlying this scheme is that the notion of *fair* shares and of *reasonable* health insurance presupposes a notion of basic health care needs it is reasonable for a *prudent* person to insure against.

This notion of prudence has a structure that reflects a concern to meet basic needs, including health care needs.[94] These are then not considered to be just one preference among many. The idea of *reasonable* insurance for a *prudent* person implies that health and health care may be objectively more important than individual choices reflect.[95] In Daniels's view, as we have seen, health resources cannot

---

[92] For the argument of fair income shares Daniels appeals to the work of Charles Fried, *Right and Wrong*, Cambridge, MA: Harvard University Press, 1978, pp. 126 ff. See: N. Daniels, 'Rights to Health Care: Programmatic Worries', in *Journal of Medicine and Philosophy* 4(1979)2, pp. 174–191; *Id.*, 'Health-Care Needs and Distributive Justice', 1981, repr. in *id.*, *Justice and Justification*, Cambridge: Cambridge University Press, 1996., p. 180–181, endn. 2, p. 202; *Id.*, *Just Health Care*, Cambridge: Cambridge University Press, 1985, pp. 20–23. Opposed to what could be expected, given the similarity of the arguments and given the fact that Dworkin also refers to the works of Fried, no references of direct interaction on the issue between Daniels and Dworkin were found.

[93] N. Daniels, *op. cit.*, 1985, p. 45.

[94] Cf. Section 2.3.4 *Health Care Needs*.

[95] *Ibid.*, pp. 21–24.

be treated as on a par with other goods that we buy out of our private resources for several reasons. Health care is comparable to education. Both play a critical role in enabling fair equality of opportunity and therefore as inheriting the special priority attributed to equality of opportunity. Furthermore, both address needs that are unequally distributed among individuals and which can be catastrophically expensive. Everyone needs food and clothing to more or less the same degree, but health care and learning needs differ greatly between people. Thus, health care, like education, is in a separate category from other basic needs, such as needs for food and shelter, which we can expect people to meet out of their fair income shares.[96]

5.5.2.3 Internal Incoherence

That the notion of prudence reflects a concern for something that goes beyond mere individual choice can also be found in Dworkin's own position. Although it is difficult for liberals to deny someone's prerogative to make trade-offs freely among goods that are just goods for him, Dworkin makes room for the fact that some goods may not be available for trade-offs on the same terms: 'We may have paternalistic reasons', he says, 'for limiting how much any individual may risk'.[97]

Although his critique on the insulation ideal is valuable, I do not believe that this requires abandoning the special status of health and health care. It can be safely argued, I believe, that health is an *important bonum*, without implying that it is the *summum bonum*, which should be protected and promoted by all possible means. Dworkin's argument that health is no more special than other goods is interesting within the framework of his hypothetical insurance mechanism, and surely has a function within this thought experiment, but as soon as we leave this position and consider the prudent insurance principle, it can no longer be defended. This is for the sole reason that the special character of health is implied in the notion of *prudent* insurance itself.

5.5.3 On Personal Responsibility and the Right to Health Care

Another issue that is related to the idea that health is no more special than any other good, concerns the role of personal responsibility in the right to health care. In view of the fact that individuals have a significant degree of control over and responsibility for their own health condition we are led to ask: does justice require society to try to restore the health of those who put their own health at risk? Why should the relatively more risk-averse pay for the risky habits of others? Should health insurance schemes charge significantly higher premiums to those who voluntarily choose to put their health at risk?[98] Although it is Dworkin's basic aim to bring into sharper focus the role of responsibility in

---

[96] Cf. Section 3.3.2.1 *The Special Moral Importance of Health Care?* See also *ibid.*, pp. 46–47.

[97] R. Dworkin, 'Equality of Resources', 1981, p. 295; 2000, p. 75; see also *op. cit.*, 2000/1994, pp. 313, endn. 6 and 7, p. 492.

[98] The issue at stake here is, first of all, the use of responsibility as a moral notion, by which society expresses approval or disapproval of certain forms of behaviour. As a moral notion, it implies that

distributive justice, he does not tackle this question.[99] Let us consider his position in this debate.

5.5.3.1 Dworkin's Cut

First of all, Dworkin argues that justice requires compensating individuals for aspects of their situation for which they are not responsible – i.e. their resources and endowments – and which hamper their achievement of whatever is valuable in life, but *only* for those aspects. Inequalities between individuals owing to acts or beliefs for which they are responsible – i.e. their tastes and preferences – are of no concern for justice. Accordingly, Dworkin sets the responsibility cut between preferences (or ambitions) and resources (or endowments): we should allow the distribution of resources to be *ambition-sensitive* and *endowment-insensitive*.

5.5.3.2 Brute Luck and Option Luck

Furthermore, Dworkin wishes to hold people responsible for *gambles* they chose to take in deciding whether or not to purchase insurance. In the same line of reasoning it might be argued that they should be held responsible for gambles they choose to take directly with their health itself, say, by smoking, bad eating habits, skiing or mountain climbing. This claim can be elucidated by means of the distinction between brute luck and option luck – the smoker who develops lung cancer has had bad option luck. The same goes for the skier with the broken leg.[100]

At first sight, these two distinctions – between preferences and resources, and between brute and option luck – give the impression that according to Dworkin's theory, people could forfeit their right to health care, due to voluntary risky behaviour, even when they wish to preserve it.[101] No one owes us anything by

---

inequalities could be legitimated as some sort of ex post punishment for mistakes and reward for good decisions. Persons should suffer the results of their own blameworthy behaviour. This use of responsibility is different from a neutral or descriptive use, which refers to the fact that something can be the result of my own doing, my own voluntary choice, that is, that some outcome can be my own fault, or the result of factors that are within my control. Yet another, again more moral interpretation of responsibility would be responsibility by delegation. Here, the crux is whether society provides the necessary conditions for the citizens to make decisions of which society decides that the citizens are responsible for further outcomes. Of course, these different interpretations of responsibility are not easily interchangeable, and there is much discussion about it. For an illuminating overview, see: K. Devooght, Essays on Responsibility-Sensitive Egalitarianism and the Measurement of Income Inequality, non-published Ph.D., Catholic University of Leuven, 2003, esp. pp. 7–36. Here, however, I will ignore the theoretical discussions on the different interpretations of responsibility and simply convey the gist of the moral use of the concept in the debate on responsibility and health care. Should we punish people for risky health behaviour?

[99] On the contrary, he explicitly puts it aside. See: *Ibid.*, 2000/1994, p. 312, endn. 4, p. 491; *op. cit.*, 2002/1993, p. 210, endn. 2, p. 222.

[100] *Ibid.*, 2000/1994, pp. 73–74. See also: N. Daniels, *op. cit.*, 1985, p. 56.

[101] Note that the forfeiture view is less about whether a person loses the full range of entitlements under the right to health care, than about whether he will have to carry the costs themselves, or will have to pay higher premiums, or will forfeit access to certain forms of care.

ways of compensation for the preferences we have and the choices we have made. Any medical problems that arise from choices count as bad option luck and should not give rise to legitimate claims for compensation at the bar of justice. On close reading, however, both Dworkin's cut and the distinction between brute and option luck turn out to be pulling in opposite directions.

5.5.3.3 Luck-Egalitarianism: Relocating Dworkin's Cut

According to the luck-egalitarian view of justice, developed by Cohen, Arneson and Roemer, Dworkin's responsibility cut between preferences and resources fails to support the distinction between brute luck and option luck because it fails to place *choice* in a central position.[102] If justice requires compensating individuals only for aspects of their situations for which they are not responsible, the cut between considerations that gives rise to issues of redistribution is not the cut between preferences and resources, as Dworkin would have it, but between choice and mere luck.[103]

What really matters, the luck-egalitarians argue, is whether people are responsible for their disadvantages or whether they are a matter of brute luck, because *justice* should tolerate only inequalities for which people are responsible, not inequalities that are a matter of luck. This view twists the previously dominating focal point around: it is concerned not just with the ways in which health conditions affect people's opportunities for other goods, but also with how people *choose* to response to their opportunities for health itself. In this view, it is unjust for someone to be poor because he was born blind and had fewer opportunities for reasonably paid employment, but it is not unjust if someone is ill because he freely chose to ruin his health for the sake of ephemeral pleasures (although

[102] G.A. Cohen, 'On the Currency of Egalitarian Justice', in *Ethics* 99(1989)4, pp. 906–944; R. Arneson, 'Equality and Equal Opportunity for Welfare', in *Philosophical Studies* 56(1989), pp. 77–93; J. Roemer, 'A Pragmatic Theory of Responsibility for the Egalitarian Planner', in *Philosophy and Public Affairs* 22(1993)2, pp. 146–166; *Id.*, 'Equality and Responsibility', in *Boston Review* 20(1995)2, pp. 3–7, 15–16; *Id.*, *Theories of Distributive Justice*, Cambridge, MA: Harvard University Press, 1996, esp. pp. 242–245, 276–279, 285–286, 308–309; *Id.*, *Equality of Opportunity*, Cambridge, MA: Harvard University Press, 1998. Their views differ in detail. Whereas Richard Arneson defends a welfarist approach, which comes down to a defence of equal opportunity for welfare, Gerald Cohen and John Roemer defend a so-called midfarist approach, that is, they defend equal opportunity for advantage irrespective of whether the advantage is in welfare or in resources. Cohen defines midfare similar to Sen's definition of functioning. Midfare contains both objectively measurable characteristics and subjective characteristics, that is, it is to be conceived as a vector, which characterises a state between having resources and enjoying welfare: It is 'posterior' to 'having goods' and 'prior' to 'having utility'. In G.A. Cohen, *op. cit.*, 1989, p. 943. Cf. Sen: a functioning is different both from (1) having goods (and the corresponding differences) to which it is posterior and (2) having utility (in the form of happiness resulting from that functioning), to which it is, in an important way, prior. In A. Sen, *Commodities and Capabilities*, Amsterdam: North-Holland, 1985, pp. 10–11. However, I will ignore the differences among the various luck-egalitarian views here and simply convey the essence of the luck-egalitarian approach.

[103] G.A. Cohen, *op. cit.*, 1989, pp. 922, 933.

of course there may be other reasons – e.g. reasons of charity – than reasons of justice to help such a person). Indeed, they argue, this conclusion does not conflict with the rule of fair equality of opportunity since those who are voluntarily risking their health have had the opportunity to be healthy but freely chose to pass it up.

On such a view, health *resources* like other advantages and disadvantages would be of concern to justice only to the extent that people are not responsible for them. Again, health, and health care per se would not have any special status. The same goes for *preferences*, since there are voluntary and involuntary preferences, like cravings or obsessions one wishes one did not have. Preferences are a concern of justice to the extent that people are not responsible for them; i.e. to the extent that they are involuntary, or 'constrained', in the words of Frankfurt.

Accordingly, Dworkin's responsibility cut between preferences and resources needs to be 'relocated', to use a term from Cohen, for it wrongly holds people responsible for the consequences of their involuntary preferences. As Cohen puts it: 'The right cut is between responsibility and bad luck, not between preferences and resources'.[104] Table 1 illustrates the difference between luck-egalitarianism and resource-egalitarianism somewhat more perspicuous.[105]

TABLE 1. The responsibility cut

		Dworkin's cut	
		Preferences	Resources
Luck-egalitarian cut	Choice (controlled)	Brian opts for a private hospital room because he wants to; because he thinks it is more comfortable	Susan's medical treatment is more expensive because she is a happy smoker
	Luck (involuntary)	Julia opts for a private room because she was raised in a family in which room-sharing with strangers is totally inconceivable	Frank's treatment is more expensive because he suffers from an unlucky natural constitution due to a geneticdefect John's treatment is more expensive because he is an unhappy smoker

[104] *Ibid.*, p. 922.

[105] I wish to thank Kurt Devooght for his illuminating suggestions in this matter. The table below is based on his writings. See: K. Devooght, Essays on Responsibility-Sensitive Egalitarianism and the Measurement of Income Inequality, non-published Ph.D., Catholic University of Leuven, 2003, esp. pp. 7–36 and pp. 42–43; E. Schokkaert; K. Devooght, 'Responsibility-Sensitive Fair Compensation in Different Cultures', in *Social Choice and Welfare* 21(2003)2, pp. 207–242, esp. pp. 212–214.

The issue of personal responsibility in health care is an important issue and one that turns up regularly in academic literature and public debate since the mid-1970s.[106] What can we say about this in the light of the debate between resource-egalitarianism and luck-egalitarianism? Note here that, contrary to most writings on the subject, I do *not* characterise Dworkin's theory as 'luck-egalitarian', for the same reason that Cohen gives in his critique: Dworkin's position is not congruent with its own underlying motivation. His master distinction between preferences and resources is less true to the motivation of his own philosophy, i.e. he does not place responsibility as choice – reflected in the distinction between brute luck and option luck – in the forefront. Let us take a closer look at the issue. Three things can be said in line with Dworkin's theory.

Health as Internal Resource. To begin with, Dworkin considers *health* as an internal *resource*. In view of his resource-egalitarianism that aims to be resource-insensitive or endowment-insensitive, health conditions are as such a concern of justice. However, in view of the aim of ambition-sensitivity, what then should we do with cases in which someone's *preferences* lead to an unhealthy or risky lifestyle (like smoking, drug abuse, lack of exercise, bad eating habits, boxing or mountain climbing), and in which one's failing health resources are due to such risky preferences? Two answers to this question can be found in his theory.

Legitimate Paternalism. Firstly, Dworkin defends, as I have said, the possibility of legitimate paternalism. This is an important addition because it allows society to *guide ex ante* personal preferences and options to some extent. Concretely, this creates the possibility of reconciling personal responsibility for one's health condition with social incentives to discourage hazardous behaviour (like consciousness-raising health campaigns, or increased taxes on risky goods like alcohol, tobacco, fatty food, or additional insurance premiums for smokers, skiers or mountain climbers).

Handicapping Tastes. Secondly, with regard to tastes and preferences, Dworkin suggests that compensation is due in the case of 'handicapping tastes'.[107] These tastes include cravings or obsessions that the person concerned *wishes he*

---

[106] R.M. Veatch, 'Voluntary Risks to Health', in *Journal of the American Medical Association* 243(1980)1, pp. 50–55; R.M. Veatch 'What is a "Just" Health Care Delivery?' in R.M. Veatch; R. Branson (eds.), *Ethics and Health Policy*, Cambridge, MA: Balinger Publishing Company, 1976, pp. 127–153; J. Butler, *The Ethics of Health Care Rationing: Principles and Practices*, London: Cassell, 1999, pp. 5–49; G. Dworkin, 'Taking Risks, Assessing Responsibility', in *Hastings Center Report* 11(1981), pp. 26–31; H. Leichter, 'Public Policy and the British Experience', in *Hastings Center Report* 11(1981), pp. 32–39, incorporated in H. Leichter, *Free to Be Foolish: Politics and Health Promotion in the United States and Great Britain*, Princeton, NJ: Princeton University Press, 1991; L.B. Russell, 'Some of the Tough Decisions Required by a National Health Plan', in *Science* 246(1989), pp. 892–896; R.L. Schwartz, 'Life Style, Health Status, and Distributive Justice', in A. Grubb; M.J. Mehlman (eds.), *Justice and Health Care: Comparative Perspectives*, New York: Wiley, 1995, pp. 244–248; Y. Denier, 'On Personal Responsibility and the Human Right to Health Care', in *Cambridge Quarterly of Healthcare Ethics* 14(2005)2, pp. 224–234.

[107] R. Dworkin, 'Equality of Resources', 1981, pp. 302–304; 2000, pp. 81–83.

did not have, because they interfere with his life plans and offer him frustration or even pain if not satisfied. In some cases, these might be tastes that have been cultivated by the person himself, but that he regrets having. He believes he would be better off without them, but nevertheless finds it painful to ignore them – an example would be John, the unhappy smoker. In such cases, these tastes are handicaps, although for other people – like Susan, the happy smoker – they can be an essential part of what gives value to their lives. The basic distinction required by equality of resources, Dworkin argues, is the distinction between those beliefs and attitudes that define what a successful life would be like, which is assigned to the person's preferences and ambitions; and those features of body or mind or personality which provide means or impediments to that success, which are assigned to the person's circumstances. Consequently, Dworkin takes the view that individuals are responsible for their preferences, as long as they *identify* with those preferences, which is *not* the case for handicapping tastes.[108] Given that handicaps count as *resource* deficiencies in Dworkin's view, they are a concern of justice. Coming back to the examples from Table 1 on the responsibility cut, it would also be correct to say for instance, that John's treatment is more expensive, because he is an unhappy smoker. His smoking behaviour is a craving, a preference that he wishes he did not have, something with which he does not identify.

All in all, however, it can be safely argued that it actually doesn't matter whether the harming preference is an identified or alienated taste, because Dworkin's responsibility cut is set between preferences and resources. To the extent that one's preferences cause resource deficiency they have a claim for compensation at the bar of justice. So although it seems at first that Dworkin's distinction between brute and option luck places *choice* in the forefront, the responsibility cut between preferences and resources pulls the issue in the opposite direction. Both John, the unhappy smoker, and Susan, the happy smoker, will receive equal access to medical care because their smoking-related disease is a health *resource* deficiency.[109] In the same line of reasoning, Robert Goodin quotes the example of the person who needs kidney dialysis because of his own botched suicide attempt.[110] This person has a perfectly *non-volitional* need for such a

---

[108] See also J. Roemer, *op. cit.*, 1996, pp. 237–238, 246–247.

[109] For critique on the inadequacy of Dworkin's typology due to such problems, see: G.A. Cohen, *op. cit.*, 1989, pp. 916–934; R. Arneson, 'Equality and Equal Opportunity for Welfare', in *Philosophical Studies*, 56(1989), pp. 77–93; J.E. Roemer, 'Equality of Talent', in *Economics and Philosophy* 1(1985)2, pp. 151–187. Cohen, for instance, argues that the idea of alienated taste is inadequate. Some people, e.g. children, may deserve compensation, even if they are not reflective enough to formulate the idea that they are alienated. In other cases, life-hampering tastes are not ones the individual really wants to renounce. Rather, the individual really just feels unlucky in having such a taste, say because it turns out to be an expensive or unhealthy one (like smoking). Ultimately, Cohen suggests that identification and disidentification with a taste matter 'only if and insofar as they indicate presence and absence of choice'. In G.A. Cohen, *op. cit.*, 1989, p. 927.

[110] Cf. R.E. Goodin, 'The Priority of Needs', in *Philosophy and Phenomenological Research* 45(1985)4, pp. 615–625. See also T.M. Scanlon, 'Preference and Urgency', 1975, esp. pp. 664–669.

machine, although this need resulted from his own undertakings, and the *objective* interpretation of *actual harm* that would result from non-fulfilment of this non-volitional need, might differ from the person's own subjective viewpoint of harm. Without entering the difficult discussion on the various interpretations of actual harm, I wish to point at the essence of what is going on here, namely at the fact that Susan's health *resource* deficiency is just as *basic*, *urgent*, *non-volitional* and *important* to fulfil as John's.

In conclusion, this means that Dworkin ultimately does not take the same path as the luck-egalitarians. His conception of responsibility does not equal voluntary choice. Dworkin's responsibility cut and his distinction between brute luck and option luck both pull in opposite directions. Perhaps this is so because he is aware of the difficulties that come with the luck-egalitarian view. Let us look at two categories of difficulties: firstly, objections that come down to difficulties in practical applicability and secondly, objections of principle based in a consistent understanding of what we owe to each other at the bar of justice.[111]

5.5.3.4 Practical Applicability

To begin with, the practical applicability of the admission of merit considerations in the instance of health care delivery appears limited. A policy of withholding societal funds cannot be justified unless several conditions are met. Firstly, it must be possible to identify and differentiate various *causal factors* in morbidity, such as natural causes, social environment and personal activities, and it must be confirmed that a pertinent disease or illness resulted from *personal activities*, rather than from some other cause. Additionally it must be shown that the personal activities in question were *autonomously undertaken* in the sense that the actors were aware of the risks and voluntarily accepted them. Furthermore, locating the autonomous risk-takers would require a rigid and complex framework of research policy. To make such a policy legitimate, considerable moral objections, for instance privacy considerations, would have to be overcome. Finally, all this would indeed have to be cost-effective.

UNAMBIGUOUS CAUSALITY. With regard to the first condition, while it is possible to define general risks from identifiable types of conduct, it is virtually impossible to draw an unambiguous link between an example of that conduct and a particular health consequence. Medical needs often result from many influences of very different kinds varying from genetic predispositions, personal actions and habits, and environmental and social conditions.[112] It is often impossible to establish the respective roles of different factors on the basis of scientific evidence. Whereas, it is often possible to determine responsibility for an injury in mountain climbing or skiing, it is not possible to determine with certainty whether a particular individual's lung cancer resulted from smoking,

[111] See also Y. Denier, *op. cit.*, 2005, pp. 224–234.
[112] See also A. Sen, 'Why Health Equity?' in *Health Economics* 11(2002), pp. 659–666.

environmental pollution, occupational conditions, heredity or some combination of these. Although we know that smoking increases the risk of lung cancer we also know that many non-smokers die of lung cancer each year while many smokers live to old age. Reality shows that the distinction between brute luck and option luck is not clear-cut. While we can identify conduct that increases the risk of illness or injury, it remains very difficult to conclude that a certain health crisis was actually caused by a certain lifestyle choice. In these cases, social policy may rest more on ignorance of causal factors than on knowledge.

Truly Voluntary? Secondly, the argument in favour of health-care rationing by responsibility shows great confidence in the free, voluntary and independent character of individual choice making. In this line of reasoning, many things, if not most, would be a matter of option luck. However, if we want to make choice central, we have to be sure that the participation in risky behaviour is *truly* voluntary. Apart from the tricky task of avoiding endless metaphysical discussions on how free the free will really is, we have to take into account that personal behaviour is to a large extent socially and culturally influenced. There are social determinants of health.[113] Nicotine is now widely recognised as a potently addictive drug, and alcoholism and eating disorders are diseases in their own right. But if many people in a cultural group or class behave similarly, this behaviour might acquire the qualities of a social or cultural norm, in which case we might wonder just how voluntary the behaviour is. A denial of a person's right to health care would be unfair if the person could not have acted otherwise, or could have acted otherwise but only with great difficulty. For instance, John might be an unhappy smoker now, for instance because he grew up in a smoking environment, and started smoking 40 years ago in a time when the health risks of smoking were not yet widely known and socially recognised. So at the very least, the proposition that individuals voluntarily bring many of their illnesses upon themselves must be challenged and tested in each situation in which it is invoked.

Rigid Policy. In addition to the previous issue, problems of rigidity in policing the system become relevant. Responsibility tests would be needed in order to determine whom we owe medical help. Thus, to locate voluntary risk-takers, officials would have to investigate the causes. In the worst-case scenario, these officials would be authorised to invade privacy, break confidentiality and keep records in order to document health abuses that could result in restriction of the right to health care. In such cases the *natural jungle*, in which morally arbitrary differences (as in race, gender or health) determine the results, makes room for a *social jungle*, in which people could be punished by society as a result of an infinite series of responsibility questions about their health behaviour and choices. This immediately raises doubts about the ethical viability of such measures. Pushing very hard on Rawls's intuitive claims about the moral arbitrariness of social and natural contingencies might lead to harsh and counterintuitive results.

---

[113] Cf. Section 3.3.2.2 *Which Health Inequalities Are Unjust?*

FREEDOM. Furthermore, as we have seen, we know that in real life people routinely trade health risks for other benefits. They do so when commuting longer distances for a better job, practicing certain sports or taking a skiing holiday. Therefore if patients needing treatment for smoking-related diseases can be fairly penalised because they smoke, we should apply the same stricture to those who drink too much alcohol, eat too much fat, drive too fast, work too hard, go out too late, go on skiing holidays or indulge themselves in sports like mountain climbing or boxing. Within such a policy only few of us might qualify for the treatment we require in our hour of need. Although there is some plausibility to the claim that rational people should refrain from trading their health for other goods, refusing *ex ante* to allow *any* trade-offs of health for other goods may seem unjustifiably paternalistic. Fundamentally, implementing the possibility of forfeiting the right to health care might entail forfeiting freedom.[114]

COST-EFFECTIVE? Furthermore, one might wonder whether health enforcement would indeed be cost-effective. One of the major reasons for the debate on responsibility in health care is the problem of increasing costs. The argument is based on the idea that those who choose to run health risks cost the rest of us money, and it is fair that they should pay it back, either by paying larger insurance premiums, or by forgoing health care for their self-induced conditions. However, there is good reason to believe that this strategy would lead to counterintuitive outcomes. In addition to the fact that the organisation of health enforcement would carry high financial costs besides its morally unattractive features, it ironically proves that some risk-taking requires less rather than more medical care, because it results in earlier and quicker deaths. Cost-effectiveness research to compare health care costs has shown that low-risk, non-smoking men with low blood pressure generate far higher health care costs per year of life than high-risk men who smoke and have high blood pressure. Ironically, it seems altogether that people with some unhealthy life styles actually save society more in overall expenditures for both health care and social security than they cost.[115]

Some of the above problems – i.e. of intrusive and demeaning policy, violating concerns of liberty and privacy, being difficult and costly to administer – might be avoided if we clarified, what counts as responsibility or fault in these cases, and found a more publicly administrable way of addressing them. John Roemer argues that we should divide a population into 'types' of people by reference to

[114] See also: E.S. Anderson, 'What Is the Point of Equality?' in *Ethics*, 109(1999), pp. 287–337; N. Daniels, 'Rawls's Complex Egalitarianism', in S. Freeman (ed.), *The Cambridge Companion to Rawls*, Cambridge: Cambridge University Press, 2003, pp. 241–276.

[115] H. Leichter, 'Public Policy and the British Experience', in *Hastings Center Report* 11(1981), pp. 32–39, incorporated in *id., Free to Be Foolish: Politics and Health Promotion in the United States and Great Britain*, Princeton, NJ: Princeton University Press, 1991; L.B. Russell, 'Some of the Tough Decisions Required by a National Health Plan', in *Science* 246(1989), pp. 892–896; R.L. Schwartz, 'Life Style, Health Status, and Distributive Justice', in A. Grubb; M.J. Mehlman (eds.), *Justice and Health Care: Comparative Perspectives*, New York: Wiley, 1995, pp. 244–248.

all their relevant biological and sociological traits (e.g. class, education, sex, and family background) which influence health behaviour and then see if a particular behaviour is more or less typical of a person belonging to this or that type.[116] If it is more typical, the person is less responsible. If it is less typical, the person is more responsible for his behaviour. Thus, we might reduce the burden for bad health of those whose ethnic or social backgrounds led to their unhealthy life style or we might find biopsychological markers for risk-takers who thus form a type that should not be held as responsible for their high-risk behaviour as non-typical people. People who are members of the same type will be similar with respect to all those factors for which we deem them not to be responsible.

Roemer suggests, for example, that members of one type might be male, black steelworkers, while members of another type might be white female college professors. Between these types, there will be differences in average smoking behaviour: black, male steelworkers might smoke on average considerably more than white female professors. People are not responsible for such *intertype* differences. Thus, such differences do make a call on justice: more health-care resources are justly allocated to black male steelworkers as a type than to white female college professors as a type, in order to cope with the former's higher overall incidence of smoking behaviour. However, within each type, there will also be variation in smoking behaviour: some black male steelworkers will smoke much less than others, some white female professors will smoke more than others. People are responsible for such *intratype* differences.

Roemer proposes that people who smoke at the median level for that type should receive the same quantity of health-care resources as people who smoke at the median level in other types. He views those at the median within each type as having exerted the same degree of effort not to smoke, as equally 'deserving'. In the case of the college professor and the steelworker, who each have smoked the median number of years for their type, it would be for example 10 and 30 years, respectively. As a result, justice and equality of opportunity with respect to health care may require allocating more health-care resources to restore the health of someone at the median in a 'less lucky', more smoking-prone type (the steelworker who has smoked for 30 years) than to someone who actually smokes less but more than the median in a 'luckier', more smoking resistant type (the professor who has smoked for 20 years).

Of course, this result conflicts with standard views about horizontal equity, which require giving two patients with equal health-care needs the same treatment. However, Roemer argues for an understanding of horizontal equity that incorporates responsibility and requires that two patients who have made the

[116] J.E. Roemer, 'A Pragmatic Theory of Responsibility for the Egalitarian Planner', in *Philosophy and Public Affairs* 22(1993)2, pp. 146–166; *Id.*, 'Equality and Responsibility', in *Boston Review* 20(1995)2, pp. 3–7, 15–16; *Id.*, *Theories of Distributive Justice*, Cambridge, MA: Harvard University Press, 1996, esp. pp. 242–245, 276–279, 285–286, 308–309; *Id.*, *Equality of Opportunity*, Cambridge, MA: Harvard University Press, 1998.

same *comparable degree of effort* not to damage their health receive the same amount of health care, even if they belong to very different types.

What should we think of all this? It is clear that the above line of reasoning has started with the worst-case scenario of whether taking the responsibility route would mean opening a Pandora's box of intrusive and demeaning policy, violating concerns of liberty and privacy, and being difficult and costly to administer. It is an argumentation that starts from the various possible *consequences* of responsibility measures. In being consequentialist in nature, this reasoning implies that if we were able to remove all the practical difficulties and make a clear distinction between individuals who are truly responsible and those who are not, like Roemer's proposal aims to do, we should implement the possibility of forfeiting health-care entitlements because of irresponsible behaviour.

However, even if all this were possible, a few objections of principle remain valid and keep us from punishing people *ex post* for the health consequences of their lifestyle choices.

5.5.3.5 Overemphasising Moral Arbitrariness

It is clear that luck-egalitarians have been most responsive to criticisms of equality based on ideals of *merit* and *responsibility*. The luck-egalitarian principle (others owe us assistance whenever we suffer through no fault of our own) emerges from pushing very hard on Rawls's intuitive claim about the moral arbitrariness of social and natural contingencies. He says:

> It seems to be one of the fixed points of our considered judgements that no one *deserves* his place in the distribution of native endowments, any more than one deserves one's initial starting place in society.[117]

Rawls is, I believe, on safe grounds if he claims that the contingencies of birth are not something we deserve. Nevertheless, it is quite another thing to claim that persons are owed compensation by others for any relative disadvantage they suffer as a result of bad brute luck, but they are owed *nothing* if they suffer as a result of bad option luck. The luck-egalitarian principle of responsibility risks undermining important protections of the human capabilities and related opportunities, i.e. it may undermine essential claims of justice.[118] Let us look at the objections of principle that keep us from punishing people for the health consequences of their lifestyle choices.

Essentially, these objections strongly relate to issues that fortify the initial basis for the moral or human right to health care, as expounded in Part I: the idea of health-care needs as *generic* characteristics of our common humanity, the importance of capability to function for *every* person's opportunities in life, the *collective moral obligation* to ensure that everyone has these opportunities,

---

[117] J. Rawls, *A Theory of Justice*, 1971, p. 104; 1999, p. 89 (my italics, YD).

[118] E.S. Anderson, *op. cit.*, 1999, pp. 288–289; N. Daniels, *op. cit.*, 1996/1990, p. 218–227; *op. cit.*, 2003, pp. 252–256.

and finally the fact that as a human right, the right to health care is granted to all individuals simply because they are human. I will not recapitulate these arguments here, as they have been analysed in detail in Part I.[119] Instead, I wish to focus on five issues that serve as counterarguments against the luck-egalitarian route. I will begin with the *non-grading*, that is, *non-meritarian* idea of respect for the person as person, as expounded in the works of Kant. Consequently, I will take this idea to the level of the just society and its basic institutions, and stress four issues that serve as counter-arguments: firstly, the primary subject of justice; secondly, a policy of inclusion; thirdly, a forward-looking basis; and fourthly, the issue of setting limits on goods, not on people. Finally, I will come back shortly to Dworkin's position.

KANT AND THE HIPPOCRATIC HERITAGE. The *non-grading* idea of respect for the person as person is to be found in its purest form in the works of Kant. Kant emphasises the idea of *respect*, which is owed to each man as a *rational moral agent*. His characteristic of man as a moral rational agent is transcendental; it is not dependent on any empirical capacities he may have.[120] Consequently the idea of respect for a person has nothing to do with grading reflection on *responsibility* for actions or actual *control* (a reflection which must have an *empirical* basis), but refers to something that goes beyond this. It refers to the value that persons have simply because they are persons. It refers to our quality of humanity, a quality that goes beyond our differences in efforts and achievements.

This idea of respect, which demands that persons should be regarded from the viewpoint of their humanity and not merely from the perspective of their different 'appearances' is part of the content that might be attached to Kant's injunction to treat each man as an end in himself, and never as a means only. This means that each man is owed the effort of understanding, and that in achieving it, each man is to be abstracted from certain conspicuous structures of inequality in which we find him.

In the field of medicine this idea returns in the Hippocratic obligation to offer appropriate care whatever the cause of the patient's condition. There is nothing in the traditional codes of medical practice exempting doctors from the care of patients who happen to be the victim of their own intemperance, weakness or simple stupidity. On the contrary, ethical opinion regards it as explicitly impermissible for doctors to be biased in their decisions by any knowledge they may possess about the causes of their patients' health crises. This reflects a long tradition of altruism, based on respect for the person as person. An interesting analogy is offered by John Butler: 'Before ministering to his needs, the Good Samaritan did not enquire about the mugged man's prudence in venturing alone

[119] Cf. Section 2.3.5.1 *What is a Human Right to Health Care?*

[120] Although Nussbaum's analysis has made clear that Kant's distinction between the rational and the animal is problematic, I here wish to lay emphasis on the Kantian idea of respect for persons, on the idea that each human being possesses an inviolable dignity, an idea that is reflected in Kant's injunction to treat each man as an end in himself, and never as a means only. Cf. Section 4.1.1.1 *A Kantian Conception of the Person*.

along a notoriously robber-infested route.'[121] In the same line of thought, it does not only seem impracticable for the doctor or health-care provider to make meritarian judgements at the moment of providing care – whether this involves emergency care, continued medical care, or long-term support of the chronically ill or disabled – it would in fact seem like a highly inappropriate thing to do. The drunk driver will and should receive medical help.[122]

If we take the *non-grading* idea of respect for the person as person to the level of *society*, commitment to the basic purpose of just health care (which is its contribution to protecting capability and opportunity), and to the *institutions* for achieving this, involves the recognition of health-care needs as they are, without going into endless responsibility research. For there are, as I have said, four things we should bear in mind.

PRIMARY SUBJECT OF JUSTICE. The first counterargument holds that in pushing very hard on Rawls's intuition that no one deserves the results of the social or natural lottery, luck egalitarianism has forgotten two things regarding the primary subject of justice.

Firstly, in defending that the fundamental aim of egalitarianism is to compensate people for undeserved bad luck – being born with poor native endowments, in a disagreeable family with bad parents, or suffering from accidents and illnesses – it has forgotten that justice or injustice does not lie in the distribution of natural or social contingencies but in the way socio-political arrangements *respond* to this distribution. 'The natural distribution', Rawls says, 'is neither just nor unjust .... These are simply natural facts. What is just and unjust is the way that institutions deal with these facts'.[123] In focusing on correcting a supposed

---

[121] J. Butler, *The Ethics of Health Care Rationing: Principles and Practices*, London: Cassell, 1999, p. 54.

[122] According to the luck-egalitarian principle, the guilty driver has no claim of justice to emergency care, to continued medical care, or to the necessary support in case of disability due to the accident. Society has no obligation of justice to assist him. See: E. Rakowski, *Equal Justice*, New York: Oxford University Press, 1991 pp. 74–75, 79; R. Arneson, 'Liberalism, Distributive Subjectivism, and Equal Opportunity for Welfare', in *Philosophy and Public Affairs* 19(1990), pp. 158–194, p. 187. Elisabeth Anderson generalises the point and indicates four problems of discrimination that follow from grading persons according to their degree of responsibility or effort: The reasons luck egalitarians offer for coming to the aid of the victims of bad brute luck, she argues, express disrespect for them. It includes and accepts abandonment of negligent victims (You reap what you sow!); discrimination among the disabled (the drunk driver who is disabled as a result has no claim of justice on society to accommodate to his disability); geographical discrimination (don't build your house at a risky spot, or in a tricky area!); occupational discrimination (don't choose dangerous occupations like being a police officer, firefighter, member of the armed forces, miner., or vulnerable occupations like dependency work at home). Of course, the latter discrimination could be remedied by arguments pointing at their use-value for society. However, this hardly reflects an expression of equal respect for all. In E.S. Anderson, *op. cit.*, 1999, pp. 295–298.

[123] J. Rawls, *A Theory of Justice*, 1971, p. 102; 1999, p. 87. See also David Miller: 'As long as a state of affairs is regarded simply as a product of natural causes, questions about its justice or injustice do not arise'. In: D. Miller, *Social Justice*, Oxford: Clarendon Press, 1976, repr. 2002, p. 18. See also Section 4.4.6 *The Tragic and the Unjust*.

cosmic injustice, luck-egalitarian writing has lost sight of the distinctive *political* aim of egalitarianism.

That is, the primary subject of justice has to do with providing basic *institutional* arrangements that generate people's *opportunities* over time. As such, we should focus on equality as a *social* relationship in which each person receives equal respect; we should conceive of justice as a matter of obligations that are not defined by the satisfaction of subjective preferences but by *objective* standards related to the *social conditions* of every person's freedom and standing as equal in society, regardless of handicap, physical appearance or intelligence; standards on which we can all agree from a public point of view.[124] Related to this, Scanlon argues that the forfeiture view 'exaggerates the importance of the fact of *choice* relative to that of the [*social*] *conditions* under which the choice was made'.[125]

Secondly, the luck-egalitarian principle has forgotten an important division of responsibility between society and the individuals. The principles of justice govern the basic structure of fundamental institutions of *society*, assuring that all individuals have the means to develop and exercise their capabilities as citizens. *Individuals* in their personal choices are free to pursue their conceptions of the good life as they see them, provided that they operate within the bounds set by the principles of justice.[126] For Rawls, this is an important division of responsibility. Society is responsible for regulating the basic structure through principles of justice; individuals are responsible for pursuing their ends in ways that comply with and support the principles of justice. Consequently, Daniels rightly wonders whether the luck-egalitarian principle does not seem more likely to be part of a particular *comprehensive* moral view than a shared feature within political theory of justice.[127] Of course there is positive evidence that in some *individualised* contexts we are concerned with actual choices and their impact. However, this does not show that it should be the target for theories of *justice* for *basic* institutions.

For purposes of justice, both Daniels and Anderson argue, egalitarian theory should not be concerned about making choice as central as the luck egalitarians wish it to be. Instead, they both defend the *democratic equality* approach, which

---

[124] Moreover, the conditions are sensitive to variations in people's circumstances, including their disabilities. People who cannot walk are entitled to accommodation in civil society (wheelchairs, ramps at public buildings ...). However, these conditions are not sensitive to people's tastes (Dworkin): Everyone is entitled to the same package of capabilities, regardless of what they would prefer to have. Thus, if a person who needs a wheelchair to get around has an expensive taste for a Stradivarius (Dworkin's example), justice does not require to substitute the wheelchair subsidy for the Stradivarius. It has to do with providing the social conditions for equal citizenship – being able to move around is much more important in functioning as a citizen than playing a Stradivarius. In R. Dworkin, *op. cit.*, 1981, pp. 240–246; 2000, pp. 59–64.

[125] T.M. Scanlon, 'The Significance of Choice', 1988, p. 196 (my italics, YD).

[126] See also: N. Daniels, *op. cit.*, 2003, p. 264.

[127] N. Daniels, *op. cit.*, 1996/1990, pp. 226–227; *Ibid.*, 2003, pp. 255–256.

is strongly inspired by the capability approach of Sen and Nussbaum. In what follows, I will gradually convey the essence of this approach.

Firstly, democratic equality guarantees not *actual* levels of functioning but effective *access* to those levels. Nussbaum puts it as follows:

> Considering health and health care, we should then distinguish between the capability or opportunity to be healthy and actual healthy functioning. A society might make the first available by providing decent health care services – among which public health is one of them – and also give individuals the freedom not to choose the relevant functioning.[128]

Individuals are free to choose how to function. Effective access to a level of functioning means that people can achieve that functioning by deploying *means* already at their disposal. Therefore, what citizens ultimately owe each other is effective access to the social conditions of the capabilities people need to function as equal citizens over the course of an entire life, regardless of how imprudently they conduct their lives. It is not a starting-gate theory, in which people could lose their access to equal standing through bad option luck. Anderson defines it as follows:

> Democratic equality guarantees effective access to the social conditions of freedom to all citizens, regardless of how imprudently they conduct their lives. It does not deprive citizens of the necessary medical care. It does not discriminate among the disabled depending on how much they can be held responsible for their disability. Under democratic equality, citizens refrain from making intrusive, moralizing judgements about how people ought to have used the opportunities open to them or about how capable they were of exercising personal responsibility. It need not make such judgements, because it does not condition citizen's enjoyment of their capabilities on whether they use them responsibly. The sole exception to this principle concerns criminal conduct. Only the commission of a crime can take away a person's basic liberties and status as an equal in civil society. Even convicted criminals, however, retain their status as equal human beings, and so are still entitled to basic human functionings such as adequate nutrition, shelter, and medical care.[129]

Policy of Inclusion. This brings us to a second point. Contrary to luck egalitarianism, democratic equality defends a policy of inclusion. Remember from the analysis in Part I that moral objections against full private pocket payment of health care in general come down to the fact that this results in a policy of exclusion (due to 'cherry-picking', only the healthy and wealthy will be able to purchase insurance and medical care) on a domain that is just *too important* to allow exclusion, i.e. the domain of guaranteeing equality of capability and opportunity for all. As Martha Nussbaum argues, certain matters are taken to be *entitlements* of citizens based upon justice. When any one of these is abridged this is an especially grave failure of the socio-political system. The abridgement is then not just a huge cost to be borne, but also a cost of a distinctive kind, involving a violation of basic justice. When, on the other hand, some matters *outside the core* are abridged, it may be a small cost – or a large cost to some persons – but it is

---

[128] M.C Nussbaum, *op. cit.*, 2000, p. 14. See also: Section 4.2.5.3 *Concerns of Pluralism*; and Y. Denier, 'Public Health, Well-Being, and Reciprocity', in *Ethical Perspectives* 12(2005)1, pp. 41–66.

[129] E.S. Anderson, *op. cit.*, 1999, pp. 326–327.

a cost of a different kind. It is not a cost that, from the viewpoint of justice, no citizen should be asked to bear.[130] In the same line of reasoning, Thomas Scanlon loosens the connection between equality and responsibility by arguing that it depends on the *objective value* of the equalisandum whether inequalities are just or unjust. Preferences and tastes are excluded from the distribution problem, not as in Rawls's proposal, because people can be held responsible for them, but because they are less *urgent*. That people can be held responsible for their preferences should only mean that they can do *without* them. They are not necessities and thus loose their urgency. Responsibility for preferences is in itself not a reason for rejecting claims; it is at most a sign of their not being very urgent. The degree of urgency depends on the *objective value* of the claim, not on the subjective stress one can lay on it.[131] Elisabeth Anderson puts it this way: some outcomes are so awful that no one deserves to suffer them, not even the imprudent. Negligent drivers don't deserve to die from a denial of health care.[132]

Accordingly, we can safely argue that a *policy of inclusion* for matters of objective, intrinsic importance, such as just health-care institutions, is a major aspect of the just and respectable society. In this line of reasoning, Elisabeth Anderson rightly argues that consistent egalitarian theory should *identify* certain types of goods to which *all* citizens must have effective access over the course of their whole lives, because they are more important from an egalitarian point of view than others.[133] A policy of inclusion is based on the idea that justice does not permit the abandonment of *anyone*, not even the imprudent, in such matters.

FORWARD-LOOKING. Thirdly, the democratic equality approach helps us to bear in mind that the notions of capability and opportunity are in itself *forward-looking* concepts. This means, in turn, that the basis of just health care is also forward-looking. To use a term from Jeremy Waldron, justice has to do with providing a 'fallback framework' that contributes to *all* persons' receiving a fair chance in life, i.e. a safety net below which *no one* would be allowed to fall.[134] As a result, it would be unfair to cut off fair equality of opportunity in the *future* because of *past* behaviour. Although it sounds paradoxical, holding people responsible for their ends means that in assuming the presence of fair

---

[130] M.C. Nussbaum, 'The Costs of Tragedy: Some Moral Limits of Cost-Benefit Analysis', 2001. See also *id.*, 'Capabilities as Fundamental Entitlements', 2003, pp. 44–48. Cf. Section 4.2.5.2 *Freedom*.

[131] T.M. Scanlon, 'Preference and Urgency', in *Journal of Philosophy* 72(1975)19, pp. 655–669; *Id.*, 'Equality of Resources and Equality of Welfare: a Forced Marriage?' in *Ethics* 97(1986), pp. 111–118; *Id.*, 'The Significance of Choice', in S. McMurrin (ed.), *The Tanner Lectures on Human Values*, Vol. 8, Salt Lake City: University of Utah Press, 1988, pp. 149–216. See also H. Frankfurt, 'Necessity and Desire,' in *id.*, *The Importance of What We Care About*, Cambridge: Cambridge University Press, 1988, pp. 104–116; Section 2.3.2 Needs and Preferences in Health Care.

[132] E.S. Anderson, *op. cit.*, 1999, p. 301.

[133] In E.S. Anderson, *op. cit.*, 1999, pp. 316, 327. Cf. Section 3.3.2.1 *The Special Moral Importance of Health Care?*

[134] J. Waldron, 'When Justice Replaces Affection: the Need for Rights', in *id.*, *Liberal Rights*, Cambridge: Cambridge University Press, 1993, pp. 370–391.

and just arrangements and institutions, we are acting as if they can exercise their underlying moral power to *form* but also to *revise* their conceptions of the good and valuable.[135] The luck-egalitarian principle, on the other hand, offers a very inadequate safety net for the victims of bad option luck. Within the hard-line version, once people risk and lose their fair share of natural wealth, they have no claims against others to stop their free fall into misery and destitution. This reflects the fact that luck-egalitarianism is in fact a 'starting-gate theory'.[136] As Elisabeth Anderson says:

> In focusing on correcting the supposed injustices of nature, luck egalitarians have forgotten that the primary subject of justice is the institutional arrangements that generate people's opportunities *over time*.[137]

So far, the net result of the discussion is that, even if there is some intuitive appeal to considering responsibility, it is not so central or important that it should not be overridden by other considerations involved in justice. Whether or not this appeal to responsibility for bad health troubles us intuitively, putting too much emphasis on it ignores egalitarian considerations central to democratic equality.

LIMITS ON GOODS, NOT ON PEOPLE. One might object that all these guarantees invite personal irresponsibility. Why act prudently? Doesn't all this imply that society is to be considered as a trough of means and services, free at everyone's disposal; that any talk of setting limits and incorporating scarcity violates central claims of justice? Should we not allow any trade-offs? Does the debate on the role of personal responsibility in the right to health care have no consequences whatsoever? No, it does not. Capabilities are diverse and resources available to

---

[135] Cf. N. Daniels, 'Equality of What? Welfare, Resources, or Capabilities?' in *Philosophy and Phenomenological Research* 50(1990), Suppl., pp. 273–296, repr. in *id., Justice and Justification: Reflective Equilibrium in Theory and Practice*, Cambridge: Cambridge University Press, 1996, pp. 208–231, esp. p. 222: 'I think Cohen misses the mark. It is not actual choice that matters but the underlying capacity for forming and revising one's ends that is at issue. If we have independent reasons to believe that a preference, whether chosen or not, whether identified with or not, cannot be eliminated and is handicapping because of a broader, underlying handicapping condition, then we have reasons to make certain resources available as compensation. It is not the unchosen taste, or the fact that the taste is unchosen, that gives rise to the claim on us. Rather, it is the underlying mental or emotional disability, and the taste, chosen or not, is but a symptom' (my italics, YD). In the same line of reasoning, Dworkin's responsibility cut implies that it is the underlying resource that raises the claim of justice.

[136] The starting-gate theory of fairness holds that justice requires equal initial resources, followed by laissez-faire thereafter. Dworkin denies that his approach is a starting-gate theory because he would allocate compensation for unequal talents over the course of a lifetime. See R. Dworkin, 'Equality of Resources', 1981, pp. 309–311; 2000, pp. 87–89. Starting-gate does apply to the luck-egalitarian approach of Cohen and Arneson: as long as people enjoy fair shares at the start of life, it does not much concern itself with he suffering and subjection generated by people's voluntary agreements in free markets.

[137] E.S. Anderson, *op. cit.*, p. 309.

provide them are scarce. Some trade-offs between capabilities must be accepted. Egalitarianism must face up to the need to uphold personal responsibility if only to avoid bankrupting the state.

There are two general strategies for doing so. One is to insure only against certain causes of loss: to distinguish between the losses for which persons are responsible and those for which they are not, and to indemnify only for the latter. This is the approach of luck-egalitarianism, which leads to intrusive and disrespectful judgements of individuals. This is also the approach of setting limits on people, excluding the bad, the irresponsible, and including the good and responsible.

The second strategy is to insure only against the losses of certain types of goods. This would be the strategy of setting limits on goods, instead of on people. Here I follow Scanlon in insisting that the weight that a citizen's claim has on others depends solely on the objective content of his interest and not on the subjective importance he places on it in his own conception of the good life.[138] In some cases, the weight of an interest can be determined by considering its impact on a person's standing as an equal in society. As Nussbaum's analysis has made clear, some deprivations of capabilities express greater disrespect than others, in ways any reasonable person can recognise. This has nothing to do with what one values privately according to one's own conception of the good, but rather with the social conditions of equal citizenship, access to public accommodation and 'the right to be in the world'.[139] Therefore, this strategy does allow trade-offs, but *not* those trade-offs that deprive *individuals* of standing as equals in society.

This means that a theory of justice has to distinguish between guaranteed and unguaranteed types of goods within the space of egalitarian concern, and to insure individuals only against the loss of the former. This is the approach of democratic equality. It does not indemnify individuals for *all* losses due to their imprudent conduct. It only guarantees a set of capabilities *necessary* to function as a free and equal citizen. Individuals must bear many *other* losses on their own. For example, the smoker would be entitled to treatment for lung cancer, regardless of his degree of responsibility by smoking. However, he would not be entitled to compensation for the loss of enjoyment of life brought about by his confinement in the hospital and reduced lung capacity, for the dread he feels upon contemplating his mortality or for the reproach of the relatives who disapprove of his lifestyle. The irresponsible have plenty to lose from their irresponsible conduct, and therefore have an incentive to behave prudently.

This means that although democratic equality first of all offers equality in the space of basic capabilities, i.e. of opportunities or freedoms, thereby setting a lower bound on responsibility below which no one should bear the consequences of one's actions, personal responsibility still remains an important value. Persons still have

---

138 T.M. Scanlon, 'Preference and Urgency', 1975, p. 659.

139 M. Nussbaum, *op. cit.*, 2002, p. 32. See also *id.*, 'The Costs of Tragedy: Some Moral Limits of Cost-Benefit Analysis', 2001.

to exercise responsible agency to achieve most of the functioning's effective access to which society guarantees. After all, people have by their behaviour an effect on their health, and society should never hesitate to underscore the importance of sensible health behaviour by generally and explicitly making people conscious of the influence they can have over their health needs. At the same time, however, personal responsibility plays a *secondary* role within a theory of just health care. In the name of fair equality of capability and of respect for the person as an end in itself, society should hold on to a policy of *inclusion* and should therefore continue to be *forward-looking*, both in providing incentives to avoid hazardous behaviour and in offering medical help.

Regarding *incentives*, consciousness-raising health campaigns show respect for individual autonomy while appealing to people's rationality to take care of their health. The same goes for cost-sharing. It is fair to require individuals who engage in certain risky actions that result in costly medical needs to pay higher premiums or taxes to absorb the extra costs of their behaviour *ex ante*. Risk-takers may be required to contribute more to particular pools such as insurance schemes or to pay a tax on their risky conduct, such as an increased tax on alcohol and tobacco, or an additional insurance premium against sports injuries. These requirements may fairly redistribute the burdens of the costs of health care, and they may deter risky conduct without disrespecting autonomy. By making risk-takers pay for the costs of their behaviour *ex ante*, democratic equality preserves their freedom and equality over the course of their whole lives. The return individuals may expect on their taxed consumption is health-care protection for themselves, although there is no requirement of compensation for loss of happiness.

As for offering medical help, it would be unjust to refuse care to a person in need, even if it is clear that he was responsible for his condition. A theory of just health care must continue to meet the most fundamental test any egalitarian theory must meet: that its principles must express equal respect and concern for *all* citizens. We must avoid a society of hierarchically positioned parties – the good and the bad, the responsible and the irresponsible – and instead hold on to a society that is determined by a forward-looking policy of inclusion.[140] Contributing to fair equality of capability and opportunity should *continue* to be one of the fundamental goals of just health care. This should not change because of past behaviour; as only then can the basic human right to health care be a truly inviolable right.

Recalling Dworkin's Position. Let us take stock. Contrary to what could be expected initially, Dworkin does not explicitly take part in this debate. This is rather surprising, for his distinction between brute and option luck initially sets the stage for the issue on personal responsibility in health care. As we have seen,

[140] For further argumentation against hierarchic positioning in health care, see: Y. Denier, 'Public Health, Well-Being, and Reciprocity', in *Ethical Perspectives* 12(2005)1, pp. 41–66.

however, it is reasonable to believe that, due to his responsibility cut between preferences and resources, Dworkin does not take the path of setting limits on care by making the issue of voluntary choice central. He does not follow the luck-egalitarian trail of Cohen and Arneson.

This is a sensible approach, I believe, given at first the problems of *practical applicability* that are attached to making choice the decisive factor within distributive theory. More importantly, however, it does seem that Dworkin's responsibility cut, his defence of legitimate paternalism, together with his proposal of universal and mandatory health insurance based on a progressive tax system, his arguments against the starting-gate approach, and his principle of equal importance, all together go into the direction of reflecting a *principled argument* against the luck-egalitarian forfeiture view.

I agree with the luck-egalitarian critique that Dworkin's master distinction between preferences and resources is not congruent with his distinction between brute luck and option luck. However, I do not agree with their alternative for the reasons I have given above, which ultimately come down to the following: although the idea of personal responsibility as it reflects in the distinction between brute luck and option luck, and in the ensuing luck-egalitarian principle (one should bear the consequences of one's own voluntary choices), generally and immediately answer to our spontaneous *intuitions*, it cannot serve as the central determining factor within a *theory of justice* in health care.

Instead, I defend Dworkin's responsibility cut between preferences and resources, and hold that for reasons of justice society should guarantee an *objective basis* – a set of basic capabilities, primary resources that are necessary *means* to lead the life of a citizen enjoying equal respect in society – to all, irrespective of whether one behaves prudently or imprudently. This however, would imply an answer to the question: what makes *some resources* so important, what makes health so important for us that we *prudently* want decent insurance in these cases and that society may not exclude people from being decently insured? In short this means that Dworkin cannot consistently maintain that health has no special status. We need to *identify* certain types of goods and explain why they are just too important to not guarantee equal access to them. With regard to health, I have argued throughout that although it is not the *summum bonum* of life, it nevertheless is an important *bonum*, which needs to be recognised as such by any just health-care system.

Throughout this part, I have taken up several positions with regard to the function and significance of health care within a theory of justice, while bearing the following question firmly in mind: how can we set limits on health care while realising an acceptable balance between *efficiency*, *justice* and *care*? Within this analysis, I have considered it as my philosophical task to obtain a clear understanding of the various arguments used, and to critically assess, explain and justify or modify the relevant categories, distinctions and their interrelations. What have we learned from this analysis? Let us briefly take the main results together, starting with Rawls.

In the first sections, I have explained the essence of a Rawlsian view on just health care. I have shown that, although various elements give the impression that Rawls's theory provides good grounds for reflections on just health care, he yet virtually remains silent on the subject. After due reflection, it turned out that the question of health care seems unanswerable from within the political conception, since it cannot fulfil the conditions of objectivity and simplicity. Consequently, Rawls considers normal health care as one of the four problems of extension. This is the point where the theory of Norman Daniels comes in.

It has been Daniels's merit to have taken up the Rawlsian challenge and ask the question: 'Why should a concern with distributive justice extend to health care at all?' Daniels's answer consists of three steps. Firstly, by fulfilling the relevant medical needs, health care seems to be necessary for maintaining *normal species-typical functioning*. Secondly, by impairing normal functioning, disease and disability restrict the *normal range of opportunities* open to individuals. Thirdly, health care thus makes a distinct but limited contribution to the protection of *fair equality of opportunity*; it helps us to preserve our status as normally functioning citizens, participating in the social, political and economic life of society. As such, health care should be subsumed under the Rawlsian fair equality of opportunity principle. In this way, it meets the condition of objectivity and simplicity since opportunity, not health care, is the primary social good. Next, I have shown that Daniels's approach incorporates the idea of *efficiency* in his conception of *just* health care in three ways.

To begin with, Daniels holds that health care meets reasonable constraints, and that meeting health-care needs should not create a bottomless pit. By concentrating on the social determinants of health, thereby illuminating that health care is not the only important good, he takes the *external* dynamics of scarcity, and the role of opportunity cost into account. Furthermore, he also takes the *internal* dynamics of scarcity into account by setting the distinction between the objective-truncated scale of well-being and the full-range scale of well-being, as well as between treatment and enhancement. The goal of health care is not to secure individual happiness, but only to provide for objective normal functioning. Similarly, basic health care is not about altering or enhancing the normal opportunity range, but about protecting individuals whose normal range is compromised.

Earlier, it had become clear that both the internal and external aspects of scarcity have shown that rationing is unavoidable. Consequently, Daniels concentrates, thirdly, on the issue of *setting limits* on health care by giving an account of a fair process for addressing critical resource allocation decisions, called 'accountability for reasonableness'. In a situation of reasonable pluralism, where we have no prior consensus on fair outcomes, Daniels holds that establishing a *fair process* is the best we can do. Overall, we may say that this has resulted in a strong and well-founded theory of just health care.

There remains, however, one major problem, which puts the value of Daniels's theory in a different perspective: the importance attached to normal functioning is not unproblematic. Following the line of reasoning of Eva Kittay and of Martha Nussbaum, I have argued that the predominant importance of normal species-typical functioning, normal opportunity range and fair equality of opportunity implies a *hierarchy* between the health-care goals of prevention, cure and care. Neither Rawls nor Daniels can place *comprehensive care* for the long-term and asymmetrically dependent within their theory of justice. In order to cover such cases, both argue that we either need another, complementary theory (like the capabilities approach), or other virtues (like beneficence or charity).

Subsequently, I have analysed Nussbaum's view on both these proposals. Making an appeal to other virtues is not an option, for it would leave the areas of long-term care up for grabs and would entail the recognition of much indeterminacy in the account of basic justice *for all* (cf. also Kittay, Waldron). If a society fails to offer a supporting framework for equal decent life chances, supporting the self-respect of both the cared-for as of the caregiver, then this is not a shortcoming of beneficence or charity or a lack of supererogatory action, but a failure of justice.

Furthermore, I have shown that Nussbaum's capabilities approach cannot be that easily added to the Rawlsian social contract theory for two reasons. Firstly, her theory is based on a different *political conception of the person*. Against a Kantian conception, which places dominant weight on having the two moral powers (reason and rationality) to a requisite minimum degree in order to be a free and equal moral person, Nussbaum's approach is based on an *Aristotelian* and *Marxian* inspired conception that considers the person as both capable and needy, both active and in need of support in many ways. Secondly, against the basic contractarian presuppositions, her theory works with a *rich and complex concept of reciprocity*, instead of one of productive reciprocity; and it holds that the benefit of cooperation is not so much that of economic efficiency and mutual advantage, but rather that of promoting the great good of doing justice itself, by including the good of vulnerable people in equal respect for the dignity of *all* human beings. In order to hold and foster such a rich concept of reciprocity between people, Nussbaum articulates a broad and substantive *list of central human basic capabilities*, which is meant to have the requisite critical force and definiteness to direct social policy. Within an *embodied liberal spirit*, the list sets the agenda for social justice.

Finally, I have shown that by arguing that the source of human dignity lies, not in the sole reference to objective normal functioning within a situation of contractual bargaining, but in the kind of complex and pluralist reciprocity in which we humanly engage, being confronted with periods of independence as well as with more or less extreme necessity and dependence, Nussbaum's theory is more apt to *integrate* the full complexity and heterogeneity of contemporary health care, e.g. prevention, cure *and* care into a theory of *just* health care.

However, I have also argued that Nussbaum's approach only partially and indirectly answers to the pressing contemporary question regarding the issue of limits to health care. Although she does say that there are obviously limits to programmes of social benefit, that there are indeed questions to be asked about *how much* the state should invest in *which* domains (cf. the *external* dynamics of scarcity), and although she holds that the capabilities list functions as a social *minimum* (cf. the *internal* dynamics of scarcity), the rich complexity of the list, and its emphasis on human flourishing, signify that this is an extensive minimum, which presupposes a realm of abundance. On closer view, it seems that her just critique on a too narrow focus on economic and productive reciprocity and her right designation that too much reliance on cost–benefit analysis as a method to deal with trade-offs can obscure the presence of a tragic situation, ultimately risk throwing out the child with the bathwater. There are no substantive or guiding answers left to deal with the fact that scarcity of resources is a condition of human life and that there are limits to what society can spend on health care, besides the proposal to leave the problem of efficiency, priority setting and cost-containment to the national democratic discussion. Are there no relevant principles conceivable that can serve as a normative guide to the democratic discussion on rendering the goals of good quality care, equal access for all, freedom of choice and economic efficiency more coherent?

In the search for a more substantive strategy in the realm of trade-offs, scarcity and limits, I have argued that Ronald Dworkin's proposal is very interesting. Within his theory of just health care, he tries to take *efficiency* into account, while saving long-term *care* by means of the hypothetical device of prudent insurance. By doing this, I have argued, his strategy occupies a middle position between scarcity and abundance, between 'too narrow' and 'too broad' a theory.

Essentially, Dworkin's *prudent insurance principle* (which is an extended example of the hypothetical insurance scheme) is based on the following question: given an equally limited amount of resources, and given that the risk of suffering brute bad luck is equal, which treatments would we prudently choose to be insured for, and which would we regard as not worth the cost of insurance? Within this analysis, I have pointed at the essential role played by the concept of *opportunity cost*. Contrary to the insulation ideal, the Dworkinian proposal provides a guiding answer to questions that refer to the *external* aspect of scarcity (How much money should a society like ours spend on health care in the aggregate?) as well as to the *internal* aspect (What should be included in the basic package of health care to which all individuals should have equal access?).

It would not be prudent to purchase no insurance at all, while at the same time it would be unwise to spend all of one's resources to insurance since then nothing would be left to pursue one's life projects. Accordingly, the hypothetical insurance market will determine an upper as well as a lower limit, which is determined by factors such as the risks people would generally be prepared to take and the premiums they would be prepared to pay given the resources available, and measured by the opportunity cost of such resources.

The point of the hypothetical device is that if *most* prudent people would buy a certain level of medical coverage in a free market if they had equal average means, then the *fairness* or unfairness of our real time society can be measured according to the amount of people that do not have such coverage now. These are essential elements in the basic package that any responsible health-care system should establish. On the other hand, if very *few* prudent people would want to buy insurance of a much higher level of coverage – including certain heroic medical technologies – it would be unjust to force everyone to have such insurance through a universal and mandatory public scheme. As such, the prudent insurance hypothesis can serve as a *workable guide* to reconciling *efficiency*, *justice* and *care* within just health-care policy. We might use our speculations about what people in the imaginary community would think prudent to provide for themselves as guides to help us determine what justice demands that should be available to everyone, at a reasonable cost, and should be supplied without charge to those who cannot carry that cost themselves. As such, his proposal serves as an interesting alternative to the dominant strategy of retreat to procedural justice.

Furthermore, I have discussed various ideas that deserve critical reflection. Here I will only mention the two most important ones. Firstly, I have pointed at the fact that although Dworkin's critique on the insulation ideal is most important and valuable, his presupposition that health and health care have no special value needs refinement for the sole reason that the special character of health and health care is implied in the notion of prudence itself. Without implying that health is the *summum bonum*, I have argued that it is reasonable to hold on to the fact that health is an *important bonum*.

Secondly, I have argued against the luck-egalitarian proposal of Arneson and Cohen to relocate Dworkin's responsibility cut between preferences and resources in order to place choice in a central position for the simple reason that the luck-egalitarian approach risks undermining important protections of the human capabilities and related opportunities, i.e. it risks undermining essential claims of justice. It is interesting to note that all three of my protagonists – Daniels, Nussbaum and Dworkin – agree in this matter. Although there are legitimate paternalistic reasons to guide personal preferences by incentive measures, refusing *ex ante* to allow *any* trades of health risks for other goods, even when the background conditions on choice are otherwise fair, may seem unjustifiably paternalistic.

Having all this in hand, i.e. having obtained a clear view, first on the various elements and mechanisms that constitute the problem of reconciling efficiency, justice and care in health care (as elaborated in Part I), and later on several positions and arguments with regard to the function and significance of health care within a theory of justice (cf. Part II), it is now time to move to the final section on *Health Care and The Limits of Human Existence*. Here, I will take general stock by formulating an answer to the following question: 'What are the desiderata for a just health-care theory?' By bringing the most essential points of the discussion together, I will try to provide a more elaborate view on what I mean by a contemporary Socratic perspective in just health care.

GENERAL CONCLUSION:
HEALTH CARE AND THE LIMITS OF HUMAN EXISTENCE

CHAPTER 6

JUST HEALTH CARE: FOUNDATIONS AND PROSPECTS

My aim in this book has been to explore the idea of just health care as it forms part of the basic structure of contemporary liberal society, i.e. as part of the structure that profoundly affects people's life prospects and chances, 'what they can expect to be and how well they can hope to do'.[1]

In this final chapter, I will take general stock by formulating an answer to the following question: what are the *desiderata* for principles of just health care? Section 6.1 starts with bringing the essential questions and positions together into a particular framework that contains the central elements of just health care. Following Nussbaum's line of reasoning, I will think of these elements as *substantive benchmarks* that will have the requisite critical force to guide and direct social policy. They will provide a *critical framework* within which various proposals for setting limits on health care and for strategies of priority setting will be justifiable or not.

In Section 6.2, I will specify the *general moral perspective* in the light of which these desiderata for principles of just health care are best understood. It is my aim to provide a more elaborate and fleshed-out view on what I mean by a contemporary Socratic perspective on just health care. After having expounded various views on the function and significance of health care within a general theory of justice, it is easier now to give somewhat more body to the general outline of the Socratic perspective as I have sketched in Section 2.4 on *Care and the Boundaries of Human Life*.

Given that it is my philosophical aim to analyse the questions, to develop and to critically assess the various categories, arguments and positions taken, the result is a framework of benchmarks that serves as a critical perspective within which the search for an acceptable balance between efficiency, justice and care *continues*. Accordingly, many prospects for further research could be suggested. In Section 6.4, I will mention four topics that I believe are particularly pressing or promising in the search for a better understanding of the problem of justice in health care.

6.1 DESIDERATA FOR PRINCIPLES OF JUST HEALTH CARE

My position can be summarised as follows. To begin with, I defend an *egalitarian* perspective. Principles of just health care have to do with guaranteeing equal access to certain forms of health care for all. I believe that this guarantee is most aptly expressed in the principle of equal access for equal need.

[1] J. Rawls, *A Theory of Justice*, 1971, p. 7; 1999, pp. 6–7.

Secondly, I hold that the principles of justice in health care must in some way *identify* the health-care goods to which all citizens must have effective access over the course of their whole lives. Earlier, I suggested that the objective and non-volitional structure of human course-of-life needs plays an important role in such identification.

Furthermore, the principles of justice must be able to specify the *moral importance* of just health care. After having investigated various approaches in Part II, I believe that Nussbaum's capabilities approach, which is based on a rich and complex idea of reciprocity and on an embodied conception of the person offers the best possibility to grasp the essence of what is going on in just health care.

Fourthly, the principles must be able to express the complex moral content of a *human right to health care*. I believe that the essence of a truly human right to health care is broader and more encompassing than the account given in Chapter 1.

Moreover, egalitarian principles of justice in health care should express a *forward-looking policy of inclusion*. This implies that the basic framework of identified health-care goods and services should be guaranteed by society on an equal basis to everyone in any case.

Next, the principles of just health care should take the *limited character* of the duties of society into account from the beginning. The human right to health care is a limited right. It cannot be a right of everyone to have access to whatever health-care services would be of any benefit to the individual. I believe that Dworkin's prudent insurance proposal offers a workable guideline for social policy to set legitimate limits on health care.

Finally, the principles should be capable to serve as a basis of agreement, i.e. they should be possible objects of collective willing within a *liberal spirit*. They should be capable of supplying sufficient reasons for citizens acting together to collectively guarantee the identified health-care goods and services to everyone on an equal basis.

6.1.1 Embodied Liberalism

Let us take up the last desideratum first. In liberal theory, the fundamental aim of the state is to equally secure the *liberty* of its members; and the fundamental obligation of citizens to one another is to secure the *social conditions* of everyone's freedom in an equal way. Different from libertarianism, liberal theory thus includes the importance of having the *means* to do what one wants in its definition of freedom. This happens, moreover, without resort to a comprehensive, perfectionist or metaphysical conception of the good life. The principles of justice should be able to serve as principles that can be endorsed for *political* purposes, as the moral basis of central constitutional guarantees, by people who have otherwise very different views of what a complete good life for a human being would be.

Principles of just health care should function in the same political-liberal spirit. In a limited way, we have seen, just health-care contributes to fair equality of opportunity. As such, equal access to health care is one of the social conditions of liberty and the opportunity to form a plan of life. Just health-care policy provides the *social basis* of the capability to healthy and independent normal functioning. As such, health care is part of the social conditions that are deeply committed to the goals of independence and liberty. This *partly* answers the question concerning the moral importance of health care: it contributes to normal, healthy, independent functioning, and as such, to one's opportunities in life.

The other part of the answer, I have argued, requires a *broader perspective*. For the moral importance of just health care is also expressed by the extent to which it truly takes care of the least-advantaged, i.e. the weak, the vulnerable and the irreversibly dependent, whose functioning is so different from functioning at the normal, independent healthy level. This implies that the principles of just health care should adopt a *rich and complex conception* of the benefits of social cooperation including 'the good of vulnerable other people' and 'respect for human dignity' because it is 'good in itself'.[2] Benefit, then, should not be understood in terms of economic efficiency and mutual advantage but rather in terms of promoting the great good of justice itself. Instead of supporting a *hierarchical relation* between prevention, cure and care, as well as a *condescending attitude* to measures of care for the irreversibly dependent, principles of just health care should include the good of vulnerable people in equal respect for the dignity of *all* human beings, i.e. in equal respect for both the temporarily and the irreversibly dependent who are in need of care.

In the line of reasoning of Nussbaum, I have suggested that this broadening of perspective requires a form of *embodied political liberalism*, i.e. a form of liberalism that is more attentive to human need and its material and institutional conditions, and which therefore can *integrate* the element of *care* as one of the major things that need to be supported and secured in a just society. This embodied perspective is necessary to avoid an exclusive focus on healthy, independent functioning and normal opportunity, as well as a condescending attitude to 'non-efficient' long-term care. It is also necessary to promote a *well-balanced relationship* between the health-care pillars of prevention, cure, and care. Finally, it implies a radical rejection of a narrow focus on productive reciprocity and efficiency, and a choice for *complex forms of reciprocity*, taking in the variety of types of dependency and independency that are generically characteristic for human life.

6.1.2 Strong Egalitarian Perspective

Included in the embodied political-liberal spirit is the defence of an egalitarian perspective. With regard to health care, this entails that the principles of just health care express a commitment to *equal respect* for all human beings by guaranteeing *equal access* to certain forms of health care for all.

---

[2] M.C. Nussbaum, *op. cit.*, 2004, pp. 436–438.

In the words of Aristotle, this entails a choice for the principle of equality in *geometrical proportion*, which demands to treat equals equally and unequals unequally. As focal variable, i.e. the criterion that defines the relevant respects under which people are equal or unequal, I defend the principle of equal *need*. This is a strong-egalitarian criterion because it is much more *inclusive* than the criteria of merit, status, or contribution.

However, the principle of equal access for equal need demands further specification since the strong normative force of the concept of need is not unambiguously valid. In order to avoid problems of opportunistic hijacking of social resources by needs-inflation on the one hand, and of illegitimate paternalism on the other, it is necessary to determine a *typology of needs*. This is especially important in health care given the internal dynamics of scarcity, which reveals health care's tendency towards infinity. This tendency towards infinity is based on the scientific and technological progress in medicine, which continuously creates new advances in clinical possibilities of diagnosis and therapy, and thus develops new needs analogously, as well as on the rise of general expectations that those advances in medicine and health care have encouraged.

6.1.3 Which Health-Care Goods and Services?

Are all health-care *needs* equally important from an egalitarian point of view? No, they are not. Analogously, some health-care *goods* and *services* are more important from an egalitarian view than others. I believe that the principles of justice in health care must in some way *identify* the health-care goods to which all citizens must have effective access over the course of their whole lives.

How can we distinguish health-care needs that can function as standards of provision for social distribution of health-care goods and services from those that cannot? In Chapter 2, I suggested that the objective and non-volitional structure of human course-of-life needs plays an important role in such identification. This has led to Braybrooke's *List of Basic Matters of Need.*

Furthermore, I have defined two strategies of further specifying this list in terms of standards of provision as an objective of social policy, one referring to the primary goods perspective of John Rawls, and the other referring to Martha Nussbaum's list of basic human capabilities. Because the primary social goods refer to things that persons need in their status as free and equal citizens, and as normal and fully cooperating members of society over a complete life, it has been criticised by many for not being sensitive to variations in need, due to variations in dependency. As such, neither Rawls nor Daniels is able to incorporate *care* from the beginning into their theories of justice. It can only provide a basis for those health-care goods and services that aim to maintain and restore normal, independent, healthy functioning, i.e. prevention and cure. I believe, however, that an adequate theory of just health care must be able to incorporate the health-care goods and services of prevention, cure, *and* care from the beginning in an equal way.

Because Nussbaum's capabilities approach is primarily based on the *generic* characteristics of human existence, which encompasses both *dependence* and *independence* as typical features of human life, it is more apt to incorporate the equal worth of the health-care goods and services of prevention, cure, and care into a general theory of justice. Her approach does not start from the things that persons need in their status as free and equal citizens and as normal and fully cooperating members of society over a complete life but from an embodied conception of human dignity, which conceives of the person as being both capable and needy; as both having activities, goals, and life projects and yet being in need of material support for the fulfilment of these many central things. Her approach takes seriously the materiality or basic dependency of human beings – their need for food, shelter, friendship, care, and so on – *in* the development of their powers and abilities. I suggested that because her approach starts from a rich and flexible sense of human dignity, including complex forms of reciprocity and social interaction, it is more apt to incorporate the *full complexity* of health-care goods and services, i.e. of prevention, cure, *and* care, into a theory of justice.

6.1.4 The Moral Importance of Just Health Care

Accordingly, I believe that the moral importance of just health care is expressed by the way in which and the extent to which it contributes to equality for all in the space of the relevant capabilities. With regard to health care, equality of access for equal need contributes to the capabilities of life, bodily health, and to the social bases of self-respect, non-humiliation and non-discrimination on the basis of disease and disability.[3] This provides a much broader interpretation of the moral importance of just health care than Daniels's approach, which focuses narrowly on the effect of health care on maintaining normal functioning.

Because capabilities measure not actually by achieved functionings but by a person's *freedom* to achieve valued functionings, it protects the reasonable pluralism of various comprehensive conceptions of the good. It leaves individuals the *choice* whether to pursue the relevant function or not. However, important exceptions are provided by cases in which we may feel that some of the capabilities are so important, so crucial to the development or maintenance of all the other capabilities, that we are sometimes justified in promoting functioning rather than capability. This, however, should happen within *limits* set by an appropriate concern for individual *liberty*. This can be done either by direct effective policy – as is mostly the case in matters of *public health* – or by providing incentives to avoid hazardous health behaviour – which is the case in matters where *personal responsibility for healthy behaviour* becomes relevant. Nussbaum's approach leaves room for the fact that we may feel that life and bodily health are so important to all the other capabilities that they are legitimate areas of interference with individual choice to some extent and up to some point.

[3] Cf. Section 4.2.4.2 *A List of the Central Human Capabilities*.

6.1.5 The Essence of a Truly Human Right to Health Care

The argument that Nussbaum's capabilities approach is better adjusted to incorporate the full complexity of health care into a theory of justice implies the necessity of offering a *broader* interpretation of the essence of the human right to health care than I have given in Chapter 2.[4]

The *moral component* of such a right remains the same; viz. from the moral viewpoint, the statement that there is a basic human right to health care means four things.[5] Firstly, it means that there is a *collective moral obligation*, i.e. an obligation on the part of society as a whole to ensure that everyone has access to some level of health-care goods and services. Secondly, it means that this obligation is a very *stringent* one. Obligations that are implied by rights have exceptional moral force in public debate. In this line of reasoning, Ronald Dworkin defines a right as a 'trump' that overrides countervailing considerations. A countervailing consideration could be, for instance, the mere fact that abandoning moral obligation could increase overall utility. Therefore, rights serve as a powerful protection of important interests every person has.[6] Thirdly, a basic right to health care implies that access to health care is *owed* to those who have the right. A right holder is not kindly asking for a favour and if society fails to fulfil this collective obligation, it does all the individuals who lack access to health care an injustice. And fourthly, as a human right, it is ascribed to all individuals because they are *human*.

The *basis* of such a right, however, should be *broadened*. Instead of founding it, as I did in Chapter 2, solely on the importance of guaranteeing fair equality of opportunity, which is based on the effect of just health care on maintaining normal functioning, the complexity of such a right is better encompassed by founding it on its contribution to equality for all in the space of *capabilities*. The other arguments remain valid, viz. the argument of the generic character of basic health-care needs, and the argument of collective social protection.

The argument of the *generic* character of basic health-care needs entails that these needs are universal in character, necessary for the fundamental projects of every person, and generically originating from human vulnerability and finitude. Although there are interpersonal differences in health-care needs to reach the same level of capability to function, enjoying equal access to health care for equal need is of fundamental value for *every* person. Securing equal access for equal need to health care is a basic moral obligation for every just society, because of health care's importance in contributing to the capabilities of life and bodily health, and to the social bases of self-respect, non-humiliation and non-discrimination on the basis of disease and disability.

---

[4] Cf. Section 2.3.5 *A Human Right to Health Care?*

[5] Partially based on: A. Buchanan, 'Philosophic Perspectives on Access to Health Care: Distributive Justice in Health Care', in *The Mount Sinai Journal of Medicine* 64(1997)2, pp. 90–95.

[6] R. Dworkin, *Taking Rights Seriously*, Duckworth: London, 1977; J. Waldron, 'When Justice Replaces Affection: the Need for Rights', in *id.*, *Liberal Rights*, Cambridge: Cambridge University Press, 1993, pp. 370–391.

The argument of *collective social protection* holds that it would be unreasonable to expect that individuals generally should be able to gain sufficient access to health care by relying solely on their own private resources for four reasons. Firstly, health-care needs are more unequally distributed than other basic needs like food, clothing, and shelter (some people need considerably more health care than others, while people's need for food and clothing is generally the same). Secondly, health-care needs can be highly unpredictable due to the element of luck. Thirdly, the fulfilment of health-care needs has an important impact on a person's capabilities to function. And finally, health care can be catastrophically expensive. If private resources could generally cover health-care needs, there would be little point in declaring entitlements to health care.

6.1.6 A Forward-Looking Policy of Inclusion

Furthermore, I strongly believe that principles of justice in health care should express a *forward-looking policy of inclusion*. This implies that the framework of identified basic health-care goods and services should be guaranteed by society on an equal basis to everyone in any case. This is part of the egalitarian aim to guarantee for everyone the social conditions of their freedom in terms of capabilities.

In this respect, I defend the *non-grading* idea of respect for the person as person as is to be found in its purest form in the Kantian injunction to treat each man as an end in himself, and never as a means only. The non-grading view of respect for the person as person is essentially different from both the *meritarian* and the *utilitarian* view. Contrary to the latter view, which derives a person's dignity from its utility value, its socio-economic worth, measured in terms of productive contribution, I argue that just health-care policy should be based on the idea of respect for persons regarded from the viewpoint of their humanity and not merely from the perspective of their different utility value (vs Gauthier). Nussbaum argues that, in order to be doing what they should be doing, states must be concerned with *all* the capabilities on the list, even when these do not seem so useful for economic growth, or even for political functioning.

Contrary to the meritarian view (cf. the luck-egalitarians), I believe that the principles of just health care should have nothing to do with grading reflection on *personal responsibility* for health outcomes *ex post*. Any starting-gate theories or any other principles that allow citizens to *lose* access to adequate levels of the identified health-care goods and services are unacceptable.

Contrary to grading viewpoints, the principles of just health care should express a *forward-looking policy of inclusion*. The framework of identified basic health-care goods and services should be guaranteed on an equal basis to everyone in any case, because it concerns a domain that is just *too important* to allow exclusion, i.e. the domain of guaranteeing for everyone the social conditions of their freedom in terms of capabilities. In the line of reasoning of Nussbaum and Scanlon, I argue that the framework of identified basic health-care goods and services concerns matters that are taken to be *entitlements* of citizens based on justice. When any one of these is abridged this is an especially grave failure of the socio-political system. The abridgement is then not just a huge cost to be borne, but also a cost of a distinctive kind, involving a violation of basic justice. When some matters *outside the core* are abridged, it may be a small cost – or a large cost to some persons – but it is a cost of a different kind. It is a cost that, from the viewpoint of justice, no citizen should be asked to bear.[7] This brings us to the final point, viz. that of limits to health care.

6.1.7 Limited Scope

The above-mentioned desiderata for just health care result in a complex, broad and inclusive basis for just health care. The principles of just health care aim to secure equal access for equal need to a framework of identified basic health-care goods and services. This framework should be guaranteed on an equal basis to everyone in any case. The perspective expresses a strong egalitarian approach of just health-care based on a policy of inclusion of all human beings in need of care.

Of course, this broad framework of entitlements is *limited*. Taking the distinct characteristics of scarcity into account, the human right to health care cannot be an unlimited right and rationing of health care has to be a fact of life. Establishing precise entitlements invariably involves trade-offs, and no right to health care will trump *all* competing claims of social utility or the common good when larger questions of macro allocation are at stake. Within this perspective, however, I believe that it is essentially important that the discussion concerning setting limits on health care should focus on setting *limits on goods not on people*. In this regard, I have argued that the *prudent insurance proposal* of Dworkin offers a workable guideline to set legitimate limits on health care. Dworkin's position combines an argument for a moral right to health care with a limitation on that right, both based on the idea of the prudential insurer.

Taking the internal and external aspects of scarcity into account, as well as the idea of opportunity costs, the prudent insurance principle expresses *a middle course*. It balances the anticipated value of medical treatment against other goods and risks, and it presupposes that people might think they lead better lives overall when they invest less in doubtful or superior quality medicine and more in making life successful or enjoyable, or in protecting themselves against other risks, like unemployment, that might also blight their lives. Health policymakers might well decide that while most prudent people would provide themselves and their family with *decent-quality basic health care* (for instance, primary medical care, hospitalisation when necessary, standard prenatal and paediatric care, routine examinations, inoculations and other preventative medicine, and finally, respectful, decent, and attentive long-term care), they would forego *expensive heroic treatment* with uncertain outcome in return for more certain benefits like education, housing, and

---

[7] M.C. Nussbaum, 'The Costs of Tragedy: Some Moral Limits of Cost-Benefit Analysis', 2001. See also *id.*, 'Capabilities as Fundamental Entitlements', 2003, pp. 44–48. Cf. Section 4.2.5.2 *Freedom*.

economic security. If this is reasonable, Dworkin argues, then justice demands that a universal health scheme should not provide such enhanced treatment.

As such, Dworkin's prudent insurance principle provides a workable *guideline* to set legitimate *limits on* the attitude of limitless *expectations* for the rights and duties within the public health-care system. Both the external and the internal aspect of *scarcity* are included in his hypothetical insurance scheme. Against the background of the idea of opportunity cost, it becomes clear that setting limits in the field of high-tech medicine produces far fewer objections of justice if this implies saving the policy of equal access to decent-quality basic health care for all.

6.2 REFINING THE CONTEMPORARY SOCRATIC PERSPECTIVE

Having brought the essential points of the discussion together, I will now try to specify the general perspective in the light of which these desiderata for principles of just health care are best understood. It is my aim to provide a somewhat more elaborate view on what I mean by a contemporary Socratic perspective in just health care than I have given in Chapter 2.

6.2.1 A General Moral Perspective

Before continuing, it is important to note that this proposal does not aim to provide specific instructions, practical solutions, and concrete cut-and-dried recipes to solve the great problems of contemporary health care. On the contrary, the philosophical aim is to offer a *general moral perspective*, which puts a different complexion on the matter. It proposes an alternative *attitude* and a renewed mentality in society's expectations for just health care. As such, it *precedes* the various rationing mechanisms and concrete proposals to enhance health care's efficiency and effectiveness. It deals with scarcity of health-care resources in a different way. It concentrates on that specific aspect of the *origin* of scarcity, which is located in various social, cultural and anthropological mechanisms that determine our general *attitudes* and *expectations* towards medicine and health care. As David Braybrooke puts it, the problem of needs-inflation can be looked upon as inherent to the needs-concept. Scarcity of resources is a case 'that may be expected to come from repeating again and again an apparently legitimate application of the concept'.[8] In view of the technological advances in the field of health care, and the rise of expectations that those advances have encouraged, it may seem that in the end 'there is no way of acknowledging that nothing already present in the concept of needs saves the need for medical care from becoming a bottomless pit'.[9] No way indeed, I believe, except for concentrating on the character of the *general expectations* that *precede* the problem of needs-inflation.

---

[8] D. Braybrooke, *Meeting Needs*, 1987, p. 295.

[9] *Ibid.*, p. 301.

The essence of the argument is captured in the idea that in order to reconcile efficiency, justice and care in health care, it is necessary to define the scope of just health care in terms of *a broad and inclusive basis of moderate expectations*, instead of a small and exclusive basis moulded by high expectations. By having formulated the desiderata for principles of just health care, I have given a concise account of what I mean by a broad and inclusive basis of decent-quality care for all. Now it is time to concentrate briefly on the desiderata of an *alternative moral perspective on reasonable attitudes and expectations* towards life, health and health care.

6.2.2 The Limits of Human Existence

The main point of a renewed Socratic attitude towards life, health and health care is that it takes the limited character of human existence into account from the beginning. As such, it expresses an approach that allows us to set limits on health care and to make the necessary choices *without* inequitable exclusion of certain groups of people on the one hand and *without* frenetically wanting to manage and control the life and health of all on the other hand. It is my viewpoint that an acceptable valuation of human life and healthy functioning that wants to be both socially efficient and equitable for all, should start from a significant and meaningful revaluation of human finitude, that is, from a renewed Socratic perspective of learning to prepare oneself for dying and death.[10]

Within the vein of contemporary health care, which is dominated by the Cartesian ideal to maintain and preserve health, prolong life and postpone death, it may seem rather strange, indeed even scandalous, to propose the Socratic perspective as leitmotiv of an alternative moral perspective of attitudes and expectations towards medicine and health care. However, as I have argued previously, it has nothing to do with an unmoved acceptance of fate and fortune, nor with an uncritical rejection of all the possibilities in medicine and health care.[11] What it does express, however, is the idea that although life and health are an important *bonum*, they are not the *summum bonum*, which needs to be protected by all means at all times and at all costs. As such, it expresses a critical attitude towards the endless possibilities of modern medicine and medical technology, founded on being able to relate meaningfully to the inevitable fact of finitude, even though we find it hard, difficult and tragic.

Against the horizon of the limits of human existence and in the light of the above-proposed framework of desiderata for principles of just health

---

[10] Cf. Plato, *Crito* and *Phaedo*, trans. H. North Fowler, Cambridge, MA: Harvard University Press, 1971. See also: O. Höffe, *Medizin ohne Ethik?* Frankfurt: Suhrkamp, 2002, esp. pp. 119–142; P. Van Parijs, 'Y a-t-il des limites à la prise en charge des soins de santé par la solidarité?' in: J. Hallet; J. Hermesse; D. Sauer (eds.), *Solidarité, Santé, Ethique*, Louvain: Garant, 1994, pp. 57–68; P. Van Parijs, *Solidariteit voor de 21e eeuw*, Leuven: Garant, 1996, pp. 35–45.

[11] Cf. Section 2.4 *Care and the Boundaries of Human Life*.

care, the contemporary Socratic perspective implies three things: a defence of a *rich and broad basis* that incorporates both the *external* and the *internal* aspect of scarcity.

6.2.2.1 Decent-Quality Basic Care

The above-proposed arguments to define the scope of just health care in terms of *a broad and inclusive basis of moderate expectations*, instead of a small and exclusive basis moulded by high expectations, should be understood against the horizon of human finitude. Taking the generic characteristics of human life, its limits and its finitude into account sets the stage for reflections on equal social valuation of respectful, decent, and attentive care for citizens' *quality* of life. It would entail breaking through the sole focus on *quantity* of life and on the quantity of independent healthy normal functioning.[12] Or as Phillippe Van Parijs puts it:

> Il s'agit de l'arbitrage que nous avons à faire entre la *longueur* de la vie et le *niveau* de vie au sens le plus large, de la question de savoir quel coût on est prêt à supporter tout au long de la vie pour avoir droit, lorsqu'on est âgé, voire très âgé, à des operations au traitements qui permettront de la prolonger.[13]

In this line of reasoning, the normative valuation of life and health becomes less strictly one-sided. It may avoid the risk of therapeutic obstinacy and install the possibility of becoming better able to deal respectfully with dependency and mortality as essential characteristics of the human condition, even though we find it tragic. In general, it avoids the risk of installing a hierarchical relation between prevention, cure and care.[14]

The main reason why Nussbaum's capabilities approach is more apt to incorporate the full complexity of just health care, resides in the fact that she starts from a *rich* and *complex* concept of human *dignity*, taking *dependence* and *finitude* into account from the beginning. The tenor of Nussbaum's rich and complex list of capabilities expresses an explicit concern for the *quality* of human life, more than for its quantity. Hence, her concern for human flourishing and truly human functioning. The ten capabilities represent basic aspirations to human *flourishing* that are universally recognisable. They express central elements of truly human functioning that can command an overlapping consensus. As I indicated before, this focus on truly human functioning and human flourishing expresses a certain logic of abundance, that may be problematic.

From the viewpoint of the contemporary Socratic perspective, however, it is reasonable to say that the *rich and complex* character of the capabilities list, its emphasis on truly human functioning and on human *flourishing*, has less to do

---

[12] Plato, *Crito* 47(d)–48(e); O. Höffe, *op. cit.*, 2002, pp. 133–136, 138–139; P. Van Parijs, *op. cit.*, 1996, p. 39.

[13] P. Van Parijs, *op. cit.*, 1994, p. 62.

[14] See also: H.A.E. Zwart, 'Ethiek, medische technologie en solidariteit', in J. Hallet; J. Hermesse; D. Sauer (eds.), *Solidariteit, gezondheid, ethiek*, Leuven: Garant, 1994, pp. 81–89.

with the problem of the infinitely rising spiral of medical possibilities and general social expectations regarding maximising health and prolonging life, than with the *abundance* that is implied in health-care policy that concentrates on *a broad and inclusive basis of decent-quality care*. The idea that decent-quality health care against the horizon of human finitude may not be colonised by a logic of scarcity is also expressed by her argument that there are moral limits to trade-offs.[15] This argument is even more valid given that scarcity is itself a problematic phenomenon, which is most aptly illuminated by its internal dynamics.

Furthermore, a contemporary Socratic perspective on health and health care incorporates both the external and the internal aspect of scarcity. From the viewpoint of the macro perspective, i.e. the perspective that concentrates on determining the scope and design of just health care, this implies that we – Dworkin-like – have to ask ourselves two questions.

6.2.2.2 The External Question

The first question asks us to take the *external aspect of scarcity* and its related idea of *opportunity costs* into account. It asks us to think about what we want to realise and what we are willing to forego by doing so. This question serves as a workable guide to set limits on what can be reasonably expected from a just health-care system, for it asks us to take *other* important *Boni* in life – like education, housing, nutrition, culture, leisure etc. – into account as well. Having discussed the issue of various distant determinants of health (climate, food and wages, working conditions, education and housing, poverty and public hygiene, the presence of hospitals, etc.) the relevance of this question becomes even clearer. The quantity and quality of life and health derive from much more than health-care alone.

In the same line of reasoning, Dworkin's *prudent insurance question* asks us which treatments would we prudently choose to be insured for, and which would we regard as not worth the cost of the insurance because their opportunity cost is too high? The answer is simple: taking the fact of opportunity costs into account, most people would probably not want to sacrifice access to decent-quality basic health care, while on the other hand, high-tech medicine is far less unproblematic in this regard.[16]

6.2.2.3 The Internal Question

In the same line of reasoning, but from a different viewpoint, the *internal question* asks us whether we truly and honestly believe that life and health have to be

---

[15] M.C. Nussbaum, 'The Cost of Tragedy: Some Moral Limits of Cost-Benefit Analysis', in M. Adler; E. Posner, *Cost-Benefit Analysis: Legal, Economic, and Philosophical Perspectives*, Chicago: Chicago University Press, 2001. Cf. Section 4.4.7 *Scarcity and the Language of Limits*.

[16] R. Dworkin, 'Justice and the High Cost of Health', 2000/1994, pp. 307–319; *Id.*, 'Justice in the Distribution of Health Care', 2002/1993, pp. 203–222.

maintained and preserved at all times, by all means and at all costs? Against this horizon, we must honestly and critically ask ourselves whether every prolongation of life, every application of medical technology is truly that important as it may seem to be at first sight. To what extent are we willing to give up certain goods and services?

Again, the relevance of a contemporary Socratic perspective becomes more clear.[17] As the gap between all that is technically possible and what should be equitably accessible increases, our moral responsibility and the answer to the question what we owe to each other at the bar of just health care has less to do with doing all we can in the name of life and health as the *Summum Bonum*, and ever more with providing equal access to decent-quality basic care for all. It has less to do with a sole focus on *quantity* of life and more with *quality*. As such, the normative valuation of health and health care becomes less strictly one-sided. It may avoid the risk of therapeutic obstinacy and install the possibility of becoming better able to deal respectfully with dependency and mortality as essential characteristics of the human condition, even though we find it tragic.

It is important to note, furthermore, that this plea for a contemporary Socratic perspective in health care certainly does not imply a romantic glorification of dependence and mortality, nor does it entail an unmoved acceptance of fate and fortune. As I have defined in the beginning, it is the true task of health care to maintain and preserve health, to restore it when possible and to care for the patients in need when cure is not or no longer possible; to support them and to ease their suffering. By this task, health-care institutions, services and measures have a major impact on our well-being. However, within this framework, the contemporary Socratic perspective aims to reduce the risk to limitless expectations.

6.2.3 *The Cardinal Virtues?*

Inspired by a suggestion made by Ottfried Höffe, I finally wish to point briefly at one further possibility to flesh out the contemporary Socratic perspective in health care. Against the problem of *pleonexia* in health care, it is fruitful, Höffe argues, to bear in mind the four classic cardinal virtues, i.e. temperance, prudence, fortitude and justice.[18] Previously, I have pointed at the Aristotelian aversion to *pleonexia*, to insatiability and endless accumulation.[19] I also argued that the criticism of the Socratic perspective is directed mainly against the dynamics of the endlessly rising spiral movement, i.e. the tendency towards infinity that characterises contemporary medicine and health care to a large degree. Let us have a look at the way in which the classic virtues may provide further refinement of this criticism.

---

[17] Cf. Section 2.4.4 *A Contemporary Socratic Perspective*.

[18] O. Höffe, *op. cit.*, 2002, pp. 119–142, 174–181, 207–232. I have mentioned the cardinal virtues earlier in Section 1.3.1.1 *Suum Cuique Tribuere* and Section 2.4.4.2 *Reassessing Scarcity in Health Care*.

[19] Cf. Section 1.3.1.1 *Suum Cuique Tribuere* and Section 2.4.4 *A Contemporary Socratic Perspective*.

6.2.3.1 *Temperance*

Against the risk of *pleonexia* in health care, Höffe argues, we need a general moral attitude of temperance or *sôphrosynê*. This entails both an attitude of scepticism about hyperactive health care and a willingness to acknowledge the value of certain limits to health care. It entails a choice for *quality* of life instead of sole concentration on quantity of health and prolongation of life. As such, it runs counter to a certain form of *hybris*, included in the contemporary tendency to limitless hopes and expectations in the field of medicine and health care.

On the level of society, the attitude of temperance implies a general modification of the normative value of life and health in the light of a positively refined valuation of the limits of human existence. It implies that reasonable expectations for contemporary health care are situated between hyperactive glorification and stoical fatalism. Within the sphere of health-care ethics, Höffe refers to the virtue of *Gelassenheit*:

> Die Gelassenheit wendet sich gegen eine Ungeduld, die sich auf eine Situation nicht einlassen kann, sie vielmehr von vornherein und ausschließlich im Sinne eigener Vorstellungen verändern will. Die Gelassenheit unterscheidet sich aber von Nachgiebigkeit und Schwäche, aufgrund deren man sich gegebenen Situationen willenlos unterwirft. Zwischen Erzwingenwollen und Gefügigkeit, zwischen Aktivität und Passivität situiert, bezeichnet sie die Bereitschaft, die natürliche Welt, die Mitmenschen aber auch die eigene Person mitsamt der dazugehörigen Geschichte anzunehmen, ohne sich deshalb als freie und schöpferisch handelnde Person aufzugeben.[20]

To the question what can we and should we reasonably expect from just health care, the answer is simple, neither everything nor nothing:

> [Die Gelassenheit] fordert hier nicht etwa zum Sparen auf, sondern setzt den Allmachtsphantasien die Einsicht in die grundsätzliche Gebrechlichkeit des Menschen und seine Sterblichkeit entgegen.[21]

However, in order to form an attitude of temperance in expectations for medicine and health care, we need another virtue, i.e. that of prudence or *phronêsis*.

6.2.3.2 *Prudence*

The virtue of prudence is present in Dworkin's prudent insurance proposal, which entails that we are willing to honestly and critically ask about the true necessity and moral importance of various health-care goods, services and treatments. As such, the virtue of prudence requires caution, circumspection, and critical distance from the idea that life and health are the *summum bonum*, which needs to be protected by all means and at all costs. Höffe puts it as follows:

> Nicht Schicksalsergebenheit herrschte früher vor, sondern ein weit geringeres medizinisches Können. Infolgedessen blieb häufig bloßer Wunsch, was im Verlauf der Neuzeit zum Machtvollen Willen wird: *eine wachsende medizinische Sorge für die Gesundheit*. Nach Auskunft von Platon und Aristoteles stellt sich die griechische Antike dabei eine Frage, die die Moderne mehr und mehr *verdrängt*, nämlich *ob man des Guten nicht manchmal zu viel tut*. Um

---

[20] O. Höffe, *op. cit.*, 2002, p. 177.

[21] *Ibid.*, pp. 222–223.

> dies zu verhindern, brauchen [wir] eine Fähigkeit und Einstellung, die den Rang einer für die Medizin unverzichtbaren Kardinaltugend hat. [Wir] bedürfen einer moralisch-praktischen Urteilskraft: der Klugheit im Sinne der griechischen *phronêsis*. Wo die Erfolge sich so rasch wie in der Moderne einstellen, wird freilich die Frage nach *Grenzen* lieber verdrängt. Statt dessen werden *Allmachtsphantasien* beschworen .... Am Ende laufen sie auf eine überwindung der menschlichen *Endlichkeit*: seiner Krankheitsanfälligkeit und Sterblichkeit hinaus. Daß den einschlägigen Ausdruck *hybris* schon die Griechen prägen, zeigt allerdings, daß erneut ein menschheitsgeschichtlich weit älteres Problem vorliegt. Man darf verallgemeinern: Zwischen der *Erwartung* an die Medizin, auch ihren Verheißungen, und der tatsächlichen *Erfüllung* tut sich eine Kluft auf, die sich auch bei größten Anstrengungen nie schließt.[22]

In health-care ethics, the essence of the virtue of prudence is that it expresses a critical attitude towards the endless possibilities of modern medicine and medical technology, founded on being able to relate meaningfully to the inevitable fact of finitude, even though we find it hard and tragic.

6.2.3.3 Fortitude

Because it is generally difficult to take the limits of human existence into account, and because it may be easier to go along with the prevailing logic of postponing the inevitable fact of finitude as much as possible, the issue of setting limits on health care requires fortitude in the classical sense of firmness of spirit in taking difficult decisions and constancy in the pursuit of the just. It requires 'die Tapferkeit in Form von Zivilcourage'.[23]

6.2.3.4 Justice

Finally, the virtue of justice in the universal sense, in the words of Aristotle, is the chief of all virtues because it combines all virtues as far as they are important to other people and to the happiness of the city.[24] The just man is honest, modest, courageous, moderate, and wise. Justice or *dikaiosyne* is the attitude to freely fulfill the demands of law and morality. It is that moral disposition 'which disposes people to do just actions, act justly, and wish for what is just'.[25] Here, Aristotle resumes the Platonian idea of justice as the perfect virtue.

Applied to health care, this means that the virtue of justice is expressed in the extent to which society allocates scarce health-care resources in a well-balanced way, i.e. in a way that harmoniously combines the other virtues. It meets the problem of just allocation in a way that combines temperance, prudence and courage; in a way that encourages temperance in expectations, prudence in application and coverage, and courage in decision-making. As such, the four cardinal virtues may be understood as giving somewhat more body to the contemporary Socratic perspective in health care.

Having developed a series of desiderata for principles of just health care, and having specified the general perspective in the light of which these desiderata are

---

[22] *Ibid.*, p. 210 (my italics, YD).
[23] *Ibid.*, p. 142; also *ibid.*, p. 211.
[24] Aristotle, *Nicomachean Ethics*, V, 1129(b).
[25] *Ibid.*, V, 1129(a).

best understood, it has been my philosophical aim to offer a general moral *perspective*, which puts a different complexion on the matter; i.e. to offer grounds for an alternative *attitude* and a change of mentality in society's expectations for just health care. The importance of this perspective resides in the fact that it *precedes* the various rationing mechanisms and concrete proposals to enhance health-care efficiency and effectiveness. It deals with scarcity of health-care resources in a different way. It concentrates on that specific aspect of the *origin* of scarcity, which is located in various social, cultural and anthropological mechanisms that determine our general *attitudes* and *expectations* towards medicine and health care. As such, it is necessarily abstract in nature. It does not aim to provide specific instructions, practical solutions, and concrete cut-and-dried recipes to solve the great problems of contemporary health care. Nevertheless, however abstract this proposal is, this need not be a problem. For it has been my philosophical task to assess, explain and justify or modify the relevant arguments, categories, distinctions and interrelations. The high level of abstraction is only a problem when the expectations are very concrete.

In the same spirit, I wish to conclude by resuming some of the questions, the questions I have posed at the very beginning.[26]

6.3 RESUMING THE CONTEMPORARY DEBATE

I started this book with a series of questions, which are all examples of topics that lie at the centre of much current debate about health-care allocation. In conclusion, I will briefly resume four of them; i.e. the questions regarding *access* to health care, regarding *rationing* and priority setting, regarding the issue of *responsibility* and finally, the question regarding the issue of *long-term care*.

6.3.1 Access to Health Care

The issue of access to health care actually contains two questions: 'Access for whom?' and 'Access to what?' The answer to the question *for whom* is simple: equal access for anyone in equal medical need. Specifically, non-medical features like race, sex, or ability to pay should not determine whether or not someone has access to health care. To avoid needs-inflation, however, we need to distinguish health-care needs that can function as standards of provision for social distribution of health-care goods and services from those that cannot. We need a typology of health-care needs, goods and services. This brings us to the second question: access *to what*?

Again, the answer is simple: we must in some way *identify* the health-care goods to which all citizens must have effective access over the course of their whole lives. Taking the complex and non-homogeneous framework of health-care goods and services, i.e. of prevention, cure *and* care, into account, I have argued that the

[26] Cf. General introduction.

objective, non-volitional structure of health-care needs, further specified in the broad and inclusive language of capabilities, provides an important abstract guideline to distinguish important goods and services from less important ones.

The moral importance of just health care is expressed by the way in which and the extent to which it contributes to equality for all in the space of the relevant capabilities. With regard to health care, equality of access for equal need contributes to the basic human capabilities of life, bodily health, the social bases of self-respect, non-humiliation and non-discrimination on the basis of disease and disability. This requires, as I have argued, equal access for equal need to a broad, inclusive and forward-looking framework of decent-quality basic health care.

6.3.2 Health-Care Rationing

The question of 'Access to what?' already contains some issues about priority setting and rationing. Even if we decide that access to health care should be based on need, which needs would we meet then when we cannot meet them all? If not all services can be provided because resources are restricted, then which are the most important to provide? Taking these questions into account is important, I have argued, for maintenance of just health care itself, i.e. not only maintenance of its actual existence in the real world, but also of its legitimacy. Two things can be said about the character of health-care rationing.

Firstly, we have the *formal* condition of fair decision-making. In order to secure the legitimacy of rationing decisions, they must be the result of a fair process for setting limits or rationing of care. In this regard, Daniels's formal condition of 'accountability for reasonableness' provides an interesting formal framework to guide rationing decisions.

Secondly, a more *substantive* strategy in dealing with trade-offs, scarcity and limits in the field of health care is provided by Dworkin's prudent insurance principle.

6.3.3 Rationing by Responsibility?

Furthermore, I have analysed the fact that the argument of rationing by responsibility often appears in contemporary debate regarding limits to health care. This argument is based on the idea that those who choose to run health risks cost the rest of us money, and it is fair that they should pay it back, either by paying larger insurance premiums, or by forgoing health care for their self-induced conditions. In this regard, three elements should be stressed.

Firstly, personal responsibility is an important value. By their behaviour, people have effect on their health, and society should not hesitate to underscore the importance of sensible health behaviour by making people conscious of the influence they have on their health needs. However, in the name of fair equality of opportunity, society should continue to be forward-looking, both in providing incentives to avoid hazardous behaviour and in offering medical help.

Regarding incentives, consciousness-raising health campaigns show respect for individual autonomy while appealing to people's rationality to take care of their health. The same goes for cost sharing. It is fair to require individuals who engage in certain risky actions that result in costly medical needs to pay higher premiums or taxes. Risk-takers may be required to contribute more to particular pools such as insurance schemes or to pay a tax on their risky conduct, such as an increased tax on alcohol and tobacco. These requirements may fairly redistribute the burdens of the costs of health care, and they may deter risky conduct without disrespecting autonomy. The return individuals may expect on their taxed consumption is health-care protection for themselves.

Thirdly, it would be unjust to refuse care to a person in need, even if it is clear that he was responsible for their condition. Contributing to fair equality of opportunity should continue to be one of the fundamental moral goals of health care. This should not change because of past behaviour; as only then can the basic human right to health care be a truly inviolable right.

6.3.4 Long-Term Care

With regard to the problem of care for the long term and irreversibly dependent, my position can be summarised in three points.

Firstly, the internal logic of the term *health* care seems to indicate that it mainly concerns public measures to maintain, restore, and possibly improve the *health* status of the citizens. Its goal is the preservation of life and health. This is only partially true. From the beginning, I have defined health care in the broad sense, including the pillars of prevention, cure *and* care. I have argued throughout that a hierarchic relation between the pillars is undesirable because it encourages a generally condescending attitude towards care for the long-term and irreversibly dependent, i.e. for the elderly, and the mentally and physically disabled. However, this kind of care is also a major part of the work that needs to be done in any society and which can be a source of great injustice.

Secondly, it is important to note that health-care derives its moral significance precisely and primarily from the horizon of finitude and our acknowledgement of the limits of human existence. Health-care services, goods, and institutions are important in the first place, precisely because people encounter dependency, illness, disease, handicap and ageing; situations in which we encounter our finitude. At the same time, these are situations that are generically characteristic for the human condition. Any adequate theory of just health care must incorporate the problem of dependency – both temporarily and irreversibly – *from the beginning.*

Thirdly, I believe that Nussbaum's capabilities approach is superior in incorporating the full complexity of health care into her theory of justice for two reasons. Firstly, the capabilities approach is sensitive to people's *variations* in needs for types of care; and secondly, her approach starts from a rich and complex concept of human dignity, incorporating both capability and neediness. As such, her approach incorporates the fact that a society's moral quality is reflected by the way in which it truly takes care for the least-advantaged, i.e. the weak and vulnerable.

BIBLIOGRAPHY

ACHTERHUIS, H., *Het rijk van de schaarste*, Baarn: Ambo, 1988.

ANAND, S.; PETER, F.; SEN, A.K. (eds.), *Public Health, Ethics, and Equity*, Oxford: Oxford University Press, 2004.

ANDERSON, E.S., 'What Is the Point of Equality?' in *Ethics* 109(1999), pp. 287–337.

ARISTOTLE, *Nichomachean Ethics*, trans. and ed. R. Crisp, Cambridge: Cambridge University Press, 2000.

ARISTOTLE, *The Politics*, trans. B. Jowett, rev. trans. J. Barnes, ed. S. Everson, Cambridge: Cambridge University Press, 1996.

ARNESON, R.J., 'Equality and Equal Opportunity for Welfare', in *Philosophical Studies* 56(1989), pp. 77–93.

ARNESON, R.J., 'Liberalism, Distributive Subjectivism, and Equal Opportunity for Welfare', in *Philosophy and Public Affairs* 19(1990), pp. 158–194.

ARNESON, R.J., 'Perfectionism and Politics', in *Ethics* 111(2000), pp. 37–63.

ARNESON, R.J., 'Primary Goods reconsidered', in *Nous* 24(1990), pp. 429–454.

ARROW, K.J., 'Some Ordinalist-Utilitarian Notes on Rawls's Theory of Justice', in *Journal of Philosophy* 70(1973)9, pp. 245–263.

ARROW, K.J., 'Uncertainty and the Welfare Economics of Medical Care', in *American Economic Review* 53(1963)5, pp. 941–973.

BAIER, A., 'The Need for More than Justice', in V. Held (ed.), *Justice and Care: Essential Readings in Feminist Ethics*, Boulder, CO: Westview, 1995, pp. 47–60.

BARRY, B., 'Justice as Reciprocity', in *id.*, *Liberty and Justice*, Oxford: Clarendon Press, 1991, pp. 211–241.

BARRY, B., *Justice as Impartiality*, Oxford: Clarendon Press, 1995.

BARRY, B., *Political Argument*, Hemel Hempstead: Harvester Wheatsheaf, 1991.

BARRY, B., *The Liberal Theory of Justice*, Oxford: Oxford University Press, 1973.

BARRY, B., *Theories of Justice*, Berkeley: University of California Press, 1989.

BAYERTZ, K., *Solidarity*, Dordrecht: Kluwer, 1999.

BEAUCHAMP, D.E., 'Public Health: Philosophy of Public Health', in W.T. Reich (ed.), *Encyclopedia of Bioethics*, 2nd edn., New York: Simon & Schuster/Macmillan, 1995, pp. 2161–2165.

BEAUCHAMP, D.E., 'The Health of the Republic: Epidemics, Medicine, Medicine and Moralism as Challenges to Democracy', Philadelphia: Temple University Press, 1988.

BEAUCHAMP, T.L.; CHILDRESS, J.F., *Principles of Biomedical Ethics*, 5th edn., New York: Oxford University Press, 2001.

BEAUCHAMP, T.L.; CHILDRESS, J.F., *Principles of Biomedical Ethics*, 4th edn., New York: Oxford University Press, 1994.

BENTHAM, J., *An Introduction to the Principles of Morals and Legislation*, eds. J.H. Burns; H.L.A. Hart, London: Methuen, 1982.

BERLIN, I., 'Two Concepts of Liberty', in *id.*, *Four Essays on Liberty*, Oxford: Oxford University Press, 1969.

BERLIN, I., *Four Essays on Liberty*, Oxford: Oxford University Press, 1969.

BÉRUBÉ, M., *Life as we Know it: a Father, a Family, and an Exceptional Child*, New York: Pantheon, 1996.

BOORSE, C., 'Health as a Theoretical Concept', in *Philosophy of Science* 44(1977), pp. 542–573.

BOORSE, C., 'On the Distinction between Health and Disease', in *Philosophy and Public Affairs* 5(1975), pp. 49–68.

Borges, J.L., 'The Immortal', in *id.*, *Labyrinths: Selected Stories and Other Writings*, Harmondsworth: Penguin Books, 1964, pp. 135–149.

Braybrooke, D., 'Let Needs Diminish That Preferences May Prosper', in N. Rescher (ed.), *American Philosophical Quarterly*, Monograph Series, No. 1, London: Basil Blackwell, 1968, pp. 86–107.

Braybrooke, D., *Meeting Needs*, Princeton, NJ: Princeton University Press, 1987.

Breslow, L., 'Public Health: Determinants of Public Health', in W.T. Reich (ed.), *Encyclopedia of Bioethics*, 2nd edn., New York: Simon & Schuster/Macmillan, 1995, pp. 2153–2156.

Brighouse, H., 'Can Justice as Fairness Accommodate the Disabled?' in *Social Theory and Practice* 27(2001)4, pp. 537–560.

Brock, D., 'Justice, Healthcare, and the Elderly', in *Philosophy and Public Affairs* 18(1989)3, pp. 297–312.

Brock, D., 'The Separability of Health and Well-Being', in C.J.L. Murray; J.A. Salomon; C.D. Mathers; A.D. Lopez (eds.), *Summary Measures of Population Health: Concepts, Ethics, Measurement and Applications*, Geneva: World Health Organization, 2002, pp. 115–120.

Buchanan, A., 'Justice as Reciprocity versus Subject-Centered Justice', in *Philosophy and Public Affairs* 19(1990), pp. 227–252.

Buchanan, A., 'Justice: a Philosophical Review', in E.E. Shelp (ed.), *Justice and Health Care*, Dordrecht: Reidel, 1981, pp. 3–21.

Buchanan, A., 'Philosophic Perspectives on Access to Health Care: Distributive Justice in Health Care', in *The Mount Sinai Journal of Medicine* 64(1997)2, pp. 90–95.

Buchanan, A., 'The Right to a Decent Minimum of Health Care', in *Philosophy and Public Affairs* 13(1984)1, pp. 55–78.

Buchanan, A.; Brock, D.; Daniels, N.; Wikler, D., *From Chance To Choice: Genetics and Justice*, Cambridge: Cambridge University Press, 2000.

Butler, J., 'The Ethics of Health Care Rationing: Principles and Practices', London: Cassell, 1999.

Calabresi, G.; Bobbitt, P., *Tragic Choices*, New York: Norton, 1978.

Callahan, D., 'Biomedical Progress and the Limits of Human Health', in R.M. Veatch; R. Branson, *Ethics and Health Policy*, Cambridge, MA: Ballinger, pp. 157–165.

Callahan, D., *Setting Limits: Medical Goals in an Aging Society*, New York: Simon & Schuster, 1987.

Caplan, A.L.; Engelhardt, H.T.; McCartney, J. (eds.), *Concepts of Health and Disease: Interdisciplinary Perspectives*, Reading, MA: Addison-Wesley, 1981.

Cassell, E.J., 'Do Justice, Love Mercy: The Inappropriateness of the Concept of Justice Applied to Bedside Decisions', in E.E. Shelp (ed.), *Justice and Health Care*, Dordrecht: Reidel, 1981, pp. 75–82.

Castoriadis, C., 'Valeur, égalité, justice, politique – De Marx à Aristote en d'Aristote à nous', in *id.*, *Les carrefours du labyrinthe*, Paris: Le Seuil, 1978, pp. 249–316.

Childress, J.F., 'Priorities in the Allocation of Health Care Resources', in E.E. Shelp (ed.), *Justice and Health Care*, Dordrecht: Reidel, 1981, pp. 139–150.

Cochrane, A.L., *Effectiveness and Efficiency*, London: The Nuffield Provincial Hospitals Trust, 1972.

Cohen, G.A., 'On the Currency of Egalitarian Justice', in *Ethics* 99(1989)4, pp. 906–944.

Copp, D., *Morality, Normativity, and Society*, Oxford: Oxford University Press, 1995.

Crocker, D., 'Functioning and Capability: The Foundation of Sen's and Nussbaum's Development Ethic, Part I', in *Political Theory* 20(1992), pp. 584–612.

Culyer, A.J., 'Commodities, Characteristics of Commodities, Characteristics of People, Utilities, and Quality of Life', in S. Baldwin; C. Godfrey; C. Proppers (eds.), *Quality of Life: Perspective and Policies*, London: Routledge, 1990, pp. 9–27.

Culyer, A.J., 'The Normative Economics of Health Care Finance and Provision', in *Oxford Review of Economic Policy* 5(1989)1, pp. 34–58.

Culyer, A.J.; Wagstaff, A., 'Equity and Equality in Health and Health Care', in *Journal of Health Economics* 12(1993), pp. 431–457.

Dahlgren, G.; Whitehead, M., *Policies and Strategies to Promote Social Equality in Health*, Stockholm: Institute of Future Studies, 1991.

Daniels, N., 'Distributive Justice and the Use of Summary Measures of Population Health Status', in *Summarizing Population Health: Directions for the Development and Application of Population Metrics*, Institute of Medicine, Washington: Academy Press, 1998, pp. 54–71.

Daniels, N., 'Enabling Democratic Deliberation: How Managed Care Organisations Ought to Make Decisions About Coverage for New Technologies', in S. Macedo (ed.), *Deliberative Politics: Essays on Democracy and Disagreement*, New York: Oxford University Press, 1999, pp. 198–210.

Daniels, N., 'Equality of What? Welfare, Resources, or Capabilities?' in *Philosophy and Phenomenological Research* 50(1990)Suppl., pp. 273–296, repr. in *id., Justice and Justification: Reflective Equilibrium in Theory and Practice*, Cambridge: Cambridge University Press, 1996, pp. 208–231.

Daniels, N., 'Four Unsolved Rationing Problems: a Challenge', in *Hastings Center Report* 24(1994)4, pp. 27–29.

Daniels, N., 'Health-care Needs and Distributive Justice', in *Philosophy and Public Affairs* 10(1981)2, pp. 146–179, repr. in *id., Justice and Justification: Reflective Equilibrium in Theory and Practice*, Cambridge: Cambridge University Press, 1996, pp. 179–207.

Daniels, N., 'Justice, Health, and Healthcare', in *American Journal of Bioethics* 1(2001)2, pp. 2–16.

Daniels, N., 'Kamm's Moral Methods', in *Philosophy and Phenomenological Research* 54(1998)4, pp. 947–954.

Daniels, N., 'Moral Theory and the Plasticity of Persons', in *Monist* 62(1979)3, pp. 265–287.

Daniels, N., 'Normal Functioning and the Treatment-Enhancement Distinction', in *Cambridge Quarterly of Healthcare Ethics* (2000)9, pp. 309–322.

Daniels, N., 'Rawls's Complex Egalitarianism', in S. Freeman (ed.), *The Cambridge Companion to Rawls*, Cambridge: Cambridge University Press, 2003, pp. 241–276.

Daniels, N., 'Rights to Health Care: Programmatic Worries', in *Journal of Medicine and Philosophy* 4(1979)2, pp. 174–191.

Daniels, N., *Am I My Parent's Keeper? An Essay on Justice Between the Young and the Old*, New York: Oxford University Press, 1988.

Daniels, N.; Brock, D., 'Ethical Foundations of the Clinton Administration's Proposed Health Care System', in *Journal of the American Medical Association*, 271(1994), pp. 1189–1196.

Daniels, N., *Just Health Care*, Cambridge: Cambridge University Press, 1985.

Daniels, N., *Justice and Justification: Reflective Equilibrium in Theory and Practice*, Cambridge: Cambridge University Press, 1996.

Daniels, N., *Seeking Fair Treatment: From the AIDS Epidemic to National Health Care Reform*, New York: Oxford University Press, 1995.

Daniels, N.; Bryant, J.; Castano, R.A.; Dantes, O.G.; Khan, K.S.; Pannarunothai, S., 'Benchmarks of Fairness for Health Care Reform: a Policy Tool for Developing Countries', in *Bulletin of the World Health Organization* 78(2000)6, pp. 740–750.

Daniels, N.; Kennedy, B.; Kawachi, I., 'Justice is Good for Our Health', in N. Daniels; B. Kennedy; I. Kawachi; A.K. Sen, *Is Inequality Bad For Our Health?* Boston: Beacon Press, 2000, pp. 3–33.

Daniels, N.; Kennedy, B.; Kawachi, I.; Sen, A.K. (eds.), *Is Inequality Bad for Our Health?* Boston: Beacon Press, 2000.

Daniels, N.; Light, D.W.; Caplan, R.L., *Benchmarks of Fairness for Health Care Reform*, New York: Oxford University Press, 1996.

Daniels, N.; Sabin, J.E., 'Limits to Health Care: Fair Procedures, Democratic Deliberation, and the Legitimacy Problem for Insurers', in *Philosophy and Public Affairs* 26(1997), pp. 303–350.

Daniels, N.; Sabin, J.E., 'The Ethics of Accountablility and the Reform of Managed Care Organisations', in *Health Affairs* 17(1998)5, pp. 50–69.

Daniels, N.; Sabin, J.E., 'Setting Limits Fairly: Can We Learn To Share Medical Resources?' Oxford: Oxford University Press, 2002.

Dasgupta, P., *An Inquiry into Well-Being and Destitution*, Oxford: Clarendon Press, 1993.

De Clercq, B.J., 'Wegen naar gerechtigheid. Een christelijk geïnspireerde ethiek van het maatschappelijk leven', Leuven: Acco, 1990.

De Dijn, H., 'Technology in Health Care: a Philosophical Ethical Appraisal', in C. Gastmans (ed.), *Between Technology and Humanity: The Impact of Technology on Health Care Ethics*, Leuven: Leuven University Press, 2002, pp. 15–34.

De Dijn, H., 'De herontdekking van de ziel. Voor een volwaardige kwaliteitszorg', Nijmegen: Valkhof Pers, 1999.

De Dijn, H., Taboes, 'monsters en loterijen. Ethiek in de laat-moderne tijd', Kapellen: Pelckmans, 2003.

Denier, Y., 'Autonomie en afhankelijkheid. Het subject in de medische ethiek', in R. Devos; A. Braeckman; B. Verdonck (eds.), *Terugkeer van het subject? Recente ontwikkelingen binnen de filosofie*, Leuven: Universitaire Pers, 2002, pp. 105–116.

Denier, Y., 'Behoefte of verlangen? De betekenisstructuur van de kinderwens', in *Ethische Perspectieven* 10(2000)3, pp. 163–174.

Denier, Y., 'Need or Desire? A Conceptual and Moral Phenomenology of the Child Wish', in *International Journal of Applied Philosophy* 20(2006)1, pp. 81–95.

Denier, Y., 'On Personal Responsibility and the Human Right to Health Care', in *Cambridge Quarterly of Healthcare Ethics* 14(2005)2, pp. 224–234.

Denier, Y., 'Public Health, Well-Being, and Reciprocity', in *Ethical Perspectives* 12(2005)1, pp. 41–66.

Denier, Y.; Meulenbergs, T., 'Health Care Needs and Distributive Justice: Philosophical Remarks on the Organisation of Health Care Systems', in R.K. Lie; P.T. Schotsmans; B. Hansen; T. Meulenbergs (eds.), *Healthy Thoughts: European Perspectives on Health Care Ethics*, Leuven: Peeters, 2002, pp. 265–297.

Descartes, R., Discours de la Méthode, Sixième Partie, in A. Bridoux (ed.), *Descartes: Oeuvres et Lettres*, Paris: Gallimard, 1952.

Descartes, R., *Discourse on the Method*, ed. and trans. G. Heffernan, Notre Dame: University of Notre Dame Press, 1994.

Devooght, K., *Essays on Responsibility-Sensitive Egalitarianism and the Measurement of Income Inequality* (nonpublished Ph.D, K.U. Leuven, Faculty of Economics and Applied Economics), Leuven, 2003.

Dierickx, K., *Genetisch gezond?* Antwerpen: Intersentia, 1999, pp. 159–166.

Duffy, J., 'Public Health: History of Public Health', in W.T. Reich (ed.), *Encyclopedia of Bioethics*, 2nd edn., New York: Simon & Schuster/Macmillan, 1995, pp. 2153–2156.

Dworkin, G., 'Paternalism', in *Monist* 56(1972)1, pp. 64–84.

Dworkin, G., 'Taking Risks, Assessing Responsibility', in *Hastings Center Report* 11(1981), pp. 26–31.

Dworkin, R., 'Equality and the Good Life', in *id.*, *Sovereign Virtue*, 2000, pp. 237–284.

Dworkin, R., 'Justice and the High Cost of Health', first appeared as 'Will Clinton's Plan Be Fair?' in *New York Review of* Books 41(1994)1–2, pp. 20–25, 1994, repr. in *id.*, *Sovereign Virtue: The Theory and Practice of Equality*, Cambridge, MA: Harvard University Press, 2000, pp. 307–319.

Dworkin, R., 'Justice in the Distribution of Health Care', in *McGill Law Journal* 38(1993), pp. 883–898, repr. in M. Clayton; A. Williams (eds.), *The Ideal of Equality*, Basingstoke: Palgrave Macmillan, 2002, pp. 203–222.

Dworkin, R., 'Playing God: Genes, Clones, and Luck', in *id.*, *Sovereign Virtue*, Cambridge, 2000, pp. 427–452.

Dworkin, R., 'The Place of Liberty', in *Iowa Law Review* 73(1987)1, pp. 1–54, repr. in *id.*, *Sovereign Virtue*, 2000, pp. 120–183.

Dworkin, R., 'What is Equality? Part 1: Equality of Welfare', in *Philosophy and Public Affairs* 10(1981)3, pp. 185–246, repr. in: R. Dworkin, *Sovereign Virtue: The Theory and Practice of Equality*, Cambridge, MA: Harvard University Press, 2000, pp. 11–64.

Dworkin, R., 'What is Equality? Part 2: Equality of Resources', in *Philosophy and Public Affairs* 10(1981)4, pp. 283–345, repr in: R. Dworkin, *Sovereign Virtue*, Cambridge, MA: Harvard University Press, 2000, pp. 65–119.

Dworkin, R., *Sovereign Virtue: The Theory and Practice of Equality*, Cambridge, MA: Harvard University Press, 2000.

Dworkin, R., *Taking Rights Seriously*, London: Duckworth, 1977.

Ellis, R.P.; McGuire, T.G., 'Supply-Side and Demand-Side Cost Sharing in Health Care', in *Journal of Economic Perspectives* 7(1993)4, pp. 135–151.

Elster, J., 'Sour Grapes – Utilitarianism and the Genesis of Wants', in A.K. Sen; B. Williams (eds.), *Utilitarianism and Beyond*, Cambridge: Cambridge University Press, 1982, pp. 219–238.

Elster, J., *Local Justice: How Institutions Allocate Scarce Goods and Necessary Burdens*, Cambridge: Cambridge University Press, 1992.

Elster, J., *Solomonic Judgements: Studies in the Limitations of Rationality*, Cambridge: Cambridge University Press, 1989.

Elster, J., *Ulysses and the Sirens: Studies in Rationality and Irrationality*, Cambridge: Cambridge University Press, 1979.

Emanuel, E., *The Ends of Human Life: Medical Ethics in a Liberal Polity*, Cambridge, MA: Harvard University Press, 1991.

Engelhardt, H.T., 'Freedom and Moral Diversity', in *Social Philosophy and Policy* 14(1997) pp. 180–196.

Engelhardt, H.T., 'Health Care Allocations: Responses to the Unjust, the Unfortunate, and the Undesirable', in E.E. Shelp (ed.), *Justice and Health Care*, Dordrecht: Reidel, 1981, pp. 121–137.

Engelhardt, H.T., 'Zielkonflikte in nationalen Gesundheitssystemen', in H.M. Sass (ed.), Ethik und öffentliches Gesundheitswesen: ordnungsethische und ordnungspolitische Einflußfaktoren im öffentlichen Gesundheitswesen, Berlin: Springer, 1988, pp. 35–43.

Engelhardt, H.T., *The Foundations of Bioethics*, Oxford: Oxford University Press, 1986.

Fedoryka, K., 'Health as a Normative Concept: Towards a New Conceptual Framework', in *Journal of Medicine and Philosophy* 22(1997), pp. 143–160.

Feinberg, J., *Social Philosophy*, Englewood Cliffs, NJ: Prentice-Hall, 1973.

Foucault, M., *Naissance de la clinique*, Paris: Presses Universitaires de France, 1963.

Frankena, W.K., 'The Concept of Social Justice', in R.B. Brandt (ed.), *Social Justice*, Englewood Cliffs, NJ: Prentice-Hall, 1962, pp. 1–29.

Frankfurt, H.G., 'Necessity and Desire', in *id.*, *The Importance of What We Care About*, Cambridge: Cambridge University Press, 1988, pp. 104–116.

Freeman, S., 'Reason and Agreement in Social Contract Views', in *Philosophy and Public Affairs* 19(1990), pp. 141–147.

Fried, C., 'Distributive Justice', in *Social Philosophy and Policy* 1(1983)1, pp. 45–59.

Fried, C., *Right and Wrong*, Cambridge, MA: Harvard University Press, 1978.

Gauthier, D., *Morals by Agreement*, Oxford: Clarendon Press, 1986.

Gehlen, A., 'Der Mensch, seine Natur, und seine Stellung in der Welt', Frankfurt: Klostermann, 1993.

George, V.; Miller, S.M., 'Social Policy towards 2000: Squaring the Welfare Circle', London: Routledge, 1994.

Gibbard, A., 'Constructing Justice', in *Philosophy and Public Affairs* 20(1991), pp. 264–279.

Gilligan, C., *In a Different Voice*, Cambridge, MA: Harvard University Press, 1982.

Gillon, R., *Philosophical Medical Ethics*, Wiley: Chichester, 1986.

Girard, R., *Des choses cachées depuis la fondation du monde*, Paris: Grasset, 1978.

Girard, R., *Mensonge romantique et vérité romanesque*, Paris: Grasset, 1961.

Golding, M., 'Justice and Rights: a Study in Relationship', in E.E. Shelp (ed.), *Justice and Health Care*, Dordrecht: Reidel, 1981, pp. 23–35.

Goodin, R.E., 'The Priority of Needs', in *Philosophy and Phenomenological Research* 45(1985)4, pp. 615–625.

Goodin, R.E., 'Protecting the Vulnerable: A Reanalysis of Our Social Responsibilities', Chicago: University of Chicago Press, 1985.

Green, R., 'Health Care and Justice in Contract Theory Perspective', in R.M. Veatch; R. Branson (eds.), *Ethics and Health Policy*, Cambridge, MA: Ballinger, 1976, pp. 111–126.

Häfner, H. (ed.), '*Gesundheit – unser höchstes Gut?*' Berlin: Springer, 1999.

Hampshire, S., 'Morality and Pessimism', in *id.*, *Morality and Conflict*, Oxford: Basil Blackwell, 1983, pp. 82–100.

Hanson, R., 'Why Health is not Special: Errors in Evolved Bioethics Institutions', in *Social Philosophy and Policy* 19(2002)2, pp. 153–179.

Harris, J., 'The Survival Lottery', in *Philosophy* 50(1975), pp. 81–87.

Harsanyi, J.C., 'Can the Maximin Principle Serve as Basis for Morality? A Critique of John Rawls's Theory', in *American Political Science Review* 69(1975), pp. 594–606.

Harsanyi, J.C., 'Cardinal Welfare, Individualist Ethics, and interpersonal Comparisons', in *Journal of Political Economy* 63(1955), pp. 309–321.

HELLSTEN, S., 'Theories of Distributive Justice', in R. Chadwick (ed.), *Encyclopedia of Applied Ethics*, 4 vols., San Diego: Academic Press, 1998, Vol. 1, pp. 815–828.

HÖFFE, O., *Gerechtigkeit: eine Philosophische Einführung*, München: Beck, 2001.

HÖFFE, O., *Medizin ohne Ethik?* Frankfurt: Suhrkamp, 2002.

HOFFMEYER, U.K.; MCCARTHY, T.R. (eds.), *Financing Health Care*, 2 vols., Dordrecht: Kluwer, 1994.

HUME, D., *A Treatise of Human Nature*, 2nd edn. ed. L.A. Selby-Bigge, rev. edn. ed. P.H. Nidditch, Oxford: Clarendon Press, 1978.

HUME, D., *Enquiries Concerning Human Understanding and Concerning the Principles of Morals*, 3rd edn. ed. L.A. Selby-Bigge, rev. edn. ed. P.H. Nidditch, Oxford: Clarendon Press, 1975.

HURLEY, J., 'An Overview of the Normative Economics of Health Care', in A.J. Culyer; J.P. Newhouse (eds.), *Handbook of Health Economics*, 2 vols., Amsterdam: Elsevier, 2000, Vol. 1, pp. 55–118.

ILLICH, I., *Medical Nemesis – The Expropriation of Health*, London: Calder & Boyars, 1975.

ILLICH, I., 'Body History', in *Lancet* 11(1986), pp. 1325–1327.

JAGGAR, A., *Feminist Politics and Human Nature*, Totowa: Rowman and Littlefield, 1983.

JAHNIGEN, D.W.; BINSTOCK, R.H., 'Economic and Clinical Realities: Health Care for Elderly People', in R.H. Binstock; S.G. Post; L.S. Mills (ed.), *Too Old for Health Care? Controversies in Medicine, Law, Economics, and Ethics*, Baltimore: Johns Hopkins University Press, 1991, pp. 13–43.

JAKUBOWSKI, E.; BUSSE, R.; CHAMBERS, G.R., *Health Care Systems in the European Union: a Comparative Study* (European Parliament Working Paper), Luxembourg: European Parliament, 1998.

KAMM, F., *Morality, Mortality*. Vol. I: *Death and Whom to Save From It*, Oxford: Oxford University Press, 1993.

KANT, I., *Grundlegung zur Metaphysik der Sitten*, Hamburg: Felix Meiner, 1999.

KARSKENS, M., 'Biopolitiek en de gezonde mens', in J. Rolies (ed.), *De gezonde burger: gezondheid als norm*, Nijmegen: Sun, 1988, pp. 71–88.

KASS, L., *Towards a More Natural Science*, New York: Free Press, 1985.

KAWACHI, I.; KENNEDY, B.P.; LOCHNER, K.; PROTHROW-STITH, D., 'Social Capital, Income Inequality, and Mortality', in *American Journal of Public Health* 87(1997), pp. 1491–1498.

KAWACHI, I.; KENNEDY, B.; WILKINSON, R.G. (ed.), *Income Inequality and Health: a Reader*, New York: New Press, 1999.

KESTELOOT, K., 'Hervormingen in de gezondheidszorg: economische achtergrond en ethische implicaties', in B. Raymaekers; A. Van de Putte; G. Van Riel; G. Verstaen (eds.), *Moeten, mogen, kunnen: ethiek en wetenschap*, Leuven: Universitaire Pers, 2001, pp. 159–185.

KESTELOOT, K.; MARCHAND, M., 'Hervormingsvoorstellen in de gezondheidszorg: een micro-economische benadering', in M. Despontin; M. Jegers (eds.), *De sociale zekerheid verzekerd?*, Brussel: VUB Press, 1995, pp. 371–412.

KITTAY, E.F., 'Can Contractualism Justify State-Supported Long-Term Care Policies? or "I'd Rather Be Some Mother's Child"', in World Health Organization, *Ethical Choices in Long-Term Care: What Does Justice Require?*, Geneva: World Health Organization, 2002, pp. 77–83.

KITTAY, E.F., *Love's Labor: Essays on Women, Equality, and Dependency*, New York: Routledge, 1999.

KLEIN, R.; DAY, P.; REDMAYNE, S., *Managing Scarcity. Priority Setting and Rationing in the National Health Service*, Buckingham: Open University Press, 1996.

KYMLICKA, W., *Contemporary Political Philosophy*, Oxford: Clarendon Press, 1990, 2nd ext. edn., 2002.

LEGRAND, J., 'Equity, Health, and Health Care', in *Social Justice Research* 1(1987)3, pp. 257–274.

LEGRAND, J., *Equity and Choice*, London: Harper Collins Academic, 1991.

LEICHTER, H., 'Public Policy and the British Experience', in *Hastings Center Report* 11(1981), pp. 32–39, repr. in *id.*, *Free to Be Foolish: Politics and Health Promotion in the United States and Great Britain*, Princeton, NJ: Princeton University Press, 1991.

LUBITZ, J.; RILEY, G., 'Trends in Medicare Payments in the Last Year of Life', in *New England Journal of Medicine* 328(1993), pp. 1092–1096.

MACCALLUM, G., 'Negative and Positive Freedom', in *Philosophical Review* 76(1967), pp. 312–334.

MACCLAIN, L., '"Atomistic Man" revisited: Liberalism, Connection and Feminist Jurisprudence', in *Southern California Law Review* 65(1992), pp. 1171–1264.

MacIntyre, A., 'The Privatization of Good: an Inaugural Lecture', in *Review of Politics* 52(1990), pp. 344–361.

MacIntyre, A., 'Dependent Rational Animals: Why Human Beings Need the Virtues', London: Duckworth, 1999.

MacIntyre, A., *Whose Justice, Which Rationality?*, Notre Dame: University of Notre Dame Press, 1988.

MacKinnon, C., *Feminism Unmodified*, Cambridge, MA: Harvard University Press, 1987.

MacKinnon, C., *Toward a Feminist Theory of the State*, Cambridge, MA: Harvard University Press, 1989.

Mann, J.M.; Gruskin, S.; Grodin, M.A.; Annas, G.J. (eds.), *Health and Human Rights*, New York: Routledge, 1999.

Marmot, M.G.; Fuhrer, R.; Ettner, S.L., Marks, N.F., Bumpass, L.L.; Ryff, C.D., 'Contribution of Psychosocial Factors to Socioeconomic Differences in Health', in *Milbank Quarterly* 76(1998), pp. 403–408.

McClosky, H.J., 'Human Needs, Rights, and Political Values', in *American Philosophical Quarterly*, 13(1976)1, pp. 1–11.

McCullough, L.B., 'Justice and Health Care: Historical Perspectives and Precedents', in E.E. Shelp (ed.), *Justice and Health Care*, Dordrecht: Reidel, 1981, pp. 51–71.

Mieth, D., 'Ethik der Gerechtigkeit. Ansätze, Prinzipien, Kriterien', in D. Mieth; P. Magino (eds.), *Vision Gerechtigkeit? Konziliarer Prozeß und Kirchliche Jugendarbeit*, Dusseldorf: Altenberg, 1992, pp. 12–32.

Miller, D., *Principles of Social Justice*, Cambridge, MA: Harvard University Press, 1999.

Miller, D., *Social Justice*, Oxford: Clarendon Press, 1976.

Mooney, G., *Key Issues in Health Economics*, New York: Harvester Wheatsheaf, 1994.

Moore, G.E., *Principia Ethica*, Cambridge: Cambridge University Press, 1903.

Mulgan, R., 'Was Aristotle an "Aristotelian Social Democrat?"', in *Ethics* 111(2000), pp. 79–101.

Nagel, T., 'Death', in *id.*, *Mortal Questions*, Cambridge: Cambridge University Press, 1979, pp. 1–11.

New, B.; LeGrand, J., *Rationing in the NHS: Principles and Pragmatism*, London: King's Fund Publishing, 1996.

Newhouse, J., 'Medical Care Cost Increases: How Much Welfare Loss?' in *Journal of Economic Perspectives* 5(1992)3, pp. 3–21.

Nord, E., 'The Person-Tradeoff Approach to Valuing Healthcare Programs', in *Medical Decision Making* 15(1995), pp. 201–208.

Nord, E., 'The Relevance of Health State After Treatment in Prioritising Between Different Patients', in *Journal of Medical Ethics* 19(1993), pp. 37–42.

Nord, E.; Richardson, J.; Street, A.; Kuhse, H.; Singer, P., 'Maximizing Health Benefits vs Egalitarianism: an Australian Survey of Health Issues', in *Social Science and Medicine* 41(1995)10, pp. 1429–1437.

Nord, E., Cost-Value Analysis in Healthcare: Making Sense out of QALY's, Cambridge: Cambridge University Press, 1999.

Nozick, R., *Anarchy, State and Utopia*, New York: Basic Books, 1974.

Nussbaum, M.C., 'Aristotelian Social Democracy', in R.B. Douglas; G.R. Mara; H.S. Richardson (eds.), *Liberalism and the Good*, New York: Routledge, 1990, pp. 203–252.

Nussbaum, M.C., 'Aristotle on Human Nature and the Foundation of Ethics', in J.E.J. Altham; R. Harrison (eds.), *World, Mind, and Ethics: Essays on the Ethical Philosophy of Bernard Williams*, Cambridge: Cambridge University Press, 1995, pp. 86–131.

Nussbaum, M.C., 'Aristotle, Politics, and Human Capabilities: a Response to Antony, Arneson, Charlesworth, and Mulgan', in *Ethics* 111(2000), pp. 102–140.

Nussbaum, M.C., 'Beyond the Social Contract: Toward Global Justice', in G.B. Peterson (ed.), *The Tanner Lectures on Human Values*, Vol. 24, Salt Lake City: University of Utah Press, 2004, pp. 413–507.

Nussbaum, M.C., 'Capabilities as Fundamental Entitlements: Sen and Social Justice', in *Feminist Economics* 9(2003)2–3, pp. 33–59.

Nussbaum, M.C., 'Disabled Lives: Who Cares?' in *The New York Review of Books* 48(2001)1, pp. 34–37.
Nussbaum, M.C., 'Human Functioning and Social Justice: In Defense of Aristotelian Essentialism', in *Political Theory* 20(1992), pp. 202–246.
Nussbaum, M.C., 'Long-Term Care and Social Justice: a Challenge to Conventional Ideas of the Social Contract', in World Health Organization, *Ethical Choices in Long-Term Care: What Does Justice Require?*, Geneva: World Health Organization, 2002, pp. 31–65.
Nussbaum, M.C., 'Nature, Function, and Capability: Aristotle on Political Distribution', in *Oxford Studies in Ancient Philosophy*, Suppl. Vol. I (1988), pp. 145–184.
Nussbaum, M.C., 'Non-Relative Virtues: an Aristotelian Approach', in M.C. Nussbaum, A.K. Sen (eds.), *The Quality of Life*, Oxford: Clarendon Press, 1993, pp. 242–269.
Nussbaum, M.C., 'Political Animals: Luck, Love, and Dignity', in *Metaphilosophy* 29(1998), pp. 273–287.
Nussbaum, M.C., 'Rawls and Feminism', in S. Freeman (ed.), *The Cambridge Companion to Rawls*, Cambridge: Cambridge University Press, 2003, pp. 488–520.
Nussbaum, M.C., 'The Cost of Tragedy: Some Moral Limits of Cost-Benefit Analysis', in M. Adler; E. Posner (eds.), *Cost-Benefit Analysis: Legal, Economic, and Philosophical Perspectives*, Chicago: Chicago University Press, 2001, pp. 169–200.
Nussbaum, M.C., 'The Feminist Critique of Liberalism', in *id.*, *Sex and Social Justice*, Oxford: Oxford University Press, 1999, pp. 55–80.
Nussbaum, M.C., 'The Future of Feminist Liberalism', in *Proceedings and Addresses of the American Philosophical Association* (2000)74, pp. 47–79.
Nussbaum, M.C., 'Victims and Agents', in *The Boston Review* 23(1998), pp. 21–24.
Nussbaum, M.C., *Women and Human Development*, Cambridge: Cambridge University Press, 2000.
Nussbaum, M.C.; Glover, J. (eds.), *Women, Culture, and Development*, Oxford: Clarendon Press, 1995.
O'Connell, J.M., 'The Relationship between Health Expenditures and the Age Structure of the Population in OECD Countries', in *Health Economics* 5(1996), pp. 573–578.
O'Neill, O., 'Public Health or Clinical Ethics: Thinking Beyond Borders', in *Ethics and International Affairs* 16(2002)2, pp. 35–45.
O'Neill, O., *Autonomy and Trust in Bioethics*, Cambridge: Cambridge University Press, 2002.
Parfit, D., 'Later Selves and Moral Principles', in A. Montefiore (ed.), *Philosophy and Personal Relations*, London: Routledge, 1973, pp. 137–169.
Parfit, D., *Reasons and Persons*, Oxford: Oxford University Press, 1984.
Pateman, C., 'The Problem of Political Obligation: a Critique of Liberal Theory', Berkeley: University of California Press, 1979.
Pateman, C., *The Sexual Contract*, Stanford: Stanford University Press, 1988.
Pauer-Studer, H., Autonom Leben. Reflexionen über Freiheit und Gleichheit, Frankfurt a.M.: Suhrkamp, 2000.
Plato, *Crito* and *Phaedo*, trans. H. North Fowler, Cambridge, MA: Harvard University Press, 1971.
Plato, *The Republic*, trans. T. Griffith, ed. G.R.F. Ferrari, Cambridge: Cambridge University Press, 2000.
Porter, R., 'The Greatest Benefit To Mankind: a Medical History of Humanity', New York: Norton, 1999.
Poterba, J.M., 'A Skeptic's View of Global Budget Caps', in *Journal of Economic Perspectives* 8(1994)3, pp. 67–74.
Powers, M., 'Hypothetical Choice Approaches to Health Care Allocation', in J. Humber; R. Almeder (eds.), *Allocating Health Care Resources, Biomedical Ethics Reviews*, Totowa: Humana Press, 1995, pp. 55–84.
Rakowski, E., *Equal Justice*, New York: Oxford University Press, 1991.
Rawls, J., 'A Kantian Conception of Equality', in *Cambridge Review* 96(1975), pp. 94–99, repr. in J. Rawls, *Collected Papers*, ed. S. Freeman, Cambridge, MA: Harvard University Press, 1999, pp. 254–266, p. 259.
Rawls, J., 'Justice as Fairness: Political not Metaphysical', in *Philosophy and Public Affairs* 14(1985), pp. 223–252, repr. in J. Rawls, *Collected Papers*, ed. S. Freeman, Cambridge, MA: Harvard University Press, 1999, pp. 388–414.

RAWLS, J., 'Kantian Constructivism in Moral Theory', in *Journal of Philosophy* 77(1980), pp. 515–572, repr. in J. Rawls, *Collected Papers*, ed. S. Freeman, pp. 303–358.

RAWLS, J., 'Social Unity and Primary Goods', in A.K. Sen; B. Williams (eds.), *Utilitarianism and Beyond*, Cambridge: Cambridge University Press, 1982, pp. 159–186, repr. in J. Rawls, *Collected Papers*, ed. S. Freeman, Cambridge, MA: Harvard University Press, 1999, pp. 359–387.

RAWLS, J., 'The Independence of Moral Theory', in *Proceedings and Addresses of the American Philosophical Association* 48(1975), pp. 5–22.

RAWLS, J., 'The Priority of Right and Ideas of the Good', in *Philosophy and Public Affairs* 17(1988), pp. 251–276, repr. in J. Rawls, *Collected Papers*, ed. S. Freeman, Cambridge, MA: Harvard University Press, 1999, pp. 449–472.

RAWLS, J., *A Theory of Justice*, Cambridge, MA: Harvard University Press, 1971.

RAWLS, J., *A Theory of Justice*, rev. edn., Cambridge, MA: The Belknap Press of Harvard University Press, 1999.

RAWLS, J., *Collected Papers*, ed. S. Freeman, Cambridge, MA: Harvard University Press, 1999.

RAWLS, J., 'Fairness to Goodness', in *Philosophical Review* 84(1975), pp. 536–555, repr. in J. Rawls, *Collected Papers*, ed. S. Freeman, Cambridge, MA: Harvard University Press, 1999, pp. 267–285.

RAWLS, J., *Justice as Fairness: a Restatement*, ed. E. Kelly, Cambridge, MA: The Belknap Press of Harvard University Press, 2001.

RAWLS, J., *Political Liberalism*, expanded paperback edition, New York: Colombia University Press, 1996.

REUTER, L., *Modern Biotechnology in Postmodern Times?* Dordrecht: Kluwer, 2003.

RICOEUR, P., *Le Juste*, Paris: Le Seuil, 1995.

RICOEUR, P., *The Just*, trans. D. Pellauer, Chicago: University of Chicago Press, 2000.

ROBBINS, L., *The Nature and Significance of Economic Science*, London: Macmillan, 1932.

ROEMER, J.E., 'A Pragmatic Theory of Responsibility for the Egalitarian Planner', in *Philosophy and Public Affairs* 22(1993)2, pp. 146–166.

ROEMER, J.E., 'Equality and Responsibility', in *Boston Review* 20(1995)2, pp. 3–7, 15–16.

ROEMER, J.E., 'Equality of Talent', in *Economics and Philosophy* 1(1985)2, pp. 151–187.

ROEMER, J.E., *Equality of Opportunity*, Cambridge: Harvard University Press, 1998.

ROEMER, J.E., *Theories of Distributive Justice*, Cambridge, MA: Harvard University Press, 1996.

ROLIES, J., 'Gezondheid: een nieuwe religie?' in *id.* (ed.), *De gezonde burger: gezondheid als norm*, Nijmegen: Sun, 1988, pp. 11–30.

ROSS, W.D., *The Foundations of Ethics*, Oxford: Clarendon Press, 1939.

ROSS, W.D., *The Right and the Good*, Oxford: Clarendon Press, 1930.

RUSSELL, L.B., 'Some of the Tough Decisions Required by a National Health Plan', in *Science* 246(1989), pp. 892–896.

SANDEL, M., *Liberalism and the Limits of Justice*, Cambridge: Cambridge University Press, 1982.

SCANLON, T.M., 'Equality of Resources and Equality of Welfare: a Forced Marriage?' in *Ethics* 97(1986), pp. 111–118.

SCANLON, T.M., 'Preference and Urgency', in *Journal of Philosophy* 72(1975)19, pp. 655–669.

SCANLON, T.M., 'The Significance of Choice', in S. McMurrin (ed.), *The Tanner Lectures on Human Values*, Vol. 8, Salt Lake City: University of Utah Press, 1988, pp. 149–216.

SCANLON, T.M., 'Value, Desire, and Quality of Life', in M.C. Nussbaum; A.K. Sen (eds.), *The Quality of Life*, Oxford: Clarendon Press, 1993, pp. 185–200.

SCANLON, T.M., *What We Owe to Each Other*, Cambridge, MA: The Belknap Press of Harvard University Press, 1998.

SCHELLING, T.C., 'The Live You Save May Be Your Own', in S.B. Chase (ed.), *Problems in Public Expenditure Analysis*, Washington: Brookings Institution, 1966, pp. 127–176.

SCHOKKAERT, E., 'Warm en koud. Solidariteit en verantwoordelijkheid in de ziekteverzekering', in *Ethische Perspectieven* 8(1998), pp. 135–146.

SCHOKKAERT, E.; DEVOOGHT, K., 'Responsibility-Sensitive Fair Compensation in Different Cultures', in *Social Choice and Welfare* 21(2003)2, pp. 207–242.

SCHOTSMANS, P., 'Is het ten laste nemen van de gezondheidszorg door de solidariteit begrensd? Een ethische verkenning', in J. Hallet; J. Hermesse; D. Sauer (eds.), *Solidariteit, gezondheid, ethiek*, Leuven: Garant, 1994, pp. 33–47.

SCHROEDER, S.A., 'Health Policy 2001: Prospects for Expanding Health Care Insurance Coverage', in *The New England Journal of Medicine* 344(2001)11, pp. 847–851.

SCHWARTZ, R.L., 'Life Style, Health Status, and Distributive Justice', in A. Grubb; M.J. Mehlman (ed.), *Justice and Health Care: Comparative Perspectives*, New York: Wiley, 1995, pp. 244–248.

SCHWARTZ, W.B., 'The Inevitable Failure of Current Cost-Containment Strategies – Why They Can Provide Only Temporary Relief', in *Journal of the American Medical Association* 257(1987)2, pp. 220–224.

SEN, A.K., 'Capability and Well-Being', in M.C. Nussbaum; A.K. Sen (eds.), *The Quality of Life*, Oxford: Clarendon Press, 1993, pp. 30–61.

SEN, A.K., 'Equality of What?' in S.M. McMurrin (ed.), *Tanner Lectures on Human Values*, Salt Lake City: University of Utah Press, Vol. 1, 1980, pp. 195–220.

SEN, A.K., 'Health in Development', in *Bulletin of the World Health Organization* 77(1999)8, pp. 619–623.

SEN, A.K., 'Justice: Means versus Freedoms', in *Philosophy and Public Affairs* 19(1990), pp. 111–121.

SEN, A.K., 'Objectivity and Position: Assessment of Health and Well-Being', in L.C. Chen; A. Kleinman; N. Ware (eds.), *Health and Social Change in International Perspective*, Boston: Harvard University Press, pp. 115–128.

SEN, A.K., 'Well-Being, Agency, and Freedom: The Dewey Lectures 1984', in *The Journal of Philosophy* 82(1985), pp. 169–221.

SEN, A.K., 'Why Health Equity?' in *Health Economics* 11(2001), pp. 659–666.

SEN, A.K., *Commodities and Capabilities*, Amsterdam: North-Holland, 1985.

SEN, A.K., *Development as Freedom*, New York: Knopf, 1999.

SEN, A.K., *Inequality Reexamined*, New York: Russell Sage, 1992.

SEN, A.K., *Resources, Values and Development*, Oxford: Basil Blackwell, 1984.

SEN, A.K.; WILLIAMS, B. (eds.), *Utilitarianism and Beyond*, Cambridge: Cambridge University Press, 1982.

SHELP, E.E. (ed.), *Justice and Health Care*, Dordrecht: Reidel, 1981.

SINGER, P., 'Utility and the Survival Lottery', in *Philosophy* 52(1977), pp. 218–222.

SKRABANEK, P., *The Death of Humane Medicine and the Rise of Coercive Healthism*, London: Social Affairs Unit, 1994.

SMITH, R., 'Being Creative About Rationing', in *British Medical Journal* 312(1996), pp. 391–392.

SOLOMON, R.C.; MURPHY, M.C. (eds.), *What is Justice? Classic and Contemporary Readings*, 2nd edn., New York: Oxford University Press, 2000.

THIELMANN, L., Ethische Grundlagen einer Prioritätensetzung im Gesundheitswesen, Bayreuth: PCO. Verlag, 2001.

TRAPPENBURG, M.J., 'Defining the Medical Sphere', in *Cambridge Quarterly of Healthcare Ethics* 6(1997), pp. 416–434.

TUGENDHAT, E., *Vorlesungen über Ethik*, Frankfurt: Suhrkamp, 1993.

ULPIANUS, *The Digest of Justinian*, Vol. I, trans. A. Watson, ed. T. Mommsen; P. Krueger, Philadelphia: University of Pennsylvania Press, 1985.

VAN DOORSLAER, E.; WAGSTAFF, A.; RUTTEN, F., *Equity in the Finance and Delivery of Health Care: an International Perspective*, Oxford: Oxford University Press, 1993.

VAN PARIJS, P., 'Assurance, solidarité, équité, les fondements éthiques de l'Etat-Providence', in *Cahiers de l'Ecole des sciences philosophiques et religieuses* 12(1993), pp. 49–72.

VAN PARIJS, P., 'Au-delà de la solidarité, les fondements éthiques de l'Etat-Providence et de son dépassement', in *Futuribles* 184(1994), pp. 5–29.

VAN PARIJS, P., *Solidariteit voor de 21e eeuw*, Leuven: Garant, 1996.

VAN PARIJS, P., 'Y a-t-il des limites à la prise en charge des soins de santé par la solidarité?' in J. Hallet; J. Hermesse; D. Sauer (eds.), *Solidarité, Santé, Ethique*, Louvain: Garant, 1994, pp. 57–68.

VAN PARIJS, P., 'ZIJN er grenzen aan de ten laste neming van de gezondheidszorg door de solidariteit?' in J. Hallet; J. Hermesse; D. Sauer (eds.), *Solidariteit, gezondheid, ethiek*, Leuven: Garant, 1994, pp. 57–67.

VAN PARIJS, P., 'Real Freedom for All. What (If Anything) Can Justify Capitalism?' Oxford: Clarendon Press, 1995.

Van Reybrouck, D., 'Die Siel van die Mier', Gent: Het Muziek Lod, 2004.

Vandevelde, T., 'Het intrestverbod bij Aristoteles', in L. Bouckaert (ed.), *Intrest en cultuur. Een ethiek van het geld*, Leuven: Acco, 1994, pp. 37–58.

Vandevelde, T., 'Ideologie en sociale zekerheid', in M. Despontin; M. Jegers (eds.), *De sociale zekerheid verzekerd?* Brussel: VUB Press, 1995, pp. 151–176.

Veatch, R.M., 'Voluntary Risks to Health', in *Journal of the American Medical Association* 243(1980)1, pp. 50–55.

Veatch, R.M., 'What is a 'Just' Health Care Delivery?' in R.M. Veatch; R. Branson (eds.), *Ethics and Health Policy*, Cambridge, MA: Balinger Publishing Company, 1976, pp. 127–153.

Veatch, R.M., 'A Theory of Medical Ethics', New York: Basic Books, 1981.

Veatch, R.M.; Branson, R. (eds.), *Ethics and Health Policy*, Cambridge: Ballinger Publishing Company, 1976.

Villey, M., Leçons d'histoire de la philosophie du droit, 2nd édn. Paris: Dalloz, 2002.

Vlastos, G., 'Justice and Equality', in J. Waldron (ed.), *Theories of Rights*, New York: Oxford University Press, 1984, pp. 41–76.

Wagstaff, A.; Van Doorslaer, E., 'Equity in Health Care Finance and Delivery', in A.J. Culyer; J.P. Newhouse (eds.), *Handbook of Health Economics*, Vol. 1, Amsterdam: Elsevier, 2000, pp. 1803–1862.

Waldron, J., 'When Justice Replaces Affection: The Need for Rights', in *id.*, *Liberal Rights*, Cambridge: Cambridge University Press, 1993, pp. 370–391.

Walzer, M., *Spheres of Justice: A Defense of Pluralism and Equality*, New York: Basic Books, 1983.

Warnock, M., *Nature and Mortality*, London: Continuum, 2003.

Weale, A., 'Rationing Health Care', in *British Medical Journal* 316(1998)7129, pp. 410–411.

Weale, A., *Equality and Social Policy*, London: Routledge, 1978.

Wiggins, D., *Needs, Values, Truth*, 3rd edn., Oxford: Oxford University Press, 1998.

Wilkinson, R.G., *Unequal Societies: The Afflictions of Inequality*, London: Routledge, 1996.

Williams, A., 'Intergenerational Equity: an Exploration of the "Fair Innings" argument', in *Health Economics* 6(1997), pp. 117–132.

Williams, A.; Cookson, R., 'Equity in Health', in A.J. Culyer; J.P. Newhouse (eds.), *Handbook of Health Economics*, Amsterdam: Elsevier, 2000, 2 vols., Vol. 1, pp. 1863–1910.

Williams, B.A.O., 'A Critique of Utilitarianism', in J. Smart; B. Williams (eds.), *Utilitarianism: For and Against*, Cambridge: Cambridge University Press, 1973, pp. 77–150.

Williams, B.A.O., 'Justice as a Virtue', in A.O. Rorty (ed.), *Essays on Aristotle's Ethics*, Berkeley: University of California Press, 1980, pp. 189–199.

Williams, B.A.O., 'Persons, Character, and Morality', in A.O. Rorty (ed.), *The Identities of Persons*, Berkeley: University of California Press, 1976, pp. 197–216, repr. in B. Williams, *Moral Luck. Philosophical Papers 1973–1980*, Cambridge: Cambridge University Press, 1981, pp. 1–19.

Williams, B.A.O., 'The Idea of Equality', in P. Laslett; G. Runciman (eds.), *Politics, Philosophy and Society*, Oxford: Blackwell, 1962, repr. in B. Williams, *Problems of the Self*, Cambridge: Cambridge University Press, 1973, pp. 230–249.

Williams, B.A.O., 'Utilitarianism and Moral Self-Indulgence', in H. Lewis (ed.), *Contemporary British Philosophy*, London: Allen and Unwin, pp. 306–321, repr. in B. Williams, *Moral Luck. Philosophical Papers 1973–1980*, Cambridge: Cambridge University Press, 1981, pp. 40–53.

Wolf, C., 'Theories of Justice: Human Needs', in R. Chadwick (ed.), *Encyclopedia of Applied Ethics*, San Diego: Academic Press, 1998, pp. 335–345.

World Health Report 2000, *Health Systems: Improving Performance*, Geneva: World Health Organization, 2000.

Xenos, N., *Scarcity and Modernity*, London: Routledge, 1989.

Zwart, H.A.E., 'Ethiek, medische technologie en solidariteit', in J. Hallet; J. Hermesse; D. Sauer (eds.), *Solidariteit, gezondheid, ethiek*, Leuven: Garant, 1994, pp. 81–89.

Zweifel, P.; Felder, S.; Meiers, M., 'Ageing of Population and Health Expenditure: a Red Herring', in *Health Economics* 8(1999), pp. 485–496.

INDEX

International Library of Ethics, Law, and the New Medicine

1. L. Nordenfelt: *Action, Ability and Health.* Essays in the Philosophy of Action and Welfare. 2000 ISBN 0–7923–6206–3
2. J. Bergsma and D.C. Thomasma: *Autonomy and Clinical Medicine.* Renewing the Health Professional Relation with the Patient. 2000 ISBN 0–7923–6207–1
3. S. Rinken: *The AIDS Crisis and the Modern Self.* Biographical Self-Construction in the Awareness of Finitude. 2000 ISBN 0–7923–6371–X
4. M. Verweij: *Preventive Medicine Between Obligation and Aspiration.* 2000 ISBN 0–7923–6691–3
5. F. Svenaeus. *The Hermeneutics of Medicine and the Phenomenology of Health.* Steps Towards a Philosophy of Medical Practice. 2001 ISBN 0–7923–6757–X
6. D.M. Vukadinovich and S.L. Krinsky: *Ethics and Law in Modern Medicine.* Hypothetical Case Studies. 2001 ISBN 1–4020–0088–X
7. D.C. Thomasma, D.N. Weisstub and C. Hervé (eds.): *Personhood and Health Care.* 2001 ISBN 1–4020–0098–7
8. H. ten Have and B. Gordijn (eds.): *Bioethics in a European Perspective.* 2001 ISBN 1–4020–0126–6
9. P.-A. Tengland: Mental Health. *A Philosophical Analysis.* 2001 ISBN 1–4020–0179–7
10. D.N. Weisstub, D.C. Thomasma, S. Gauthier and G.F. Tomossy (eds.): *Aging: Culture, Health, and Social Change.* 2001 ISBN 1–4020–0180–0
11. D.N. Weisstub, D.C. Thomasma, S. Gauthier and G.F. Tomossy (eds.) *Aging: Caring for our Elders.* 2001 ISBN 1–4020–0181–9
12. D.N. Weisstub, D.C. Thomasma, S. Gauthier and G.F. Tomossy (eds.): *Aging: Decisions at the End of Life.* 2001 ISBN 1–4020–0182–7
    (Set ISBN for vols. 10–12: 1–4020–0183–5)
13. M.J. Commers: *Determinants of Health: Theory, Understanding, Portrayal, Policy.* 2002 ISBN 1–4020–0809–0
14. I.N. Olver: Is Death Ever Preferable to Life? 2002 ISBN 1–4020–1029–X
15. C. Kopp: *The New Era of AIDS.* HIV and Medicine in Times of Transition. 2003 ISBN 1–4020–1048–6
16. R.L. Sturman: *Six Lives in Jerusalem.* End-of-Life Decisions in Jerusalem-Cultural, Medical, Ethical and Legal Considerations. 2003 ISBN 1–4020–1725–1
17. D.C. Wertz and J.C. Fletcher: *Genetics and Ethics in Global Perspective.* 2004 ISBN 1–4020–1768–5
18. J.B.R. Gaie: *The Ethics of Medical Involvement in Capital Punishment.* A Philosophical Discussion. 2004 ISBN 1–4020–1764–2
19. M. Boylan (ed.): *Public Health Policy and Ethics.* 2004 ISBN 1–4020–1762–6; Pb 1–4020–1763–4
20. R. Cohen-Almagor: *Euthanasia in the Netherlands.* The Policy and Practice of Mercy Killing. 2004 ISBN 1–4020–2250–6
21. D.C. Thomasma and D.N. Weisstub (eds.): *The Variables of Moral Capacity.* 2004 ISBN 1–4020–2551–3
22. D.R. Waring: *Medical Benefit and the Human Lottery.* An Egalitarian Approach. 2004 ISBN 1–4020–2970–5

International Library of Ethics, Law, and the New Medicine

23. P. McCullagh: *Conscious in a Vegetative State? A Critique of the PVS Concept*. 2004 ISBN 1-4020-2629-3
24. L. Romanucci-Ross and L.R. Tancredi: *When Law and Medicine Meet*: A Cultural View. 2004 ISBN 1-4020-2756-7
25. G.P. Smith II: *The Christian Religion and Biotechnology*. A Search for Principled Decision-making. 2005 ISBN 1-4020-3146-7
26. C. Viafora (ed.): *Clinical Bioethics*. A Search for the Foundations. 2005 ISBN 1-4020-3592-6
27. B. Bennett and G.F. Tomossy: *Globalization and Health*. Challenges for health law and bioethics. 2005 ISBN 1-4020-4195-0
28. C. Rehmann-Sutter, M. Düwell and D. Mieth (eds.): *Bioethics in Cultural Contexts*. Reflections on Methods and Finitude. 2006 ISBN 1-4020-4240-X
29. S.E. Sytsma, Ph.D.: Ethics and Intersex. 2006 ISBN 1-4020-4313-9
30. M. Betta (ed.): *The Moral, Social, and Commercial Imperatives of Genetic Testing and Screening*. The Australian Case. 2006 ISBN 1-4020-4618-9
31. D. Atighetchi: *Islamic Bioethics: Problems and Perspectives*. 2007 ISBN 1-4020-4961-7
32. Rispler-Chaim: *Disability in Islamic Law*. 2007 ISBN 1-4020-5052-6
33. Y. Denier: *Efficiency, Justice and Care*. Philosophical Reflections on Scarcity in Health Care. 2007 ISBN 1-4020-5213-8

The manufacturer's authorised representative in the EU is Springer Nature Customer Service Centre GmbH, Europaplatz 3, 69115 Heidelberg, Germany. If you have any concerns regarding our products, please contact ProductSafety@springernature.com

Printed and bound by CPI Group (UK) Ltd, Croydon, CR0 4YY
10/10/2026
02231227-0007